Holistic Plastic Pollution Solutions and Science-Based Policies

Editors

Sandhya Babel
School of Biochemical Engineering and Technology
Sirindhorn International Institute of Technology
Thammasat University
Pathum Thani, Thailand

Velma I Grover
McMaster University
Hamilton, ON, Canada

Amila Abeynayaka
Institute for Global Environmental Strategies (IGES)
Japan
and
Quantitative Sustainability Assessment
Department of Environmental and Resource Engineering
Technical University of Denmark (DTU)
Kongens Lyngby, Denmark

CRC Press
Taylor & Francis Group
Boca Raton London New York

CRC Press is an imprint of the
Taylor & Francis Group, an **informa** business
A SCIENCE PUBLISHERS BOOK

Cover credit: Original image by Dr. Sandhya Babel. Reproduced with her kind permission.

First edition published 2027
by CRC Press
2385 NW Executive Center Drive, Suite 320, Boca Raton FL 33431

and by CRC Press
4 Park Square, Milton Park, Abingdon, Oxon, OX14 4RN

CRC Press is an imprint of Taylor & Francis Group, LLC

Library of Congress Cataloging-in-Publication Data (applied for)

ISBN: 978-1-032-84575-3 (hbk)
ISBN: 978-1-032-87214-8 (pbk)
ISBN: 978-1-003-53146-3 (ebk)

DOI: 10.1201/9781003531463

Typeset in Times New Roman
by Prime Publishing Services

Preface

Plastic is a comparatively cheaper and versatile material used daily in many things including toothpaste, cosmetics, clothes, computers, mobiles, electronics, cars and is used in construction, healthcare, agriculture, and the list just goes on. This has driven the plastic production from just two million in 1950 to over 450 million tonnes in 2023 (Ritchie et al. 2023). Unfortunately, plastic does not breakdown that easily or quickly, it stays around for decades and are now detected even in air we breathe and water we drink (even in the deepest parts of ocean and Arctic Sea ice).

Whenever we talk about plastic waste, the first thing that comes to mind is plastic bottles, but the list is longer which includes synthetic fibres, disposable cups, plates, cutlery, straws, cigarette buts, plastic bags, packaging, protective mesh, and disposable health care instruments such as syringes and gloves. This book focuses on plastic and microplastics (MPs) pollution in riverine, mangrove, coastal, and marine environments. It also highlights detection techniques for MPs in aquatic systems, and various methods for microplastic removal. Although several strategies exist to reduce plastic pollution, prevention at the source remains the most effective approach to stopping plastic and MP contamination. The book also discusses recycling and the reuse of plastics as efficient, sustainable, and eco-friendly solutions to this long-term issue.

Taking a holistic approach, this book examines the environmental impacts of MPs, evaluates sustainable alternatives, promotes recycling through circular economy principles, analyzes relevant policy frameworks, and presents case studies. It serves as an essential guide for understanding, addressing, and mitigating global plastic pollution.

Velma I Grover
Sandhya Babel
Amila Abeynayaka

REFERENCE

Ritchie, H., V. Samborska and M. Roser. 2023. Plastic Pollution. Published online at OurWorldinData.org. Retrieved from: 'https://ourworldindata.org/plastic-pollution' [Online Resource]

Contents

Chapter 1

Introduction and Road Map to the Book

Velma I Grover*[1] and Sandhya Babel[2]

[1]Associate Adjunct Professor, McMaster University, Canada

[2]Professor, Thammasat University, Thailand

Plastics were introduced as a revolutionary polymer in the early 1900s due to their lightweight, malleability, durability, and longevity. Today, plastics are used in nearly every sector and have become a major contributor to riverine and coastal pollution. Recently, the widespread presence of microplastics (MPs) has raised significant concern, as studies have shown their harmful effects on organisms when ingested. This book focuses on plastic and MP pollution in riverine, mangrove, coastal, and marine environments. It also highlights detection techniques for MPs in aquatic systems and various methods to remove microplastics. Although several strategies exist to reduce plastic pollution, prevention at the source remains the most effective approach for stopping plastic and MP contamination. The book also discusses recycling and the reuse of plastics as efficient, sustainable, and eco-friendly solutions to this long-term issue.

Taking a holistic approach, this book examines the environmental impacts of MPs, evaluates sustainable alternatives, promotes recycling through circular economy principles, analyzes relevant policy frameworks, and presents case studies. It serves as an essential guide for understanding, addressing, and mitigating global plastic pollution.

In "Plastic Pollution in Coastal and Marine Environments: Sources, Impacts, and Solutions," Mahagamage and Gamage examine the origins of plastic

*For Correspondence: velmaigrover@yahoo.com

DOI: 10.1201/9781003531463-1

pollution in coastal and marine habitats. A major source of this pollution is the mismanagement of plastic waste disposal on both land and sea. This includes littering, improper landfill practices, and the urban plastic runoff into water bodies. Additional sources include industrial discharges, as well as fishing and shipping activities that release plastic particles and waste directly into the marine environment. Single-use plastics, such as bottles, bags, and packaging, are among the most common contributors. These items often end up in the ocean due to their durability and improper disposal. Marine life is heavily affected, with species becoming entangled in discarded plastics or ingesting MPs. This disrupts marine ecosystems, harms wildlife, and affects the food chain. MPs have spread throughout the ocean, posing a threat to organisms at all trophic levels. They also have the potential to carry and transfer harmful chemical compounds. Tackling plastic waste in coastal and marine environments requires a range of strategies. These include improved waste collection systems, advances in recycling technologies, circular economy approaches, awareness campaigns, and incentives that encourage responsible plastic use and disposal. The authors conclude that effective global legislation and international agreements are essential. These tools can help limit plastic production, reduce pollution, and promote sustainable practices. Together, they offer a path toward a cleaner and healthier future for coastal and marine environments.

It is not only the oceans that are polluted by plastic waste; many rivers are also severely affected. A major cause of this pollution is municipal waste. In "Holistic Analysis of Floating Debris in Thailand's Chao Phraya Tributary: From Community Source to Sink," Hawangchu et al. examine the sources and pathways of debris entering major water bodies, while also providing insights into society's waste management practices. According to the authors, understanding the composition of the collected waste helps reveal people's behavior toward municipal waste management at both household and community levels, which can support the development of more effective policies. As seen globally, the study conducted over 20 months at the Lat Luang Canal, which flows through a densely populated community in Bangkok and discharges into the Chao Phraya River, identified plastic as the main pollutant in rivers near urban areas in Thailand. Single-use plastics were the largest contributor to this problem. The study found that the debris collected at the canal's sluice gates was primarily composed of rigid single-use containers such as bottles, boxes, and cups made from polyethylene terephthalate (PET), high-density polyethylene (HDPE), and polypropylene (PP), along with thin plastic films used in food packaging and multi-material items. The researchers argue that one of the most effective ways to address plastic pollution is by reducing plastic use at the early stages of product design, which would help decrease river waste and its associated carbon footprint. The study presented three important findings. First, canals can act as major sources and transport routes for plastic debris moving toward the ocean. Second, take-out consumer items significantly contribute to plastic pollution in urban waterways. Third, unintentional disposal of waste, such as overflow from collection points and hotspots, plays a major role in canal pollution. The study also noted that the distance between residential areas or hotspots and the canals is a key factor influencing the likelihood of waste leakage into water bodies. Furthermore, socio-economic factors were found to be strongly linked to the amount of plastic litter present in these aquatic environments.

After examining plastic pollution in marine ecosystems and rivers, the next chapter shifts focus to the plastic pollution issues in Vietnam. Plastic production and consumption in Vietnam have been increasing alongside socio-economic development. However, this growth has not been matched by improvements in management and mitigation efforts to prevent environmental contamination by plastic waste. As a result, plastic waste remains a major component of environmental pollution in the country. In the chapter titled "Current Status of Plastic Waste Pollution and Management in Vietnam: Challenges and Opportunities Toward a Circular Economy," Lap and Nguyen Lan analyze the current situation of plastic waste generation and management in Vietnam. Their analysis is based on data and information from government reports, academic research articles, public media, on-site surveys, and a review of Vietnam's legal framework and related policies. The authors aim to highlight critical issues while assessing the challenges and opportunities in transitioning to a circular economy. The study found that plastic waste accounts for approximately 12% of domestic solid waste in Vietnam. The volume of plastic waste generated has been increasing in recent years. In 2021, the Red River Delta produced the highest amount, with 1,914 tons per day, while the Central Highlands recorded the lowest, at 473 tons per day. In urban areas, the average volume of plastic waste generated per capita was about 43 kg per person per year, more than twice the amount produced in rural areas, at 20.33 kg per person per year. Single-use plastic waste accounted for between 38% and 63% of total plastic waste. Overall, coastal areas generated more plastic waste than inland regions, and urban areas produced more than rural ones. Plastic waste has polluted the soil, air, water, and sediments of Vietnam through its degradation into fragments, MPs, and nano-plastics. Although Vietnam has made progress in plastic waste management, particularly in collection, sorting, and treatment, several challenges remain. These include outdated technologies, insufficient funding, and low levels of public awareness. Vietnam has introduced legal frameworks, strategies, and policies focusing on plastic production and consumption, as well as on the collection, sorting, and disposal of plastic waste.

This is followed by the chapter "Microplastic Pollution in the Water Column and Sediment of Cochin Estuary, India: Abundance, Characteristics and Potential Ecological Risk Assessment." In this chapter, Viji and Ramasami focus on the impact of MPs in the Cochin Estuary, which is part of Vembanad Lake, a Ramsar site located in Kerala, India. The lake receives discharges from six rivers, along with input from several canals and streams. The artificial Thannermukkam bund separates Vembanad Lake into a saltwater-dominated northern part and a freshwater-dominated southern part. Water and sediment samples were collected from eight selected locations in the Cochin Estuary using a depth water sampler and a sediment grab sampler, both from Hydro-Bios. The findings of this study reveal the widespread distribution of MPs in the estuary. The average abundance was recorded as 35.34 ± 7.68 particles per liter in sub-surface water (SSW), 26.87 ± 3.36 particles per liter in bottom water (BW), and 1367.5 ± 403.75 particles per kilogram (dry weight) in sediment samples. Five morphotypes of MPs were identified: fiber, fragment, film, foam, and beads. Among these, fragments were the most dominant in both SSW and BW, while fibers were the major morphotype in the sediment samples. The shape inventory of the MPs suggests that they originate from pre-existing plastic materials, classified as secondary MPs, formed through the breakdown of larger plastic debris. More than 50% of the MPs in the SSW and BW samples were in the size range of 100 to 500 μm, while about 46% of the MPs in the sediment fell within the same range. These smaller particles are more bioavailable to aquatic organisms, increasing the risk of their entry into the food web of the estuarine ecosystem and eventual transfer to higher trophic levels. The accidental or intentional ingestion of MPs by aquatic organisms can result in various harmful effects, including internal injuries, reduced growth rates, false satiation, reproductive failure, enzyme inhibition, and eventually death. MPs can also be transferred to humans through the consumption of contaminated seafood, such as fish, clams, prawns, and crabs. The smaller the size of the MPs, the greater the likelihood of ingestion by aquatic organisms, making them more hazardous. The study area is well known for tourism, boating, fishing, and industrial activities. These human activities are considered major sources of microplastic pollution in the region. To control the problem at its source, strict regulations are needed to prevent plastic debris from entering the Cochin Estuary. In addition, proper waste management practices should be implemented, and awareness campaigns should be conducted to educate the local population about the issue.

The next chapter by Gautama et al. focuses on "Global Microplastics Contamination in Bottled and Tap Water: Prevalence and Sources." MPs, ranging in size from 1 to 5,000 μm, have been detected in various food products, such as seafood, table salt, honey, beverages, and drinking water. In 2018, the World Health Organization (WHO) initiated a review to assess the potential risks associated with plastic particles in drinking water. The variation in MP concentration and type between bottled and tap water highlights the widespread nature of plastic pollution and the need for comprehensive measures to address this issue across all water sources. MP contamination in bottled water varies globally, with the highest levels reported in China and the lowest in India. This indicates the ubiquitous presence of MPs worldwide. The most dominant size range of MPs in bottled water is less than 50 μm. Fragments and fibers are the most prevalent morphologies, with fragments being more common. These particles are typically transparent, blue, black, or green. PET is the most dominant polymer found, followed by PP, polyethylene (PE), polyamide (PA), polystyrene (PS), and polyvinyl chloride (PVC). PET contamination primarily originates from the bottle itself, due to mechanical stress, squeezing, and ultraviolet (UV) exposure, while bottle caps often cause PE contamination through repeated opening and closing. Interestingly, single-use plastic bottles generally show lower MP contamination levels compared with reusable ones. MPs have also been detected in tap water sources. The highest levels of MP contamination in tap water have been reported in China, while the lowest levels have been recorded in Iran, with particle sizes ranging from 3 to 5,000 μm. The presence of MPs in tap water is a global concern, as they are found throughout the water supply chain, from sources to water treatment plants and ultimately to household taps. In tap water, the detected MP morphologies include fibers, fragments, films, and spheres, with fibers being the most common, followed by fragments. These particles are generally transparent, black, or blue. The most dominant size range is less than 60 μm, with most particles measuring less than or equal to 500 μm. The most common polymers are PE, PP, PET, PVC, and PS. Additionally, polyacrylamide (PAM), copolymers of PE and PP, and various forms of PVC have also been identified. These findings suggest that MP generation may increase through processes within water treatment facilities and distribution pipelines. Given the widespread detection of MPs in both bottled and tap water, the chapter concludes that more rigorous investigations are necessary and that clear guidelines should be established to mitigate the impact of MP contamination on public health and the environment.

Mangrove forests around the world are suffocating from the accumulation of plastic waste and persistent MPs on the forest floor. Both abiotic and biotic degradation processes drive the generation of secondary MPs from accumulated plastic waste. Initially, plastic waste undergoes weathering and breaks down into smaller pieces, which microorganisms can then utilize as a source of carbon and energy. Mangrove forests, with their unique features, such as mudflats and pneumatophores, present significant challenges for implementing mitigation strategies against plastic and MP pollution. In the chapter "Microplastics Intercepted in Mangrove Forests: Sources, Generation and Mitigation," Jittalerk and Babel explore MP pollution in mangrove environments, with a focus on potential sources, mechanisms of generation, and feasible mitigation approaches. The authors highlight bioremediation by microorganisms as one of the most promising solutions to combat this form of pollution in mangrove forests. However, to be effective, mitigation strategies must be integrated from the point of origin to the destination, and from the stage of generation to the degradation of MPs. Waste trapping is identified as a practical approach to protect mangrove forests from floating debris carried by seawater and to support cleanup efforts. Alongside this, long-term strategies are necessary to address the deeply rooted issue of MP pollution in these ecosystems. Producer responsibility is essential at the starting point of the plastic life cycle. Incentives within a circular economy framework, along with better choices of raw materials and the implementation of strict legislation, can help steer plastic production toward more sustainable practices. Consumers also play a critical role in ensuring that plastic waste is directed toward proper disposal or recycling pathways. Raising awareness, improving environmental literacy, and fostering a sense of responsibility among consumers are all indispensable elements of the solution. Finally, plastic waste should be effectively managed both within mangrove forests and at their upstream sources. Efforts should be made to sustainably recycle this waste and reintegrate it into the production cycle, supporting a circular and environmentally responsible approach to plastic use.

MPs have been increasingly detected in drinking water, raising concerns about potential risks to human health. Accurate assessment of microplastic contamination requires robust sampling and analysis methodologies tailored to the complexities of drinking water and biological matrices. The chapter

"Comprehensive Approaches to Microplastic Sampling and Analysis in Drinking Water," by Tuan Ta and Babel, provides comprehensive recommendations for the sampling and analysis of microplastics in drinking water. Various sampling techniques, including grab and passive sampling, are presented, each with its advantages and considerations. In addition, the chapter outlines advanced analytical techniques, such as microscopy, spectroscopy, and chromatography, which are effective in detecting and characterizing microplastics across different sample types. Quality assurance and quality control measures are also discussed to ensure the reliability and reproducibility of analytical results. By integrating these methodologies, researchers and practitioners can improve their ability to assess MP contamination accurately in drinking water. This, in turn, supports informed decision-making aimed at reducing potential risks to human health.

Shooshtarian et al., in "Application of Recycled Plastic Waste in the Greenest Freeway in Australia," describe the Mordialloc Freeway Project (MFP) in Victoria. This project was designed to deliver a world-first freeway that incorporates technological advancements in sustainability. It aimed not only to construct a much-needed roadway but also to contribute to the state's waste reduction goals. The project has delivered several social, economic, and environmental benefits. Recycled materials were integrated into the construction process, including resources derived from plastic waste such as rotationally molded panels (RMP), PP, and HDPE. The chapter presents findings from an analysis of this case project, which was informed by a series of interviews with representatives of key stakeholder groups, including the client, designer, recycled material suppliers, and builders. The authors explore the key challenges and motivations behind using these recycled resources in the project.

Improper disposal of single-use plastics and heavy metals contributes significantly to environmental pollution by contaminating soil and water, posing serious risks to both ecosystems and human health. One innovative approach to addressing this problem is the development of highly porous metal-organic frameworks (MOFs) using both organic and inorganic waste materials. This strategy has the potential to eliminate waste from the environment while conserving resources and recycling hazardous materials. PET, a common plastic waste, can be depolymerized into benzene-dicarboxylic acid (BDC) and used to produce waste-derived MOFs. Metal precursors required for MOF synthesis can be obtained from various inorganic waste sources such as electroplating sludge, spent batteries, aluminum dross, industrial wastewater, bones, and eggshells. These sources provide the metal ion clusters needed to construct the MOF structure. Using waste-derived MOFs helps divert plastic and metal waste from the environment by converting it into valuable materials. In "Beyond Disposal: Unleashing the Potential of Plastic and Metal Waste in Producing Metal Organic Frameworks (MOFs) for Environmental Revitalization," Waribam et al. describe different waste sources, synthesis protocols, and types of waste-derived MOFs, along with their applications in environmental remediation. These MOFs provide a platform for creating value-added porous materials that align with the principles of the 3Rs, namely Reduce, Reuse, and Recycle, thereby supporting a cleaner and more sustainable environment. Waste-derived MOFs have demonstrated promising properties as adsorbents for removing pollutants such as dyes, antibiotics, heavy metals, and MPs, and for capturing toxic gases. Additionally, they have shown good catalytic performance and have been used as heterogeneous catalysts in the conversion of CO_2 into valuable products. This chapter highlights the potential of transforming plastic and inorganic wastes, which are typically viewed as pollutants, into functional materials with diverse environmental applications. It presents this transformation as part of a holistic circular economy approach, in which waste streams are converted into useful and sustainable materials.

The global share of medical plastics accounts for approximately 2% of total plastic production and is growing at a rate of 6.1% per annum (Rizan et al. 2020). This market is expected to expand further as both developed and developing countries continue to enhance their healthcare infrastructure. The United States is currently the largest consumer of medical plastic devices, while healthcare facilities are also expanding rapidly in developing countries, particularly in Brazil, India, and China. Although most nations, including India, support restrictions or bans on single-use plastics, such materials remain essential in sectors like healthcare. The chapter **Plastic in Healthcare Waste – Pathways for Sustainable Strategies in the Indian Scenario** by Suneja and Kaur highlights the challenges posed by plastic waste in India's healthcare system. With a population exceeding 1.44 billion, India has become the world's most populous country. As pointed out by the authors, the Indian medical plastic market is expected to grow at a compound annual growth rate of 15.2% from 2020 to 2030. Biomedical waste generation in India ranges from 0.11 to 0.87 kg per patient per day, with the country producing approximately 774 metric tonnes of biomedical waste daily. The growth of the healthcare sector, increased government support, the rise of private healthcare providers, and greater public awareness have collectively contributed to the increasing use of plastics in line with societal demands. To address these challenges, the Government of India has implemented the Plastic Waste Management Rules, which include specific provisions for the use of plastics in the medical and healthcare sectors. In 2023, the Central Pollution Control Board (CPCB) issued an update regarding plastic packaging and related products for manufacturers and importers of medical devices. The update introduced fees for managing plastic waste. It also mandated that manufacturers and importers obtain Extended Producer Responsibility (EPR) registration before importing any medical product containing plastic or plastic packaging. These entities must quantify their plastic consumption and report the amount of waste generated to the State Pollution Control Board. Furthermore, the CPCB has established targets for minimum recycling levels of plastic packaging to be achieved within the plastic waste stream. The global movement today emphasizes banning or minimizing disposable plastics. In the medical sector, comprehensive and stringent steps are necessary to manage plastic waste effectively. The 2024 United Nations pollution treaty highlighted the risks associated with healthcare plastics, particularly for women and informal sector workers. The treaty also recommended the development of fact sheets on healthcare waste to support policy and action. Promoting safe reuse and recycling, encouraging the adoption of biodegradable plastics, and improving the recovery of energy or materials from waste can make the healthcare sector more cost-effective and environmentally friendly. Measures supporting a circular economy and fostering international collaboration can significantly reduce the carbon and water footprints of medical plastics. It is essential not only to provide quality healthcare but to provide it sustainably, in alignment with the broader goals of achieving the Sustainable Development Goals (SDGs).

The next section of the book explores some of the solutions to the growing problem of MPs, which are now found in almost every aspect of the environment. Artificial Intelligence (AI) has been playing a crucial role in combating plastic pollution by offering innovative solutions across various domains, including detection, monitoring, recycling, cleanup, and public engagement. AI-powered technologies are increasingly used to detect and monitor plastic pollution in diverse environments. For example, drones equipped with AI-driven image recognition systems can survey and monitor hard-to-reach areas, capturing real-time images of plastic debris. These images from the sea surface can then be analyzed using AI algorithms to identify and map regions with high concentrations of plastic waste in oceans, rivers, and coastal zones. This detailed classification and quantification of plastic litter provides essential benchmark data that informs decision-making for citizens, policymakers, and stakeholders across both public and private sectors. AI also contributes to the cleanup of plastic waste in natural environments. Drones and robots powered by AI can detect and collect plastic debris from water bodies, making cleanup efforts more efficient and targeted. In the next chapter, "AI-Powered Solutions for Combating Plastic Pollution," Nagpal et al. examine the use of AI to address plastic pollution. This chapter explores the complex and urgent nature of the plastic crisis, with a particular focus on the environmental damage caused by increasing plastic production, especially in marine ecosystems. The authors present promising technological innovations designed to mitigate plastic pollution. The chapter highlights the development of AI- and robotics-based waste sorting systems and their potential to revolutionize waste management practices. It also explores AI applications in 3D printing, materials modeling, and sustainable design, which contribute to innovative waste reduction strategies and more sustainable material choices. Furthermore, the authors discuss emerging technologies for detecting sources of plastic waste and describe the use of AI-powered robots and vehicles to address pollution. AI algorithms are capable of identifying and tracking plastic waste in oceans, rivers, and on land using input from drones, satellites, or underwater robots equipped with cameras and sensors. The chapter also outlines the benefits of AI and technology in addressing plastic pollution and discusses future perspectives in this area. In addition to the advantages, the authors identify several challenges associated with implementing AI systems on a large scale. One key issue is that AI models require large, high-quality datasets to function effectively, and acquiring and maintaining these datasets can be difficult. Furthermore, scaling up AI solutions globally involves significant logistical and financial challenges. The effective use of AI in combating plastic pollution requires collaboration among technologists, environmental scientists, policymakers, and the public. As

AI continues to evolve, its role in addressing plastic pollution is expected to grow, offering more sophisticated tools and solutions to this critical environmental issue.

The Basel Convention regulates the transboundary movement of hazardous waste. Recent amendments to Annexes II, VIII, and IX of the Convention have been adopted to strengthen the management and control of plastic waste movement across borders, aiming to protect both human health and the environment. As a signatory, Sri Lanka is obligated to align its national legal frameworks with these amendments. In the chapter "Assessing the Impact and Awareness of Plastic Waste Amendments to the Basel Convention in Sri Lanka," Meetiyagoda et al. evaluate the current status and implications of the recent Basel Convention amendments in the country. A survey was conducted among key stakeholders from diverse sectors, including policymakers, enforcement agencies, businesses, academia, civil society, and the general public. The objective was to gain insights into implementation challenges and potential opportunities. The study found that nearly half of the respondents were unaware of the plastic waste amendments or the Prior Informed Consent (PIC) procedure. However, policymakers and enforcement authorities were generally well informed due to recent administrative updates and regulatory changes. Sri Lanka currently prohibits the import of plastic waste listed under Annex VIII (List A), while the import of waste under Annex IX (List B) is permitted only after thorough screening. Government and academic stakeholders identified several positive outcomes associated with the amendments. These include improved plastic waste management practices, a cleaner environment, reduced health risks, and economic benefits such as job creation and growth in tourism. The study also confirmed that Sri Lanka has banned the import of hazardous waste listed under Annex VIII, due to limited capacity to manage such materials. In contrast, non-hazardous plastic waste under Annex IX can be imported with formal approval from the Technical Expert Committee of the Central Environmental Authority (CEA) and the Ministry of Environment. Despite these regulatory advancements, awareness about the Basel Convention amendments remains low among government and academic entities outside the Ministry of Environment and the CEA. This highlights the need for focused awareness programs aimed at building understanding and support across all relevant institutions. The authors recommend several actions to strengthen implementation. These include conducting targeted workshops to improve awareness among academics and government officials, enhancing the monitoring of plastic waste activities, improving coordination among government authorities, and educating the public through outreach initiatives. Overall, the chapter emphasizes the importance of carefully considering the Basel Convention's plastic waste amendments to ensure that they not only protect the environment but also align with and support the economic interests of Sri Lanka.

Plastic pollution and global warming are often considered separate issues, yet they are closely interconnected in multiple ways. Both problems impact terrestrial and marine environments, ultimately affecting ecosystems and species. GHG emissions from plastic pollution contribute to climate change. The processing and use of plastics generate substantial amounts of GHGs, and reducing plastic consumption is a critical step in addressing the climate crisis.

Tackling plastic pollution and global warming effectively requires coordinated and synergistic policy interventions at local, national, and international levels. Such efforts are essential for promoting sustainable development and building resilience in vulnerable communities. While the links between global warming and plastic pollution have been studied, a holistic approach to managing both issues is necessary to achieve long-term sustainability. The final chapter by Bari, Parvez, and Rahman, titled "Plastic Pollution and Global Warming: Integrated Approach to Develop Policy-Based Solution," identifies the current challenges and barriers to addressing both problems simultaneously. It offers recommendations for an integrated approach aimed at policymakers, stakeholders, and scientists. This integrated approach includes a range of strategies, such as implementing legislation and regulations to reduce plastic consumption and promote recycling, investing in innovative technologies for plastic management, and advancing public awareness and education initiatives.

This chapter illustrates how a synergistic approach in policy design can effectively influence the success or failure of environmental strategies. When implemented together, policies aimed at reducing plastic use and those addressing climate change can interact to create synergistic effects, leading to a reduction in GHG emissions from plastics. There is no single policy model that guarantees success, as different situations require tailored responses. Therefore, policies focused solely on reducing GHG emissions cannot serve as substitutes for those targeting plastic waste reduction. Likewise, policies addressing plastic waste cannot replace the specific actions needed to mitigate GHG emissions. Only through the aggressive implementation of both types of policies—such as promoting renewable energy in plastic production and ensuring widespread plastic recycling—combined with comprehensive strategies to reduce plastic consumption, can emissions be significantly reduced to help mitigate global warming. It is also clear that demand control, improved recycling infrastructure, the adoption of bio-based plastics, and the decarbonization of energy systems are all essential components in achieving a substantial decrease in GHG emissions across the plastic life cycle. Therefore, extraordinary efforts are required from the international community to promote sustainable plastic use, minimize plastic waste, and reduce the associated emissions.

REFERENCE

Rizan, C., F. Mortimer, R. Stancliffe and M.F. Bhutta. 2021. Plastics in healthcare: Time for a re-evaluation. Journal of the Royal Society of Medicine. 113(2): 49–53. DOI: 10.1177/0141076819890554

Chapter **2**

Plastic Pollution in Coastal and Marine Environments: Sources, Impacts, and Solutions

M.G.Y.L. Mahagamage*[1] and S.G. Gamage[2]

[1]Centre for Environmental Studies and Sustainable Development
The Open University of Sri Lanka
Nugegoda, Sri Lanka-10250

[2]Centre for Environmental Studies and Sustainable Development
The Open University of Sri Lanka
Nugegoda, Sri Lanka-10250

2.1 PLASTIC POLLUTION IN THE COASTAL AND MARINE ENVIRONMENT: AN EMERGING CONTAMINANT

Over the past few decades, the global demand for plastics has increased drastically. For instance, in the last 50 years alone, an estimated 9,200 million metric tons (Mt) of plastics have been produced, and alarmingly, more than 6,900 Mt of the produced plastics have been mismanaged (Walker and Fequet 2023). As Geyer et al. (2017) discuss, approximately 75% of the produced virgin plastics end up as plastic waste in the environment. Furthermore, as a direct result of the modern throwaway society, plastic production has drastically shifted from durable products to single-use plastics. Thus, this has created the global plastic problem that we experience today (Chen et al. 2021). Generally, the sources for marine

*For Correspondence: mgyma@ou.ac.lk

DOI: 10.1201/9781003531463-2

plastic pollution can be divided into two categories: land-based sources and ocean-based sources. Urban industrialization, poor waste management, and coastal development activities are only a few examples of land-based sources.

On the other hand, the commercial fishery industry can be identified as the dominant ocean-based plastic source (Napper and Thompson 2018). Once introduced into the environment, the plastics are categorized by size into three groups: Macroplastics, Mesoplastics, and Microplastics (Figure 2.1). As documented, many marine animals, ranging from large mammals and birds to turtles, fish, and crabs, face life-threatening incidents due to such plastic particles in the environment. Around 700 species encounter various plastic debris, and of these, at least 17% of the species are recorded in the IUCN red list as near threatened or above (Napper and Thompson 2018; Gall and Thompson 2015). The persistence of mismanaged plastics in the environment is unknown. However, their long durability combined with external physical, chemical, and biological factors causes parent plastics to degrade into smaller daughter plastic particles known as microplastics (less than 5 mm in diameter), which can cause unforeseen impacts on the environment (Wayman and Niemann 2021).

Figure 2.1 Macroplastics, mesoplastics, and microplastics on the beach surfaces.

Through entanglement, suffocation, ingestion, transportation of invasive alien species (IAS), and accumulation of toxic substances, plastic waste threatens humans, wildlife, and the environment. Many countries around the world have now recognized the severity of this catastrophe and are implementing national- and regional-level management measures. Yet, in light of such concerns, plastic production in the next 20 years is expected to double, raising many uncertainties (Walker and Fequet 2023).

2.2 SOURCES OF MARINE PLASTIC POLLUTION

2.2.1 Macro and Mesoplastic Sources

Marine plastic pollutants are outcomes of various anthropogenic activities, and as discussed in the introduction, plastic pollutants can either have a terrestrial origin or an ocean-based origin. It is widely accepted that around 80% of marine plastics originate from land, and through numerous transport mechanisms, they end up on beaches, ocean bottoms, or in the form of floating and suspended particles (Napper and Thompson 2018).

2.2.1.1 Inadequate Waste Management Systems

One of the major land-based plastic pollutant sources is inadequate waste management facilities. When the waste management infrastructure or facilities are deficient or inadequate, proper disposal of the pollutants is not practiced. Such cases often lead to open dumping, unsanitary landfills, and uncontrolled incineration. In most developed countries, landfills are regularly covered with soil or synthetic material and fenced to prevent accidental emissions. However, in developing countries, such steps are not always practiced. Thus, open dumping leads to plastic ending up in aquatic systems through water or air (Siddiqua et al. 2022; Napper and Thompson 2020). Indonesia is suspected to be the second largest plastic polluter in the world after China, and its solid waste contains 14% plastic. Even though 70% of the total solid waste management facilities in Indonesia are designed as sanitary landfills, around 99% of them currently function as open dumping sites. Such cases have influenced plastic pollution in the country (Lestari and Trihadiningrum 2019). Out of all the solid waste collected in Bangladesh, 8% represents plastic waste. Since up to 80–90% of the allocated budget for plastic management is spent only to collect plastics from households, open dumps, and open burning have become the primary plastic disposal methods (Mourshed et al. 2017). Through such cases, most macro- and meso plastics end up in internal resources and rivers. The estimated plastic waste generation in Sri Lanka was recorded as 938 MT/day, and around 638 MT/day of plastics were mismanaged (Ministry of Environment 2021). Although several management techniques have been initiated to control plastic waste generation and reduce the amount of plastics in the environment, around 260 MT were managed through open dumping in selected areas (Figure 2.2). Bandara et al. (2023) further revealed that most public communities practiced burning as a plastic waste disposal method since they had no proper collection from local authorities.

Figure 2.2 Mismanagement of plastic waste in Sri Lanka.

Commercial production of plastics started around the 1930s (Jambeck et al. 2015), and it has increased from 1.5 million metric tons in 1950 to 368 million metric tons in 2019 (Tiseo 2020). While billions of metric tons of plastic were produced during this period, around 76% of the total production ended up as waste

(Ali et al. 2021). Total plastic waste contribution to overall waste composition in the United States, the United Kingdom, China, and the European Union was 13%, 11%, 11%, and 8% respectively (Idumah and Nwuzor 2019). While production is skyrocketing, the plastic waste disposal process is lagging. Geyer et al. (2017) reported that only 19.5% of globally disposed plastics were recycled in 2015, while 25.5% were incinerated and 55.5% were discarded into the environment. When introduced into the environment, large plastic litter undergoes processes such as photo-degradation, oxidation, or weathering (physical, chemical, or biological), creating secondary plastics (Browne 2015; Boucher and Friot 2017). The packaging industry is the main contributor to plastic waste worldwide. In 2015, 141 million tons of plastic waste were generated from plastic packaging (Geyer et al. 2017). In addition, plastic bottles are a major source of microplastics (MPs). While MPs can develop inside the bottles due to various processes, this facilitates their ingestion, and when discarded, they undergo various erosion processes that create secondary plastics. Similarly, plastic bags, straws, and plastic pallets are among other plastic litter types that provide macro and mesoplastics (Antunes et al. 2018; Oßmann et al. 2018; Schymanski et al. 2018).

On the other hand, uncontrolled burning of solid wastes is generally practiced in many developing countries. In any burning process, particle matter (PM) is generated, and most of the time, the generated PM is smaller in size than either $PM_{2.5}$ or PM_{10}. Through air circulation, these plastic particles will ultimately end up in rivers or oceans. As scientific literature indicates, 90% of ocean plastic is generated from only 10 rivers. Of these, eight rivers are in Asia, while the other two are in the African continent. The Great Pacific garbage patch, which contains 79,000 tons of plastic debris, mostly macro- and mesoplastics, is one of the grim reminders of the severity of the problem (Rajmohan et al. 2019).

2.2.1.2 Urbanization and Development Activities

Urbanization and various development activities along the coastline and inland produce large amounts of plastic debris. In coastal areas, plastic disposal through tourism hot spots contributes rapidly to beach plastics and floating/submerged plastics in the ocean. On Santa Marta beaches in the Colombian Caribbean, plastic accounted for 30–77% of the total macro litter during the high tourism season (Garcés-Ordóñez et al. 2020). In Australia, marine debris generation through tourism and its effect on the Southern Great Barrier Reef and the surrounding region were studied. Heron Island and Northwest Island are two tourism hotspots on the east coast of Australia, and their tourism activities were directly affected by two isolated and restricted islands: Wreck Island and Tryon Island. On the latter two islands, plastic accounted for 68–98% of the total marine debris and was found on both windward and leeward beaches of the Islands. On Wreck Island, plastic caps, bottles, drinking straws, lollypop sticks, balloon fragments, and cigarette lighters were found. On the Northwest Island, tourism activities are more focused on the leeward side. Correlating to such activities, most plastic debris, such as sheets (e.g., wrappers), and fibrous plastics (e.g., cigarette butts) were found on the leeward side of the island rather than the windward side. (Wilson and Verlis 2017).

2.2.1.3 Fisheries Activities

Plastic generated through fisheries activities represents a large portion of the global marine debris and could be identified as a major ocean-based plastic source. Due to the low cost and durability, most of the fishing gear is now made out of plastic. However, due to various stresses, the plastic material could crack and fragment, creating secondary plastics (Figure 2.3). In a similar manner, abandoned, lost, and discarded fishing gear highly contributes to plastic pollution in many marine environments (Napper and Thompson 2020). As evidence suggests, 17% of the total litter in the northeast Atlantic region was generated through the fisheries and aquaculture activities in the area, while an estimated 18–20% of the global marine litter is generated through marine activities. Fisheries activities such as Bottom trawling and demersal seining are known contributors of plastic waste in the marine environment. According to statistics, at least 25 MT of dolly ropes (threads), which are used in bottom trawling, end up in the North Sea annually. Furthermore, around 40 MT of dolly rope particles are generated by Norwegian trawlers per year. At the same time, Danish, Scottish, and pair seine fishing gears have been recognized as the most plastic-polluting methods in the sea (Syversen et al. 2022).

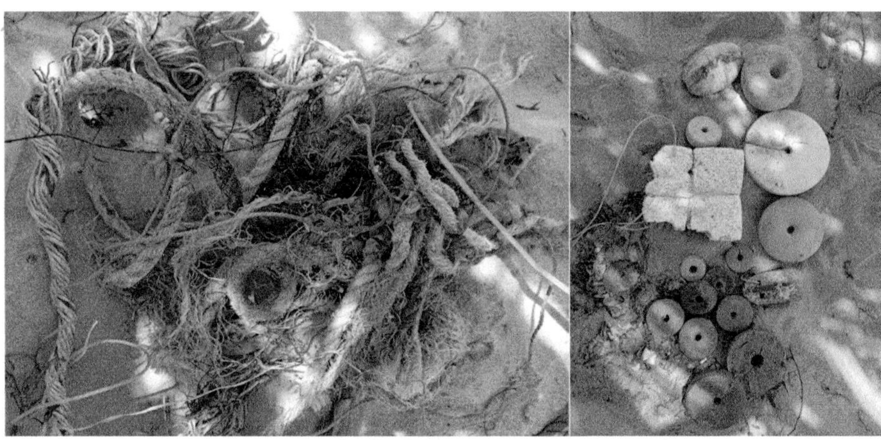

Figure 2.3 Plastic fishing gear parts on beaches.

In the present day, plastics are widely used in aquaculture as construction materials (e.g., ropes), buoys, cages, packaging material, and more. In approximately 25% of all aquaculture practices, mollusk aquaculture (e.g., oysters, mussels, and clams) is prominent (Moore 2013). In French Polynesia, as an outcome of pearl farming, around 1,603 tons of plastic waste were estimated to be emitted annually. Moreover, 75% of the generated plastic waste was found to be mismanaged, raising significant environmental concerns (Crusot et al. 2023).

2.2.2 Microplastic Sources

Since MP pollution causes devastating effects on the environment, it is crucial to understand the different sources of MPs to control and manage this ever-increasing global disaster successfully (Lv et al. 2019). Universally, MPs are classified into two categories: primary and secondary MPs. Primary MPs are intentionally manufactured plastics for various domestic and industrial applications, while secondary MPs result from the fragmentation of larger plastic waste into smaller particles (Auta et al. 2017). Even though these two categories provide a basis for terrestrial sources of MPs, Browne (2015) argues that the vague nature of these two terms may confuse the general public, scientists, industries, and governments, as they do not identify the specific sources of MPs. Therefore, the following specific origins are examined as MP emission sources into the natural environment.

2.2.2.1 Personal Care Products and Cosmetics

Microbeads (MBs) are a type of primary MPs that can be found in many personal care products and cosmetics such as toothpaste, face wash gels, skin creams, facial scrubs, etc (Figure 2.4) (Auta et al. 2017; Miraj et al. 2019; Gamage and Mahagamage 2024a). The occurrence of MBs/MPs in different types of personal care and cosmetic products worldwide is represented in Table 2.1. MBs are intentionally added to products to act as a bulking agent (Conkle et al. 2018), abrasive scrubbers (Rochman et al. 2015), or to replace natural exfoliating materials (e.g., oatmeal, walnut husk) (Xanthos and Walker 2017; Rochman et al. 2015).

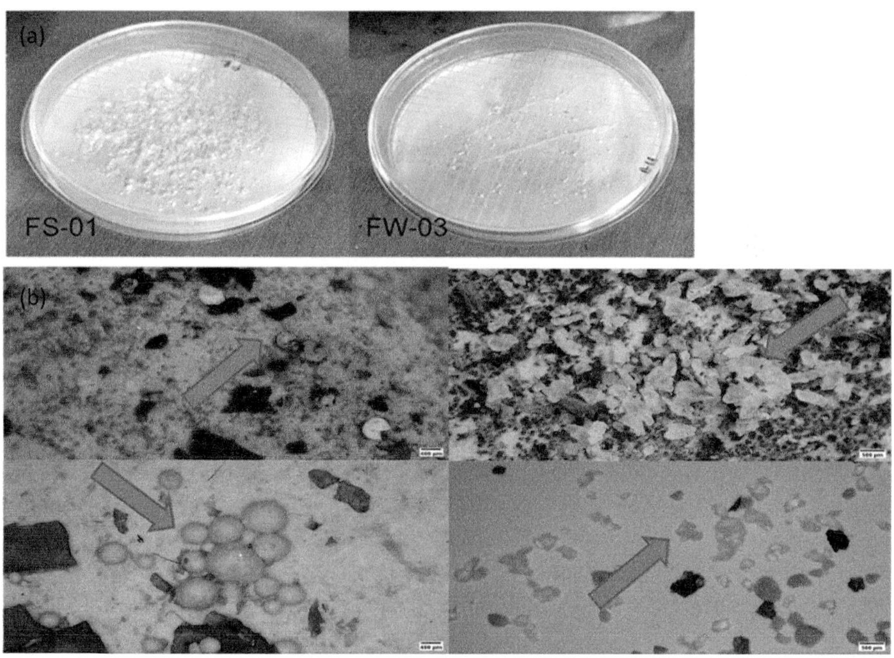

Figure 2.4 MBs in cosmetic products. (a) Extracted MBs from cosmetic products. (b) Stereo microscopic images of MBs.

Table 2.1 Occurrence of MBs/MPs in various personal care and cosmetic products worldwide

Product Type	Type of Plastic	Country	Reference
Toothpaste	Polyethylene (PE)	Turkey	Ustabasi and Baysal 2019
	Low-density polyethylene (LDPE)	Malaysia	Praveena et al. 2018
Facial cleansers	PE	New Zealand	Fendall and Sewell 2009
	PE	United States of America	Chang 2015
	PE	China	Lei et al. 2017
	LDPE, Cellophane, Polypropylene	Malaysia	Praveena et al. 2018
Facial scrubs	PE. Polyvinyl Chloride (PVC)	China	Cheung and Fok 2017
	PE	United Arab Emirates (UAE)	Habib et al. 2020
	PE	United Kingdom	Napper et al. 2015
	PE	Spain	Godoy et al. 2019
	PE	Sri Lanka	Gamage and Mahagamage 2024a
Body scrubs	PE	UAE	Habib et al. 2020
	PE	Spain	Godoy et al. 2019
Shower gel	PE	China	Lei et al. 2017
	PE	Spain	Godoy et al. 2019
Foot scrubs	PE	Spain	Godoy et al. 2019

Since MBs in these products are intended to wash off, they end up in domestic sewage, wastewater, or are directly emitted into water bodies (Gamage and Mahagamage 2024b; Cheung and Fok 2017). Napper et al. (2015) estimated that between 4,594 and 94,500 MBs could be released from each use of MB-containing products. Unfortunately, MBs may escape from the wastewater treatment process due to incomplete removal, causing the final effluent released from the wastewater treatment plants (WWTP) (Mason et al. 2016; Kalčíková et al. 2017; Cheung and Fok 2017). Rochment et al. (2015) calculated that around 95–99.9% of MBs would settle into sludge and leave the rest in the final effluent during the treatment process. Since treated effluent is released directly into aquatic habitats most of the time, it has been widely recognized as a source of MP pollution (Mason et al. 2016). In many countries, sludge generated during the water treatment process is used as a landfill material or applied to agricultural lands as fertilizers (Bratina et al. 2016). When applied, many MBs accumulate in the soil, but surface runoff, stormwater, or floodwater can carry them into natural water bodies such as lakes, rivers, and oceans (Rochman et al. 2015).

Considering the emissions of MBs worldwide, the United States of America alone discharges 8 trillion MBs into water bodies per day (Rochman et al. 2015). Similarly, China emits 209.7 trillion microbeads per year, consisting of 80% from the final effluent of wastewater treatment plants and 20% from direct emissions (Cheung and Fok 2017). In Ljubljana, Slovenia, the average of 15.2 mg of MBs is released per person, making 112.5 million MBs emissions per day (Kalčíková et al. 2017). Praveena et al. (2018) estimated that a total of 0.199 trillion MPs from cosmetics and personal care products were emitted annually from Malaysia. According to a survey conducted in 2012, all European Union countries, including Norway and Switzerland, used 4,360 tons of MBs in 2012. Further, it was estimated that around 0.1–1.5% of that total has been emitted into the North Sea through the usage of cosmetic products (Gouin et al. 2015).

2.2.2.2 Textile Industry

With the mass usage of synthetic fibers such as Polyester and Polyamide, the textile industry worldwide could be identified as a major source of MP pollution (Ngo et al. 2019). In 2019, 74 million tons of synthetic fibers were produced worldwide, which is a staggering growth from 14 million tons in 1980 (The Fiber Year 2020). According to Henry et al. (2019), approximately two-thirds of all textiles in the world are considered made with synthetic fibers (Figure 2.5). Fragmentation of synthetic fibers creates microfibers (MFs)/MPs, and it has been estimated that 20–35% of all primary plastics found in oceans belong to plastic microfibers (Liu et al. 2019; Laitala et al. 2018). Geyer and colleagues have estimated that the textile industry generated 42 million tons of plastic waste in 2015 (Geyer et al. 2017).

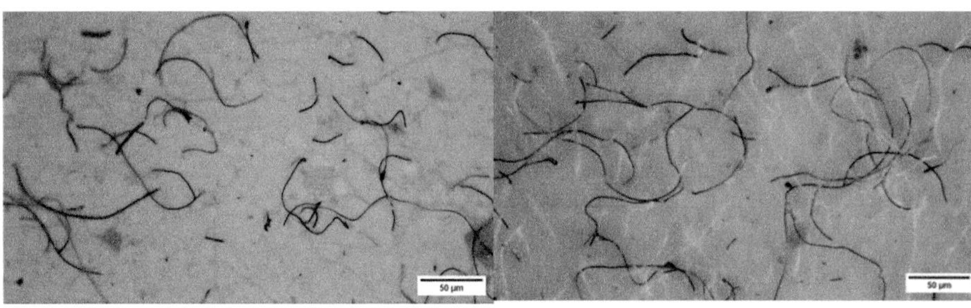

Figure 2.5 Microfibers (MFs) in the textile industry.

There are three potential microfiber emission points during the life cycle of textiles: production, usage, and disposal (Liu et al. 2019; Henry et al. 2019). Microfibers are emitted either during the washing process or as airborne particles during production and usage, and when disposed of, textiles degrade, creating more microfibers (Henry et al. 2019). If wastewater reached the WWTP, most of the microfibers would settle in the sludge during water treatment. But a significant amount will still be released with final effluent (Peller et al. 2019; Yang et al. 2019a).

In a study conducted by Carney Almroth et al. (2018), the shedding of microfibers during the washing process was measured for acrylic, nylon, and polyester polymers. While all tested synthetic fibers were shedding microfibers, textiles with loose textile constructions and worn material caused more shedding. In the same year, De Falco et al. (2018) compared microfiber emission from woven polyester, knitted polyester, and woven polypropylene. Mahagamage et al. (2024) studied the generation of MFs during washing after a typical 6.5 kg wash load; around 2 million microfibers were found in the effluent water. Further, over 700,000 microfibers were found to be released from an average 6 Kg wash load that included polyester, polyester-cotton blend, and acrylic fabrics (Napper and Thompson 2016). A comparison of polyester, polyamide, and acetate fabrics showed the highest discharge of fibers from acetate fabrics at a rate of 74,816 ± 10,656 microfibers/m^2 per wash (Yang et al. 2019b).

In line with many studies all fabrics shed microfibers. Wastewater analysis of Keqioa Industrial Park, one of the largest textile industrial parks in China, demonstrated releasing 54,100 microfibers per liter. Even though this wastewater undergoes treatment, it was estimated that 537.5 microfibers per liter were still present in the final effluent of WWTP, releasing 430 billion microfibers per day (Zhou et al. 2020). Microfiber emission during the washing process depends on factors such as temperature (Yang et al. 2019b), usage of detergent (Yang et al. 2019b), age of the garment (Carney Almroth et al. 2018), type of washing machine (Hartline et al. 2016), etc. In Sri Lanka, Mahagamage et al. (2024) found that 10,028.7 ± 1,420.8 microfibers per liter were emitted by doing 12 kg of laundry during a cold water cycle. However, it was reduced to 191.5 ± 109.4 by applying a filtration technique.

Throughout the global shorelines, microfiber contamination has been widely reported. Browne et al. (2011) discuss that in 18 shorelines (Australia (Port Douglas, Busselton Beach), Japan (Kyushu), Oman, United Arab Emirates (Dubai), Chile (Vina Del Mar, Punta Arenas), Philippines (Malapascua Island), Portugal (Faro), Azores (Ponta Delgado), USA (Virginia, California), South Africa (Western Cape), Mozambique(Pemba), and the United Kingdom (Sennon Cove)) representing six continents shows high microplatic pollution. One of the major methods for these pollution sites was found to be the sewage contaminated with microfibers from washing clothes. An estimated 35% of all the microplastics in the global oceans originate from synthetic textiles. Simultaneously, it has been calculated that 0.19 million tons of textile microplastics enter the global seas per year (Boucher and Friot 2017).

2.2.2.3 Agriculture

Plastics have many applications in the agriculture sector, such as greenhouse/tunnel building, mulching, irrigation system development, shading, and bird nets (Scarascia et al. 2012), among others. Plastic usage in agriculture has significantly improved productivity. For example, plastic film mulching in China has helped to increase grain and cash crop yields to 20–35% and 20–60% respectively (Liu et al. 2014). In 2017, 6.96 million tons of plastic were used for agricultural practices worldwide (Jansen et al. 2019). With such huge usage, the absence of capable recycling facilities directly influences the generation of MPs in agricultural lands (Qi et al. 2020). Further, tillage, UV radiation, and biodegradation in agricultural lands break macro plastics into MPs, making agricultural soil a major sink for MPs (Igalavithana et al. 2022). While macroplastic usage highly contributes to MPs in agricultural soil, accumulation of MPs in large amounts was demonstrated after repeated usage of MPs containing fertilizers (Kumar et al. 2020), sewage sludge (Corradini et al. 2019), plastic-coated fertilizers and pesticides (Qi et al. 2020). Soil-accumulated MPs could be transported into aquatic habitats through surface runoff (Piehl et al. 2018) or wind erosion (Razaei et al. 2019). Apart from that, MPs could also transport through soil animals, livestock, and crop offtake (Qi et al. 2020).

Harms et al. (2021) found MPs in 15 farmlands across northern Germany, with the majority being black polyethylene particles, followed by polypropylene particles. The mean MP abundance of each site was found to be 3.7 ± 11.9 particles per kg dry weight. Agricultural farmlands in southeast Germany had 0.34 ± 0.36 particles of MPs per kg dry weight of soil with films and fragments of polyethylene, polystyrene, and polypropylene (Piehl et al. 2018). Interestingly, plastic-related agricultural practices or MPs containing fertilizers were not used in this study site. Thus, it highlights the MP's contamination of conventional agricultural lands. The concentration of MPs in agricultural soil in western China was significantly greater than in eastern China (Huang et al. 2020). An average MP concentration of 83.6 Kg per hectare was found in the considered sampling sites, and Huang et al (2020) found a significant linear relationship (R^2 = 0.61) between the consumption of mulching films and MPs in agricultural lands.

2.2.2.4 Tire and Road Wear

Two types of rubber are used when manufacturing tires: natural and synthetic (Kole et al. 2017). In a light vehicle tire, 19% natural and 24% synthetic rubbers, such as Butadiene rubber, Styrene-Butadiene rubber, and Halogenated Polyisobutylene rubber, are present (U.S Tire Manufacturers Association 2019). The application of synthetic rubbers ensures wear resistance, impermeability, and other performance improvements, but it causes MPs to emit when driving due to tire and road friction (Andersson-Sköld et al. 2020). Global data suggests tire wear emissions are nearly 6 million tons annually, with a global average of 0.81 Kg per year per capita (Kole et al. 2017). In their review, Kole et al. have discussed the wear and tear emissions of tires in 13 countries. The United States emits 4.7 kg per capita per year, which is the highest emission of the considered countries, while India has the lowest: 0.23 kg per capita per year. IVL Swedish Environmental Research Organization reported approximately 13,000 tons of MPs were released in Sweden because of road wear and tire aberrations (Magnusson et al. 2016). Released MPs may then be transported into soil or water bodies through rain or wind (Boucher and Friot 2017). In, stormwater floating treatment wetland on Queensland's Gold Coast between 15–38% of total MPs were identified as derived from car tires (Ziajahromi et al. 2020) while airborne particle analysis of frequented industrial, agricultural, and urban roads in Germany has revealed 90% of volume dominated by composites of tire, road, and brake-abrasion materials (Sommer et al. 2018).

Thermoplastic and polymer-based paints are used worldwide for road markings. Weathering and aberrations due to tire friction may remove particles of these paints as MPs. In European countries, thermoplastics are most dominantly used for road markings (Boucher and Friot 2017). According to Sundt et al. (2014), 12,476 tons of thermoplastic were used for road marking in Norway in 2014. When road dust was analyzed in Kumamoto, Okinawa, and Tokyo, Japan, poly (diallyl phthalate), polyvinyl chloride, polymethyl methacrylate, and polyester were present in the dust. In addition, road marking was found to be a major source for MPs in study sites (Kitahara and Nakata 2020). On 4 sample sites in the Thames River basin, sediment analysis revealed an average of 66 MP particles per 100 g of sediment. Interestingly, most of the MPs of the four sites were considered to originate from thermoplastic road surface marking paints (Horton et al. 2017). Plastics are also used in road construction materials. For instance, Styrene-Butadiene and Styrene Ethylene Butylene Styrene Copolymer are added to asphalt concrete to increase its viscosity (Sundt et al. 2014). Thus, degradation of road construction material is also a potential MP emission source into the environment (Norén and Naustvoll 2010). In all these scenarios, the residue microplastics will ultimately end up in terrestrial and aquatic bodies, shorelines, and the oceans. The following illustration (Figure 2.6) summarizes the above land-based sources and their ultimate fates.

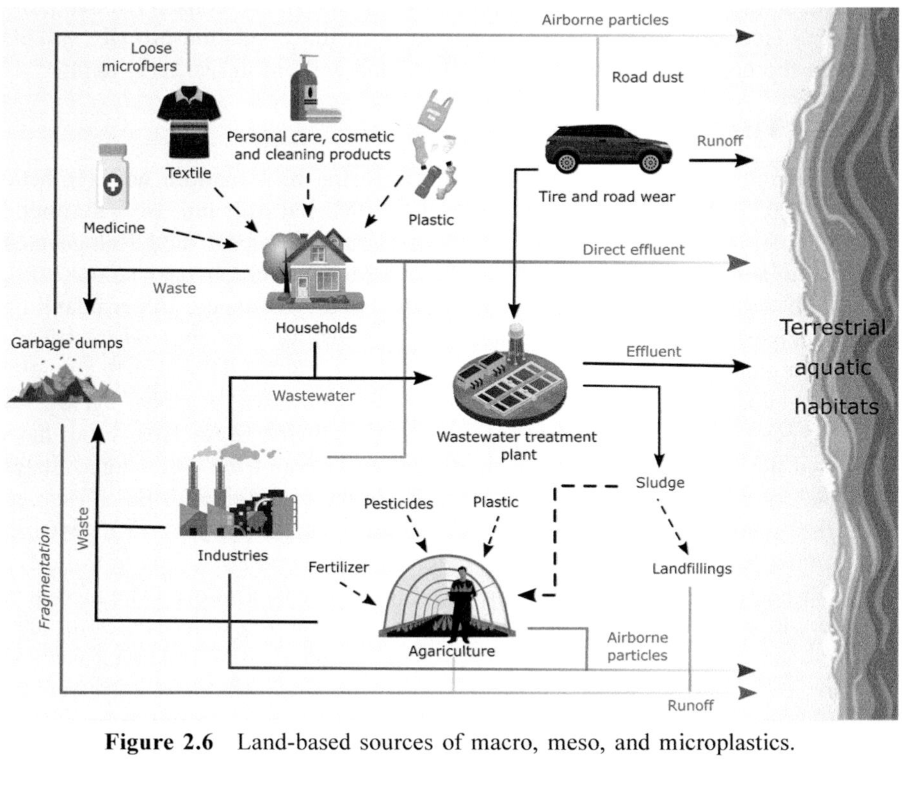

Figure 2.6 Land-based sources of macro, meso, and microplastics.

2.3 IMPACT OF PLASTIC POLLUTION ON THE MARINE ENVIRONMENT

Marine plastic pollution affects every living thing in the world. From physical and behavioral effects to ecological impacts, the presence of plastics in the natural environment has created a significant environmental problem. In the earlier section, we discussed the different macro, meso, and microplastic sources for the marine environment. Once aggregated, plastic pollutants create numerous consequences, and this section discusses several harmful impacts of plastics in detail.

2.3.1 Entanglement

The entanglement of marine animals occurs when certain plastic debris (e.g., fishing gear and marine debris), such as ropes, nets, and buoys, wrap around the animal, which could lead to the animal's death or certain physical injuries. According to Sul et al. (2011), in the Antarctic environment, the most identified impact of plastics during 1982–2010 is the entanglement of marine mammals and birds. Moreover, it has been reviewed that a considerable number of fur seals (*Arctocephalus gazella*) were entangled mainly with nylon ropes, packaging bands, rubber rings, and other types of fishing gear. The entanglement of animals is typically considered a type of bycatch. It has been identified that the bycatch of seabirds in longline fishing for Patagonian toothfish is around 6,000 in South Georgia, while it is 13,000–14,000 in Crozet and Kerguelen Islands each year (Bestley et al. 2020). Furthermore, large plastic-based items can negatively impact the habitats through ghost fishing on marine wildlife (Stelfox et al. 2016). Macroplastics can develop anoxic conditions within the sediment by affecting the survival of the benthic community as well. Entanglement is the visible problem that leads to other associated impacts, such as macroplastic ingestion (Mordecai et al. 2011). While plastics could strangle any animal species, the most affected species around the world are birds, turtles, mammals, fish, and crabs.

Tiger sharks (*Galeocerdo cuvier*), a top predator in marine food webs, were found to be affected by the circular plastic debris in the Western South Atlantic Ocean. The aftermath of the entanglement showed severe trauma and anatomical deformities in these species. Circular plastic straps in this environment proved to be causing severe trauma in 3% of all entangled animals. Similarly, the European hedgehog (*Erinaceus europaeus*) in Great Britain has an estimated mortality of 4,000–7,000 due to plastic entanglement. Plastic accumulation zones in the world's oceans are considered to be hotspots for entanglement. Moreover, coastlines such as the Atlantic coast off the Bahamas and several locations in Southeast Asia, and marginal seas (e.g., Mediterranean) are considered as high plastic impact zones (Thrift et al. 2023; Afonso and Fidelis 2023)

2.3.2 Ingestion

While there are many records of plastic ingestion by birds, mammals, and turtles, with the emerging advances in the field of microplastic analysis, more attention is now given to studying plastic ingestion by fish and invertebrates (Thompson et al. 2004). As studies point out, ingestion of plastics by organisms could happen due to intentional ingestion or accidental ingestion. Intentional ingestion occurs when plastic becomes a target, typically during foraging, whereas accidental digestion is linked to feeding mechanisms, such as Baleen whales filtering large volumes of water that contain both microplastics and plankton (Lusher 2015). Scavenging could be identified as an important pathway for plastic digestion in marine animals. Green turtles (*Chelonia mydas*) were found to scavenge on dead squids; due to this scavenging behavior, such animals had a high plastic ingestion rate. While scavenging is an adaptive behavior in the marine ecosystem, this could be positively correlated with plastic ingestion by marine animals (Andrades et al. 2019). Savoca, McInturf, and Hazen (2021) documented plastic ingestion of 177,774 individual marine fish from 555 species. An alarming 386 species have ingested plastic debris, and 210 species from this lot are considered commercially important. In all research that considered microplastics, fish consumed plastic at a rate of 26%. However, this incidence has doubled in the last ten years, rising by 2.4 ± 0.4% annually. According to Thayer's law, the color of the plastic debris plays an important role in the ingestion. Thus, dark-colored plastic debris is dominantly consumed by animals that perceive food from below, and light-colored particles are consumed by animals that perceive food from above (Santos et al. 2016).

2.3.3 Transportation of Invasive Species

Invasive alien species (IAS) are any species that occupy habitats outside their natural areas and cause damage to the native species and ecosystems. By this definition alone, we could comprehend the IAS pose on the environment and the biodiversity. It has been widely reported that the IAS species uses plastics as vectors to distribute throughout the world. As plastics, pollution is more abundant in the environment, durable and slow-moving, it provides optimal conditions to spread IAS from one place to another. In real-life scenarios like *Membranipora tuberculata*, which was spread from Australia to New Zealand through plastic substrates, *Thalamoporella evelinae* was spread from Brazil to Florida shores via attaching to plastic substrates (Gregory 2009). Similarly, the transportation of non-native and pathogenic organisms to new habitats on floating plastic litter has been identified as an emerging impact of microplastic pollution. As a consequence of such incidents, there is a propensity to introduce IAS into the food chains (Barnes 2002).

Plastics have increased the possibility of marine organisms dispersing through rafting material. Organisms that range from algae to reptiles have been observed to raft on floating objects. Among them, the most common species include barnacles, Polychaeta worms, bryozoans, hydroids, and mollusks. On small plastic fragments collected by plankton nets, the presence of diatoms, hydroids, and bacteria was first reported in the early 1970s. In a way, anthropogenic plastic debris has given marine organisms, including microbes and invertebrates, a new space in the pelagic zone. (Barnes and Milner 2005; Barnes 2002)

2.3.4 Accumulation of Toxic Compounds

The adsorption of several organic and inorganic pollutants on microplastic particles has been identified as another considerable effect of plastics on the environment. According to the findings of Ogata et al. (2009), the concentrations of environmental chemicals on microplastics are frequently greater than those found in seawater, owing to the higher specific surface area and the affinity of organic pollutants for the hydrophobic characteristics of polymers. Furthermore, the potential for plastics to absorb chemical contaminants from their surrounding has been extensively studied in laboratory conditions. Based on such experiments, it has been found that certain polymers, including polyvinyl chloride (PVC), polyethylene (PE), polypropylene (PP), and polystyrene (PS/EPS), have a high sorption capacity for DDTs, polycyclic aromatic hydrocarbons (PAHs), hexachlorocyclohexanes ($C_6H_6Cl_6$), and chlorinated benzenes (Bakir et al. 2012). Besides that, metals also show a greater adsorption capacity in plastics. As found by Holmes et al. (2012) in beach pellets, metals have shown an adsorption ability with measured quantities of up to 300 mg/g for Al, Fe, Cu, Pb, and Zn and up to 80 ng/g for Cd, Cr, Co, and Ni.

2.3.5 Physical, Behavioral, and Ecological Changes

Ecological impacts of microplastics concerning marine ecosystems have been well investigated in invertebrates. In sessile organisms such as mussels, microplastics disrupt the filtering mechanism (Thompson et al. 2004; von Moos et al. 2012). Moreover, there is a tendency for monomers and plastic additives to cause carcinogenesis and endocrine disruption. Likewise, it was found that upon exposure to microplastics, reduced feeding rate and fecundity were observed in some copepod species, which could negatively affect copepod populations in the long term. Furthermore, exposure to microplastics can cause behavioral changes that could lead to sub-optimal conditions for individual organisms and wider effects on a system. As observed in beach hoppers (*Platorchestia smithi*), short-term ingestion of microplastics has shown an impact on their survival and behavior (Tosetto et al. 2017). Consequently, there is a considerable impact of microplastics on the organisms' behavior and physical mechanisms.

2.4 SOLUTIONS

As discussed in the earlier sections, the consequences and the scale of plastic pollution are dire and far-reaching. However, with the correct approaches and management methods, there is still hope for a sustainable future. Before providing solutions, it is crucial to grasp the scope and the complexity of plastic pollution. Since many methods exist for plastics to accumulate in the environment, understanding and mapping each source is essential in developing management practices. One of the alarming obstacles to managing plastic pollution in the marine environment is the large quantity of plastic entering the marine ecosystems. Such quantities exhaust the existing cleanup efforts and the infrastructure.

One approach to minimize the plastic quantity is the implementation of circular economy approaches. One of the inherent flaws in the linear economy is the disposability of the products. As discussed in the earlier stages of this chapter, the usage of unsustainable practices leads to the accumulation of vast amounts of plastic in the environment. However, in a circular economy scenario, resources are kept in use for as long as possible through reducing, recycling, and reusing (3R). Thus, the circular economy is considered the spearhead of sustainability, and the 3Rs govern this process (Manickam and Duraisamy 2018).

Therefore, priority should be given to the 3Rs of plastic products before implementing other management practices. For example, Wichai-utcha and Chavalparit (2019), discussed that with the implementation of the "Clean technology concept," "3Rs," and "Waste exchange" concepts in Thailand, industrial waste generation has decreased, and 10.88 million tons of solid waste have been recycled, including plastics. Furthermore, some large factories adopted "Zero waste to landfills" concepts to further reduce the quantity of solid waste that ends up in the environment. To strengthen the 3R concept, the following measurements could be applied (Singh et al. 2017).

(i) **Production Stage:** During the plastic production stage, a closed-loop process could be adopted to minimize waste generation and promote recycling efforts. For example, plastic products from a single type of plastic could be used as the raw materials for generating plastic resins.

(ii) **Consumption Stage:** At the post-consumption stage, segregation of plastic waste according to the type of plastic encourages the plastic recycling process. Accomplishing the segregation process at the source rather than the waste management facilities is important since it is easier to segregate the plastics at the source. To facilitate this, regulated labeling practices should be introduced to identify the plastic product easily.

(iii) **End of Life Stage:** Once plastic products reach their end-of-life stage or are contaminated beyond recovery, such products could be used to generate energy. As the literature indicates, plastic products have a high calorific value of about 37 MJ/kg. So they could be incinerated to produce electricity or heat energy. One other end-of-life application of plastic is the extraction of crude oil through the pyrolysis process.

Central to the circular economy approach is designing out waste and pollution. The major objective of this concept is to eliminate waste generation or at least minimize it (Keys et al. 2000). This entails rethinking the way plastic is produced by promoting biodegradable materials and eco-friendly alternatives. Simultaneously, enforcing extended producer responsibility (EPR) will make the manufacturers responsible for the entire lifecycle of their products. Thus promoting innovative solutions for plastics (Leal Filho et al. 2019). EPR has been used as one of the major instruments in managing plastic waste in the European Union. As discussed by Andreasi Bassi et al. (2020), an investigation of the role of EPR in plastic waste management and its efficiency in Italy revealed that packaging waste management is an environmental burden. However, environmental benefits could be obtained if plastic waste collection rates were maximized and by minimizing the impurities through the segregation of plastics.

As illustrated by Leal Filho et al. (2019), the recyclability of plastic varies. Polyethylene Terephthalate (PET), High-density polyethylene (HDPE), Low-density polyethylene (LDPE), and Polypropylene (PP) are known to be easy to recycle, while Polyvinyl chloride (PVC), Polystyrene (PS/EPS), and miscellaneous plastics are difficult to recycle. However, due to the collection level and the technological advances of each nation, the recyclability of these could be a problem. In light of such circumstances, biodegradable plastics are often considered an alternative to conventional plastics, and they already play an important role in many industries. Currently, the production of bioplastics accounts for approximately 1% of the 300 million tons of conventional plastics. While biodegradability is a great feature, it also requires a specific environment and a period for biodegradation. Although it may not be a long period compared to traditional plastics, in the meantime, the same issues, such as entanglement, ingestion, and aesthetic problems of conventional plastics, could arise (Shen et al. 2020; Rujnić-Sokele and Pilipović 2017).

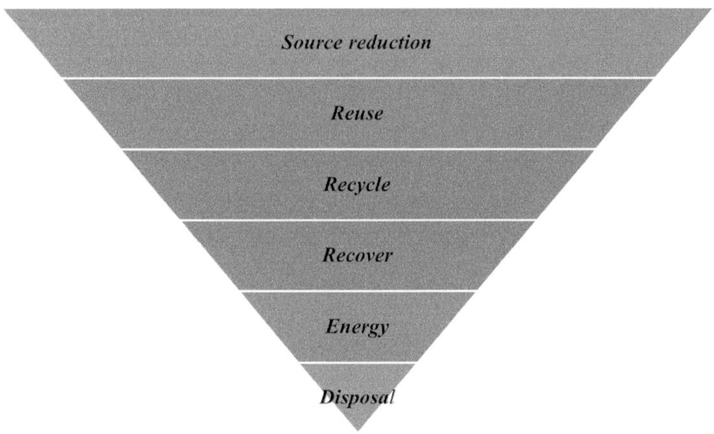

Figure 2.7 Plastic waste management hierarchy.

As illustrated by Figure 2.7, source reduction should be considered the primary plastic management step to reduce and manage plastics in the coastal environment. Apart from the 3Rs, the incineration process for energy generation and disposal is the next three favored management methods. To manage the existing plastic pollutants in the environment, awareness campaigns, technological advances in garbage collection, transportation, and processing are much needed, and by educating the general public and actively participating. Furthermore, new and improved local, regional, and international plastic waste management policies and regulations should be enforced.

REFERENCES

Afonso, A.S. and L. Fidelis. 2023. The fate of plastic-wearing sharks: Entanglement of an iconic top predator in marine debris. Mar. Pollut. Bull. 194. 115326. https://doi.org/10.1016/j.marpolbul.2023.115326

Ali, S.S., T. Elsamahy, E. Koutra, M. Kornaros, M. El-Sheekh, E.A. Abdelkarim, et al. 2021. Degradation of conventional plastic wastes in the environment: A review on current status of knowledge and future perspectives of disposal. Sci. Total Environ. https://doi.org/10.1016/j.scitotenv.2020.144719

Andersson-Sköld, Y., M. Johanesson, M. Gustafsson, I. Järlskog, D. Lithner, M. Polukarova, et al. 2020. Microplastics from tyre and road wear—A literature review. Retrieved from www.vti.se

Andrades, R., R.A. Dos Santos, A.S. Martins, D. Teles and R.G. Santos. 2019. Scavenging as a pathway for plastic ingestion by marine animals. Environ. Pollut. 248: 159–165. https://doi.org/10.1016/j.envpol.2019.02.010

Andreasi Bassi, S., A. Boldrin, G. Faraca and T.F. Astrup. 2020. Extended producer responsibility: How to unlock the environmental and economic potential of plastic packaging waste? Res. Conserv. Recycl. 162: 105030. https://doi.org/10.1016/j.resconrec.2020.105030

Antunes, J., J. Frias and P. Sobral. 2018. Microplastics on the Portuguese coast. Mar. Pollut. Bull. 131: 294–302. https://doi.org/10.1016/j.marpolbul.2018.04.025

Auta, H.S., C.U. Emenike and S.H. Fauziah. 2017. Distribution and importance of microplastics in the marine environment: A review of the sources, fate, effects and potential solutions. Environ. Int. 102: 165–176. https://doi.org/10.1016/j.envint.2017.02.013

Bakir, A., S.J. Rowland and R.C. Thompson. 2012. Competitive sorption of persistent organic pollutants onto microplastics in the marine environment. Mar. Pollut. Bull. 6412: 2782–2789.

Bandara, L.D., K.I. Marasinghe and Y.L. Mahagamage. 2023. The usage of household polythene and its management approaches. Environ. Qual. Manage. 1–11. https://doi.org/10.1002/tqem.22086

Barnes, D.K.A. 2002. Invasions by marine life on plastic debris. Nature. 4166883, 808–809.

Barnes, D.K.A. and P. Milner. 2005. Drifting plastic and its consequences for sessile organism dispersal in the Atlantic Ocean. Mar. Biol. 146: 815–825.

Bestley, S., S. Bengtson Nash, C.M. Brooks, C. Cotté, M. Dewar, A.S. Friedlaender, et al. 2020. Marine Ecosystem Assessment for the Southern Ocean: Birds and Marine Mammals in a Changing Climate. Front. Ecol. Evol. 8: 566936.

Boucher, J. and D. Friot. 2017. Primary microplastics in the oceans: A global evaluation of sources. https://doi.org/10.2305/iucn.ch.2017.01.en

Bratina, B., A. Šorgo, J. Kramberger, U. Ajdnik, L.F. Zemljič, J. Ekart, et al. 2016. From municipal/industrial wastewater sludge and FOG to fertilizer: A proposal for economic sustainable sludge management. J. Environ. Manage. 183: 1009–1025. https://doi.org/10.1016/j.jenvman.2016.09.063

Browne, M.A. 2015. Sources and pathways of microplastics to habitats. *In*: Marine Anthropogenic Litter. 334: 229–244. Springer International Publishing. https://doi.org/10.1007/978-3-319-16510-3_9

Browne, M.A., P. Crump, S.J. Niven, E. Teuten, A. Tonkin, T. Galloway, et al. 2011. Accumulation of Microplastic on Shorelines Worldwide: Sources and Sinks. Environ. Sci. Technol. 4521: 9175–9179. doi:10.1021/es201811s

Carney Almroth, B.M., L. Åström, S. Roslund, H. Petersson, M. Johansson and N.K. Persson. 2018. Quantifying shedding of synthetic fibers from textiles: A source of microplastics released into the environment. Environ. Sci. Pollut. Res. 252: 1191–1199. https://doi.org/10.1007/s11356-017-0528-7

Chang, M. 2015. Reducing microplastics from facial exfoliating cleansers in wastewater through treatment versus consumer product decisions. Mar. Pollut. Bull. 1011: 330–333. https://doi.org/10.1016/j.marpolbul.2015.10.074

Chen, Y., A.K. Awasthi, F. Wei, Q. Tan and J. Li. 2021. Single-use plastics: Production, usage, disposal and adverse impacts. Sci. Total Environ. 752: 141772. https://doi.org/10.1016/j.scitotenv.2020.141772

Cheung, P.K. and L. Fok. 2017. Characterisation of plastic microbeads in facial scrubs and their estimated emissions in Mainland China. Water Res. 122: 53–61. https://doi.org/10.1016/j.watres.2017.05.053

Conkle, J.L., C.D. Báez Del Valle and J.W. Turner. 2018. Are we Underestimating Microplastic Contamination in Aquatic Environments? Environ. Manage. 611. https://doi.org/10.1007/s00267-017-0947-8

Corradini, F., P. Meza, R. Eguiluz, F. Casado, E. Huerta-Lwanga and V. Geissen. 2019. Evidence of microplastic accumulation in agricultural soils from sewage sludge disposal. Sci. Total Environ. 671: 411–420. https://doi.org/10.1016/j.scitotenv.2019.03.368

Crusot, M., J. Gaertner, T. Rodriguez, C. Lo and N. Gaertner-Mazouni. 2023. Assessment of plastic waste generated by the aquaculture industry: The case study of pearl farming in French Polynesia. J. Clean. Prod. 425: 138902. https://doi.org/10.1016/j.jclepro.2023.138902

De Falco, F., M.P. Gullo, G. Gentile, E. Di Pace, M. Cocca, L. Gelabert, et al. 2018. Evaluation of microplastic release caused by textile washing processes of synthetic fabrics. Environ. Pollut. 236: 916–925. https://doi.org/10.1016/j.envpol.2017.10.057

Fendall, L.S. and M.A. Sewell. 2009. Contributing to marine pollution by washing your face: Microplastics in facial cleansers. Mar. Pollut. Bull. 588: 1225–1228. https://doi.org/10.1016/j.marpolbul.2009.04.025

Gall, S. and R. Thompson. 2015. The impact of debris on marine life. Mar. Pollut. Bull. 92(1–2): 170–179. https://doi.org/10.1016/j.marpolbul.2014.12.041

Gamage, S. and Y. Mahagamage. 2024a. Microplastics in personal care products and cosmetics in Sri Lanka. Heliyon. 108. 10.1016/j.heliyon.2024.e29393

Gamage, S.G. and M.G.Y.L. Mahagamage. 2024b. Consumer perception of personal care products and cosmetics on health and environmental effects. Environ. Qual. Manag. 34(2): 22339. https://doi.org/10.1002/tqem.22339

Garcés-Ordóñez, O., L.F. Espinosa Díaz, R. Pereira Cardoso and M. Costa Muniz. 2020. The impact of tourism on marine litter pollution on Santa Marta beaches, Colombian Caribbean. Mar. Pollut. Bull. 160: 111558. https://doi.org/10.1016/j.marpolbul.2020.111558

Geyer, R., J.R. Jambeck and K.L. Law. 2017. Production, use and fate of all plastics ever made. Sci. Adv. 37: e1700782. https://doi.org/10.1126/sciadv.1700782

Godoy, V., M.A. Martín-Lara, M. Calero and G. Blázquez. 2019. Physical-chemical characterization of microplastics present in some exfoliating products from Spain. Mar. Pollut. Bull. 139: 91–99. https://doi.org/10.1016/j.marpolbul.2018.12.026

Gouin, T., J. Avalos, I. Brunning, K. Brzuska, J. Graaf, J. De Kaumanns, et al. 2015. Use of micro-plastic beads in cosmetic products in Europe and their estimated emissions to the North Sea environment. SOFW Journal. 141: 40–46. Retrieved from https://www.ikw.org/fileadmin/ikw/downloads/Schoenheitspflege/SOFW_Micro-Plastic_beads_in_Cosmetic_Products.pdf

Gregory, M.R. 2009. Environmental implications of plastic debris in marine settings—entanglement, ingestion, smothering, hangers-on, hitch-hiking and alien invasions. Philos. Trans. R. Soc. B 364: 2013–2025. https://doi.org/10.1098/rstb.2008.0265

Habib, R.Z., M.M. Salim Abdoon, R.M. Al Meqbaali, F. Ghebremedhin, M. Elkashlan, W.F. Kittaneh, et al. 2020. Analysis of microbeads in cosmetic products in the

United Arab Emirates. Environ. Pollut. 258: 113831. https://doi.org/10.1016/j.envpol.2019.113831

Harms, I.K., T. Diekötter, S. Troegel and M. Lenz. 2021. Amount, distribution and composition of large microplastics in typical agricultural soils in Northern Germany. Sci. Total Environ. 758: 143615. https://doi.org/10.1016/j.scitotenv.2020.143615

Hartline, N.L., N.J. Bruce, S.N. Karba, E.O. Ruff, S.U. Sonar and P.A. Holden. 2016. Microfiber masses recovered from conventional machine washing of new or aged garments. Environ. Sci. Technol. 5021: 11532–11538. https://doi.org/10.1021/acs.est.6b03045

Henry, B., K. Laitala and I.G. Klepp. 2019. Microfibres from apparel and home textiles: Prospects for including microplastics in environmental sustainability assessment. Sci. Total Environ. 652: 483–494. https://doi.org/10.1016/j.scitotenv.2018.10.166

Holmes, L.A., A. Turner and R.C. Thompson. 2012. Adsorption of trace metals to plastic resin pellets in the marine environment. Environ. Pollut. 160: 42–48.

Horton, A.A., C. Svendsen, R.J. Williams, D.J. Spurgeon and E. Lahive. 2017. Large microplastic particles in sediments of tributaries of the River Thames, UK – Abundance, sources and methods for effective quantification. Mar. Pollut. Bull. 1141: 218–226. https://doi.org/10.1016/j.marpolbul.2016.09.004

Huang, Y., Q. Liu, W. Jia, C. Yan and J. Wang. 2020. Agricultural plastic mulching as a source of microplastics in the terrestrial environment. Environ. Pollut. 260: 114096. https://doi.org/10.1016/j.envpol.2020.114096

Idumah, C.I. and Nwuzor, I.C. 2019. Novel trends in plastic waste management. SN Applied Sciences. Springer Nature. https://doi.org/10.1007/s42452-019-1468-2

Igalavithana, A.D., Mahagama Gedara Y.L. Mahagamage, P. Gajanayake, A. Abeynayaka, P.J.D. Gamaralalage, M. Ohgaki, et al. 2022. Microplastics and potentially toxic elements: Potential human exposure pathways through agricultural lands and policy based countermeasures. Microplastics. 11: 102–120.

Jambeck, J.R., Q. Ji, Y.-G. Zhang, D. Liu, D.M. Grossnickle and Z.-X. Luo. 2015. Plastic waste inputs from land into the ocean. Science. 3476223: 764–768. Retrieved from http://www.sciencemag.org/cgi/doi/10.1126/science.1260879

Jansen, L., M. Henskens and F. Hiemstra. 2019. Report on use of plastics in agriculture. Retrieved from www.schuttelaar.nl

Kalčíková, G., B. Alič, T. Skalar, M. Bundschuh and A.Ž. Gotvajn. 2017. Wastewater treatment plant effluents as source of cosmetic polyethylene microbeads to freshwater. Chemosphere. 188: 25–31. https://doi.org/10.1016/j.chemosphere.2017.08.131

Keys, A., A. Baldwin and S. Austin. 2000. Designing to encourage waste minimisation in the construction industry Version 1. Loughborough University. https://hdl.handle.net/2134/4945

Kitahara, K.I. and H. Nakata. 2020. Plastic additives as tracers of microplastic sources in Japanese road dusts. Sci. Total Environ. 736: 139694. https://doi.org/10.1016/j.scitotenv.2020.139694

Kole, P., A.J. Löhr, F.G.A.J. Van Belleghem and A.M.J. Ragas. 2017. Wear and tear of tyres: A stealthy source of microplastics in the environment. Int. J. Environ. Res. Public Health. 1410. https://doi.org/10.3390/ijerph14101265

Kumar, M., X. Xiong, M. He, D.C.W. Tsang, J. Gupta, E. Khan, et al. 2020. Microplastics as pollutants in agricultural soils. Environ. Pollut. 265: 114980. https://doi.org/10.1016/j.envpol.2020.114980

Laitala, K., I.G. Klepp and B. Henry. 2018. Does use matter? Comparison of environmental impacts of clothing based on fiber type. Sustainability Switzerland. 107: 1–25. https://doi.org/10.3390/su10072524

Leal Filho, W., U. Saari, M. Fedoruk, A. Iital, H. Moora, M. Klöga, et al. 2019. An overview of the problems posed by plastic products and the role of extended producer responsibility in Europe. J. Clean. Prod. 214: 550–558. https://doi.org/10.1016/j.jclepro.2018.12.256

Lei, K., F. Qiao, Q. Liu, Z. Wei, H. Qi, S. Cui, et al. 2017. Microplastics releasing from personal care and cosmetic products in China. Mar. Pollut. Bull. 1231–1232: 122–126. https://doi.org/10.1016/j.marpolbul.2017.09.016

Lestari, P. and Y. Trihadiningrum. 2019. The impact of improper solid waste management to plastic pollution in Indonesian coast and marine environment. Mar. Pollut. Bull. 149: 110505. https://doi.org/10.1016/j.marpolbul.2019.110505

Liu, E.K., W.Q. He and C.R. Yan. 2014. "White revolution" to "white pollution"—Agricultural plastic film mulch in China. Environ. Res. Lett. Institute of Physics Publishing. https://doi.org/10.1088/1748-9326/9/9/091001

Liu, J., Y. Yang, J. Ding, B. Zhu and W. Gao. 2019. Microfibers: A preliminary discussion on their definition and sources. Environ. Sci. Pollut. Res. 2628: 29497–29501. https://doi.org/10.1007/s11356-019-06265-w

Lusher, A. 2015. Microplastics in the marine environment: Distribution, interactions and effects. *In*: Bergmann, M., L. Gutow, M. Klages. (Eds). Mar. Anthropog. Litter. Springer, Cham.

Lv, X., Q. Dong, Z. Zuo, Y. Liu, X. Huang and W.M. Wu. 2019. Microplastics in a municipal wastewater treatment plant: Fate, dynamic distribution, removal efficiencies and control strategies. J. Clean. Prod. 225: 579–586. https://doi.org/10.1016/j.jclepro.2019.03.321

Magnusson, K., K. Eliasson, A. Fråne, K. Haikonen, J. Hultén, M. Olshammar, et al. 2016. Swedish sources and pathways for microplastics to the marine environment. A Review of Existing Data. IVL, C, 183C. 183: 1–89. Retrieved from www.ivl.se

Mahagamage, M.G.Y.L., S.G. Gamage, R.M.S.K. Rathnayake, P.J.D. Gamaralalage, M. Hengesbugh, T. Abeynayaka, et al. 2024. Mitigating microfiber pollution in laundry wastewater: Insights from a filtration system case study in Galle, Sri Lanka. Microplastics. 34: 599–613.

Manickam, P. and G. Duraisamy. 2018. 3Rs and circular economy. Circular Economy in Textiles and Apparel. 77–93. https://doi.org/10.1016/B978-0-08-102630-4.00004-2

Mason, S.A., D. Garneau, R. Sutton, Y. Chu, K. Ehmann, J. Barnes, et al. 2016. Microplastic pollution is widely detected in US municipal wastewater treatment plant effluent. Environ. Pollut. 218: 1045–1054. https://doi.org/10.1016/j.envpol.2016.08.056

Ministry of Environment 2021. The National Action Plan on Plastic Waste Management 2021–2030, "Sobadam Piyasa", 416/C/1, Robert Gunawardana Mawatha, Battaramulla, Sri Lanka.

Miraj, S.S., N. Parveen and H.S. Zedan. 2019. Plastic microbeads: small yet mighty concerning. Int. J. Environ. Health Res. 0000: 1–17. https://doi.org/10.1080/09603123.2019.1689233

Moore, C. 2013. Rapidly increasing plastic pollution aquaculture threatens marine life. Tul. Envtl. LJ. 27: 205.

Mordecai, G., P.A. Tyler, D.G. Masson and V.A. Huvenne. 2011. Litter in submarine canyons off the west coast of Portugal. Deep Sea Research Part II: Topical Studies in Oceanography. 58(23–24): 2489–2496.

Mourshed, M., M.H. Masud, F. Rashid and Md U.H. Joardder. 2017. Towards the effective plastic waste management in Bangladesh: A review. Environ. Sci. Pollut. Res. 24: 27021–27046. https://doi.org/10.1007/s11356-017-0429-9

Napper, I.E., A. Bakir, S.J. Rowland and R.C. Thompson. 2015. Characterisation, quantity and sorptive properties of microplastics extracted from cosmetics. Mar. Pollut. Bull. 99(1–2): 178–185. https://doi.org/10.1016/j.marpolbul.2015.07.029

Napper, I.E. and R.C. Thompson. 2016. Release of synthetic microplastic plastic fibres from domestic washing machines: Effects of fabric type and washing conditions. Mar. Pollut. Bull. 112(1–2): 39–45. https://doi.org/10.1016/j.marpolbul.2016.09.025

Napper, I.E. and R.C. Thompson. 2018. Marine plastic pollution: Other than microplastic. pp. 425–442. *In*: T.M. Letcher and D.A. Vallero (eds). Waste, 2nd Ed., Elsevier B.V. https://doi.org/10.1016/B978-0-12-815060-3.00022-0

Napper, I.E. and R.C. Thompson. 2020. Plastic debris in the marine environment: History and future challenges. Global Challenges. 46: 1900081. https://doi.org/10.1002/gch2.201900081

Ngo, P.L., B.K. Pramanik, K. Shah and R. Roychand. 2019. Pathway, classification and removal efficiency of microplastics in wastewater treatment plants. Environ. Pollut. Elsevier Ltd. https://doi.org/10.1016/j.envpol.2019.113326

Norén, F. and L.J. Naustvoll. 2010. Survey of microscopic anthropogenic particles in Skagerrak. Pilot study October-November 2010, March, 19.

Ogata, Y., H. Takada, K. Mizukawa, H. Hirai, S. Iwasa, S. Endo, et al. 2009. International Pellet Watch: Global monitoring of persistent organic pollutants POPs in coastal waters. 1. Initial phase data on PCBs, DDTs and HCHs. Mar. Pollut. Bull. 5810: 1437–1446. https://doi.org/10.1016/j.marpolbul.2009.06.014

Oßmann, B.E., G. Sarau, H. Holtmannspötter, M. Pischetsrieder, S.H. Christiansen and W. Dicke. 2018. Small-sized microplastics and pigmented particles in bottled mineral water. Water Res. 141: 307–316. https://doi.org/10.1016/j.watres.2018.05.027

Peller, J.R., L. Eberhardt, R. Clark, C. Nelson, E. Kostelnik and C. Iceman. 2019. Tracking the distribution of microfiber pollution in a southern Lake Michigan watershed through the analysis of water, sediment and air. Environ. Sci.: Process. Impacts. 219: 1549–1559. https://doi.org/10.1039/c9em00193j

Piehl, S., A. Leibner, M.G.J. Löder, R. Dris, C. Bogner and C. Laforsch. 2018. Identification and quantification of macro-and microplastics on an agricultural farmland. Sci. Rep. 8: 17950. https://doi.org/10.1038/s41598-018-36172-y

Praveena, S.M., S.N.M. Shaifuddin and S. Akizuki. 2018. Exploration of microplastics from personal care and cosmetic products and its estimated emissions to marine environment: An evidence from Malaysia. Mar. Pollut. Bull. 136: 135–140. https://doi.org/10.1016/j.marpolbul.2018.09.012

Qi, R., D.L. Jones, Z. Li, Q. Liu and C. Yan. 2020. Behavior of microplastics and plastic film residues in the soil environment: A critical review. Sci. Total Environ. 703: 134722. https://doi.org/10.1016/j.scitotenv.2019.134722

Rajmohan, K.V.S., C. Ramya, M. Raja Viswanathan and S. Varjani. 2019. Plastic pollutants: Effective waste management for pollution control and abatement. Curr Opin Environ Sci Health. 12: 72–84. https://doi.org/10.1016/j.coesh.2019.08.006

Rezaei, M., M.J.P.M. Riksen, E. Sirjani, A. Sameni and V. Geissen. 2019. Wind erosion as a driver for transport of light-density microplastics. Sci. Total Environ. 669: 273–281. https://doi.org/10.1016/j.scitotenv.2019.02.382

Rochman, C.M., S.M. Kross, J.B. Armstrong, M.T. Bogan, E.S. Darling, S.J. Green. 2015. Scientific Evidence Supports a Ban on Microbeads. Environ. Sci. Technol. 4918: 10759–10761.

Rujnić-Sokele, M. and A. Pilipović. 2017. Challenges and opportunities of biodegradable plastics: A mini review. Waste Manage. Res. 352: 132–140.

Santos, R.G., R. Andrades, L.M. Fardim and A.S. Martins. 2016. Marine debris ingestion and Thayer's law – The importance of plastic color. Environ. Pollut. 214: 585–588. https://doi.org/10.1016/j.envpol.2016.04.024

Savoca, M.S., A.G. McInturf and E.L. Hazen. 2021. Plastic ingestion by marine fish is widespread and increasing. Glob. Change Biol. 2710: 2188–2199. https://doi.org/10.1111/gcb.15533

Scarascia-Mugnozza, G., C. Sica and G. Russo. 2012. Plastic Materials in European Agriculture: Actual Use and Perspectives. J. Agric. Eng. 423: 15. https://doi.org/10.4081/jae.2011.3.15

Schymanski, D., C. Goldbeck, H.U. Humpf and P. Fürst. 2018. Analysis of microplastics in water by micro-Raman spectroscopy: Release of plastic particles from different packaging into mineral water. Water Res. 129: 154–162. https://doi.org/10.1016/j.watres.2017.11.011

Shen, M., B. Song, G. Zeng, Y. Zhang, W. Huang, X. Wen, et al. 2020. Are biodegradable plastics a promising solution to solve the global plastic pollution? Environ. Pollut. 263: 114469. https://doi.org/10.1016/j.envpol.2020.114469

Siddiqua, A., J.N. Hahladakis and W.A.K.A. Al-Attiya. 2022. An overview of the environmental pollution and health effects associated with waste landfilling and open dumping. Environ. Sci. Pollut. Res. 29: 58514–58536. https://doi.org/10.1007/s11356-022-21578-z

Singh, N., D. Hui, R. Singh, I. Ahuja, L. Feo and F. Fraternali. 2017. Recycling of plastic solid waste: A state of art review and future applications. Compos. Part B Eng. 115: 409–422. https://doi.org/10.1016/j.compositesb.2016.09.013

Sommer, F., V. Dietze, A. Baum, J. Sauer, S. Gilge, C. Maschowski, et al. 2018. Tire abrasion as a major source of microplastics in the environment. Aerosol Air Qual. Res. 188: 2014–2028. https://doi.org/10.4209/aaqr.2018.03.0099

Stelfox, M., J. Hudgins and M. Sweet. 2016. A review of ghost gear entanglement amongst marine mammals, reptiles and elasmobranchs. Mar. Pollut. Bull. 111(1–2): 6–17.

Sul, J.A.I., D.K.A. Barnes, M.F. Costa, P. Convey, E.S. Costa and L.S. Campos. 2011. Plastics in the Antarctic environment: Are we looking only at the tip of the Iceberg? Oecol. Australis. 1501: 150–170. https://doi.org/10.4257/oeco.2011.1501.11

Sundt, P., P. Schultze and F. Syversen. 2014. Sources of microplastic-pollution to the marine environment. Mepex, Norwegian Environment Agency.

Syversen, T., G. Lilleng, J. Vollstad, B.J. Hanssen and S.A. Sønvisen. 2022. Oceanic plastic pollution caused by Danish seine fishing in Norway. Mar. Pollut. Bull. 179: 113711. https://doi.org/10.1016/j.marpolbul.2022.113711

The Fiber Year. 2020. Deceleration along the textile chain | International Fiber Journal. n.d. Retrieved May 21, 2021, from https://fiberjournal.com/the-fiber-year-2020-deceleration-along-the-textile-chain/

Thompson, R.C., Y. Olsen, R.P. Mitchell, A. Davis, S.J.G. Rowland, A.W. John, et al. 2004. Lost at Sea: Where Is All the Plastic? Science.

Thrift, E., P. Nouvellet and F. Mathews. 2023. Plastic entanglement poses a potential hazard to European Hedgehogs Erinaceus Europaeus in Great Britain. Animals. 1315: 2448. https://doi.org/10.3390/ani13152448

Tiseo, I. 2020. Production of plastics worldwide from 1950 to 2019 in million metric tons. Retrieved from https://www.statista.com/statistics/282732/global-production-of-plastics-since-1950/

Tosetto, L., J.E. Williamson and C. Brown. 2017. Trophic transfer of microplastics does not affect fish personality. Animal Behaviour. 123: 159–167.

U.S Tire Manufacturers Association. 2019. What's in a tire. Retrieved from https://www.ustires.org/whats-tire-0

Ustabasi, G.S. and A. Baysal. 2019. Occurrence and risk assessment of microplastics from various toothpastes. Environ. Monit. Assess. 1917: 1–8. https://doi.org/10.1007/s10661-019-7574-1

Von Moos, N., P. Burkhardt-Holm and A. Köhler. 2012. Uptake and effects of microplastics on cells and tissue of the blue mussel *Mytilus edulis* L. after an experimental exposure. Environ. Sci. Technol. 4620: 11327–11335. doi:10.1021/es302332w

Walker, T.R. and L. Fequet. 2023. Current trends of unsustainable plastic production and micronanoplastic pollution. TrAC Trends Anal. Chem. 160: 116984. https://doi.org/10.1016/j.trac.2023.116984

Wayman, C. and Niemann, H. 2021. The fate of plastic in the ocean environment: A minireview. Environ. Sci.: Processes. Impacts. 232: 198–212. https://doi.org/10.1039/d0em00446d

Wichai-utcha, N. and O. Chavalparit. 2019. 3Rs Policy and plastic waste management in Thailand. J. Mater. Cycles Waste Manag. 21: 10–22. https://doi.org/10.1007/s10163-018-0781-y

Wilson, S.P. and K.M. Verlis. 2017. The ugly face of tourism: Marine debris pollution linked to visitation in the southern Great Barrier Reef, Australia. Mar. Pollut. Bull. 117(1–2): 239–246. https://doi.org/10.1016/j.marpolbul.2017.01.036

Xanthos, D. and T.R. Walker. 2017. International policies to reduce plastic marine pollution from single-use plastics plastic bags and microbeads: A review. Mar. Pollut. Bull. Elsevier Ltd. https://doi.org/10.1016/j.marpolbul.2017.02.048

Yang, L., K. Li, S. Cui, Y. Kang, L. An and K. Lei. 2019a. Removal of microplastics in municipal sewage from China's largest water reclamation plant. Water Res. 155: 175–181. https://doi.org/10.1016/j.watres.2019.02.046

Yang, L., F. Qiao, K. Lei, H. Li, Y. Kang, S. Cui, et al. 2019b. Microfiber release from different fabrics during washing. Environ. Pollut. 249: 136–143. https://doi.org/10.1016/j.envpol.2019.03.011

Zhou, H., L. Zhou and K. Ma. 2020. Microfiber from textile dyeing and printing wastewater of a typical industrial park in China: Occurrence, removal and release. Sci. Total Environ. 739: 140329. https://doi.org/10.1016/j.scitotenv.2020.140329

Ziajahromi, S., D. Drapper, A. Hornbuckle, L. Rintoul and F.D.L. Leusch. 2020. Microplastic pollution in a stormwater floating treatment wetland: Detection of tyre particles in sediment. Sci. Total Environ. 713: 136356. https://doi.org/10.1016/j.scitotenv.2019.136356

Chapter 3

Holistic Analysis of Floating Debris in Thailand's Chao Phraya Tributary: From Community Source to Sink

Yotwadee Hawangchu[1], Chatpong Na Pombhejara[2], Kasidij Peantham[2], Abdul Shobar[2], Chadalak Samritthinanta[3], Dao Janjaroen[2], Unruan Leknoi[3], Thomas Mani[4] and Jenyuk Lohwacharin*[2]

[1]Aquatic Resources Research Institute, Chulalongkorn University, Bangkok, Thailand

[2]Department of Environmental and Sustainable Engineering, Faculty of Engineering, Chulalongkorn University, Bangkok, Thailand

[3]Social Innovation and Sustainability for Vulnerable People Research Unit, Chulalongkorn University Social Research Institute, Chulalongkorn University, Bangkok, Thailand

[4]The Ocean Cleanup, Rotterdam, The Netherlands

3.1 SOURCE OF MACROPLASTICS IN RIVER SYSTEMS

Due to rapid population growth and urbanization, waste reduction measures have been overwhelmed, resulting in the peak global production of municipal solid waste (MSW) after the year 2100 (Hoornweg et al. 2013). Concurrently, growing plastic

*For Correspondence: Jenyuk.L@chula.ac.th

DOI: 10.1201/9781003531463-3

production markedly increases the plastics' share of MSW (Ryan 2015). Poor management of non-recyclable and non-recycled plastics results in increased waste leakage into river networks (Morales-Caselles et al. 2021). Specifically, plastic debris may commute from land via aeolian transport, as well as from untreated waste and storm water to water bodies (Lebreton et al. 2022; Mellink et al. 2024). Mainly buoyant macroplastics were found to be highly mobile, leading to their widespread dispersal (Lebreton et al. 2022). Plastic debris can find its way to the ocean via canal and river systems (Figure 3.1) (Mani et al. 2025). It may be trapped temporarily as floating debris (Figure 3.1c and 3.1d) or deposited with sediment in the water systems (Figure 3.1b) until re-mobilized during high flow and flooding events. Previous studies reported that high flow and flood events cause a significant increase of up to 14 times the average in terms of the number of microplastics (Gundogdu et al. 2018). Globally, about 1,000 rivers carry 80% of all rivers' plastic load toward the oceans (Meijer et al. 2021), highlighting the importance of understanding freshwater plastic pollution to mitigate marine plastic pollution.

Figure 3.1 Macroplastic debris in canals. Canal hotspot sources (a), aggregation of plastic litter (b), and (c), blockage of floating plastic litter on the water surface (d).

He et al. (2024) reported hotspots in global river plastic waste, indicating that they are located in densely populated urban environments. The authors report that, based on the numbers, transparent (40.4%), film (46.9%), mesoplastic (52.5%), and polyethylene (PE) (36.9%) dominated the color, shape, size, and polymer of plastic litter. They found that the top three plastic litter types include plastic food wrappers, synthetic ropes, and plastic bags. Wrappers are widely used as packaging for products in shops (e.g., snacks, tobacco sachets, plastic bags, etc.).

This plastic type has little to no recycling value; therefore, it is more likely to be discarded and accumulate in the environment (He et al. 2024). In terms of litter origins, take-out consumer items made up the largest contribution to urban water plastic pollution. This finding was in line with recent surveys, which have shown that plastic bags and wrappers are the main contributors to riverine plastic litter (Weideman et al. 2020; Blondel and Buschman 2022). The abundance of plastic debris was positively correlated with various socio-economic factors, e.g., population density, per capita disposable income, and passenger traffic. Waterway freight and passenger traffic, one of the socio-economic factors, were the most significantly correlated with the abundance of plastic litter in waterbodies (He et al. 2024). It has been emphasized that the socio-economic factors play a major role in driving the spatial distribution of plastic pollution in freshwater systems (Yin et al. 2020; Zhou et al. 2020). Several studies demonstrate the role of riparian vegetation in the plastic debris transport along rivers. Previous studies, for example, report that mangroves retain plastic litter and microplastics in coastal habitats (Jiao et al. 2022).

Among meteorological factors, air temperature, air pressure, and wind speed were positively correlated with the abundance of plastic litter in waters, while relative humidity showed a negative correlation. The abundance of plastic debris had a relatively strong positive correlation with chemical oxygen demand (COD), but a weak correlation with other hydrological factors (He et al. 2024).

3.2 CHARACTERISTICS OF FLOATING WASTE

3.2.1 Characterization of Riverine Waste in Urbanized Community Areas

The characteristics of riverine waste in areas where both riverbanks are utilized for urban activities can reflect the lifestyle of the surrounding community. For the case study in this chapter, we focus on the Lat Luang canal that flows through a densely populated community in the metropolitan region of Bangkok. The surface water from this canal is discharged through a system of sluice gates into the Chao Phraya River, a principal river of Thailand and its capital city. This river emits floating waste to the Gulf of Thailand, ultimately contributing to marine debris. These processes are well-documented by models that demonstrate the significance of small rivers as pathways for transporting waste to the sea (González-Fernández et al. 2021).

The waste collected for this study was gathered from debris trapped at the sluice gates, ready to flow into the Chao Phraya River once the gates are opened. Over a period of 20 months, encompassing all local seasons, riverine debris was recorded by item count and mass. Here we present the item count.

The study found that the majority of plastic waste falls into three categories as shown in Figure 3.2:

(i) Rigid single-use containers (66.4%): This includes items such as bottles, boxes, and cups.

(ii) Thin films (15.9%): This includes soft plastic sheets used as wrappers and food packaging.
(iii) Multimaterial items (17.6%): This category comprises everyday items such as blister foil, cartons, shoes, refilled packaging, and cigarette lighters.

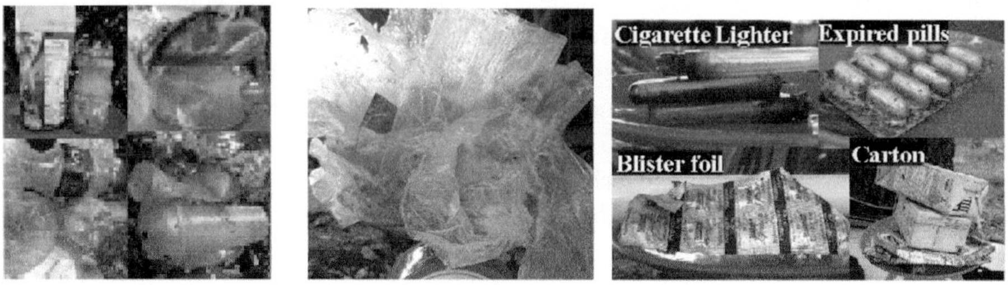

Rigid containers Thin films Multimaterials

Figure 3.2 Three categories represent the majority of plastic waste found in the study area.

3.2.2 Recyclable Plastics

When examining the number of riverine waste items collected for this study, it was found that the most prevalent type of rigid container is the single-use polyethylene terephthalate (PET) bottle (37%), primarily used for drinking water. The second most common type of rigid container is made from high-density polyethylene (HDPE) (14%), which is found in two main product categories: ready-to-drink milk and household products such as talcum powder containers and shampoo bottles. The third most common category is food containers made from polypropylene (PP) (9%), which are readily found in large convenience stores throughout the country. Notably, over 90% of these materials are single-use plastics, primarily in the form of containers for consumption. This finding aligns with existing research on floating litter within an European monitoring programme, which shows that 82% of floating litter consists of plastic waste, predominantly fragments and single-use items such as bottles, packaging, and bags (González-Fernández et al. 2021). Additionally, this finding is consistent with research in Georgia, USA, which used a similar methodology to examine a creek in Atlanta. The study found that single-use plastics, such as food and beverage packaging, were the most prevalent items. Polystyrene (PS) was the most frequently collected material, accounting for 56.4% of the waste (Poletti and Landberg 2021).

3.2.3 Non-Recyclable Polymers in Riverine Waste

As illustrated in Figure 3.3, which compiles data on recyclable polymers, recycling symbols play a crucial role in identifying and separating recyclable materials, such as 1 for PET, 2 for HDPE, and 3 for polyvinyl chloride (PVC). Beyond the polymers marked with these recycling symbols, foam or expanded polystyrene (EPS) emerges as the second most commonly found material after PET, comparable to HDPE in this study. EPS waste was collected in two primary forms: more than 80% were fragments of foam crates (Figure 3.4a) used for temporary cold storage, and styrofoam take-out containers commonly used for food packaging from street vendors or takeout services (Figure 3.4b). Additionally, food containers in this context were also found to be made from PP and PS materials.

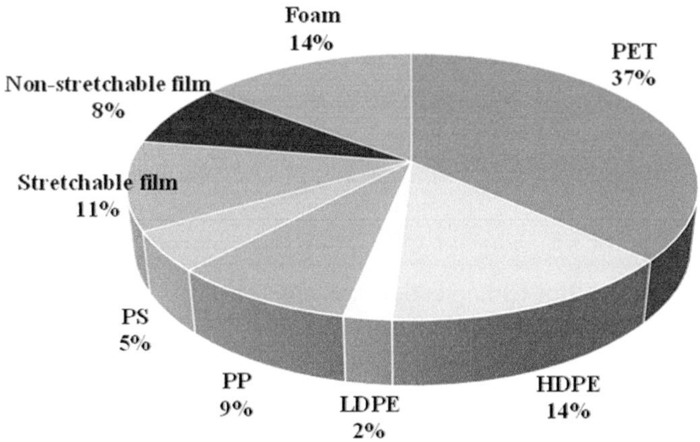

Figure 3.3 Proportion of polymers found in the floating riverine waste study area.

(a) EPS fragment (b) Foam container

Figure 3.4 Expanded polystyrene (EPS) small pieces and foam containers.

3.2.4 Impact of Urban Lifestyles on Riverine Plastic Waste

Furthermore, the urban lifestyle, which often involves purchasing takeaway food rather than cooking at home, contributes significantly to the presence of food contact plastic bags (Figure 3.5a) in riverine waste. As shown in Figure 3.5, these bags are typically made from three main types of polymers: low-density polyethylene (LDPE), PP, and HDPE.

In this study, soft plastic or film plastic waste was categorized into two types: stretchable (Figure 3.5b) and non-stretchable (Figure 3.5c). The first type, stretchable plastics, mainly includes linear low-density polyethylene (LLDPE) and LDPE, which are used for food containers and shrink film for bottle packaging. These were found to be slightly more prevalent than non-stretchable plastics. Non-stretchable plastics predominantly consist of HDPE and PP, used for food bags, shopping bags, and consumer product wrappers. Regarding polymer types and plastic waste, this aligns with research indicating PE, PP, PET, and PS are predominantly used in packaging, while EPS is primarily for food packaging (van Emmerik and Schwarz 2020).

(a) Food contact plastic bags (b) Stretchable films (c) Non-stretchable films

Figure 3.5 Films or soft plastics found in the study area.

Although these thin-film plastics are not typically recyclable, they can be incinerated to generate thermal energy (Chen et al. 2021; Dihan et al. 2024). However, to support this process, designated collection points are necessary. Nonetheless, a more effective strategy would be to reduce the use of unnecessary plastics through upstream design innovations (Ubando et al. 2024).

3.2.5 From Riverine Plastic Waste to Marine Debris

When analyzing the types of plastics found in the study area, which drains into the Chao Phraya River, and comparing them with beach litter data from recreational areas (similar to the residential and recreational uses of the riverbanks in the study), notable similarities emerge. According to the Department of Marine and Coastal Resources (DMCR) data from 2022, beach litter comprises 25% thin plastic (8% non-stretchable, 5% food wrappers/bags, 5% fragments/pieces, and 7% other plastic), 11% beverage bottles, and 3% each of foam and plastic food containers. This composition closely mirrors the waste found in our study area, where drinking bottles and thin plastics are among the most prevalent items. However, when considering the floating riverine debris at the mouth of the Chao Phraya River, which DMCR reports as the leading source of floating waste into the upper Gulf of Thailand, the composition shifts slightly. According to DMCR's 2022 data (DMCR 2022), the primary types of floating plastic waste entering the upper Gulf of Thailand include fast food packaging (33%), food contact plastic bags (18%), shopping bags 11%), and elastic bands (other polymers, 11%). Additionally, similar to the waste gathered at our study points, foam containers and HDPE plastics for household products are also commonly found. This may provide clear evidence suggesting that anthropogenic litter, including that originating in upstream urban areas, is transported through creeks and rivers to downstream ecosystems, such as the Gulf of Thailand, and eventually to the ocean (Poletti and Landberg 2021).

The alignment between the types of plastics found in different environments underscores the interconnected nature of waste management challenges and the importance of coordinated efforts to mitigate plastic pollution from source to sea.

3.2.6 Multimaterial Riverine Waste—Items of Daily Life

In addition to classifying waste by polymer type, we can also categorize it by function, especially multimaterials commonly used in daily life. As shown in Figure 3.6, blister foil and Tetra Pak cartons are the most frequently found items. This raises concerns about the disposal of unused medications in such forms. However, Tetra Pak cartons can be sent for upcycling into durable materials such as furniture, gypsum boards, notebooks, and Braille media through campaigns promoting the donation of milk cartons. Recently, Wongpanit, a major waste buyer with nationwide branches, began purchasing Tetra Pak cartons at approximately 4.50 baht per kilogram, indicating a promising future for this type of waste and suggesting a potential decrease in its volume. This development indicates improved waste management by promoting the collection and recycling of multimaterial waste, reducing its environmental impact.

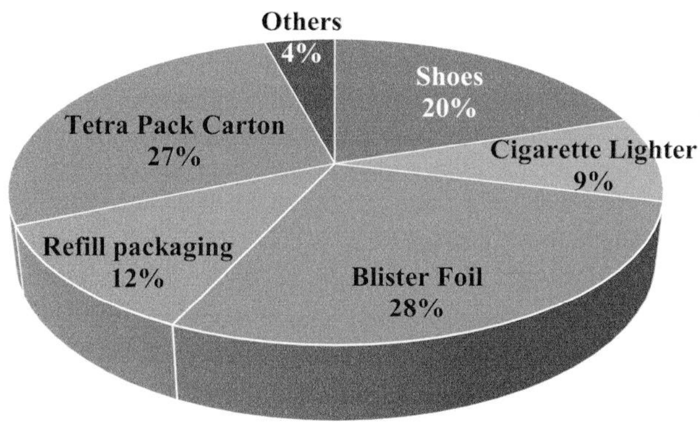

Figure 3.6 Multimaterial items found in daily life.

Additionally, footwear consistently found in the waste at the study point indicated former use in recreational activities like sports and fishing (Figure 3.7). The recorded proportions reflect the local population's lifestyle, particularly their reliance on single-use plastics, as waste from surface water sources (Lat Luang canal) flows through community areas.

Blister foil Tetra Pak cartons Shoes Refilled bags

Figure 3.7 Multimaterial waste in the canal in the daily life of an urbanized area.

The most common type of waste found is PET beverage bottles. It is striking that these bottles are not being recycled, possibly due to their low market value.

Other types of plastics found in the study face similar issues, with some not being purchased at all. Therefore, waste management efforts should focus on PET, HDPE, and foam materials.

In Thailand, some operators can convert stretchable plastic film waste into energy and recycled plastic, though they are few and concentrated in specific areas. Despite this capability, additional support is necessary to drive the reduction of this type of waste. Effective collection remains challenging, particularly in urban areas with limited space. To address this issue, it is essential to establish designated collection points and implement incentives for proper disposal, as this type of waste cannot be sold. Moreover, the placement of adequate trash bins in public recreational areas is crucial to prevent littering. Providing designated storage areas for footwear can also help prevent these items from falling into the water, thereby reducing riverine waste.

3.3 MATERIAL FLOW ANALYSIS OF PLASTIC WASTE

Plastic wastes posed a risk to aquatic and human health. Lithner et al. (2011) identified that by-products and additives released from degraded plastics could leach into the marine environment. These chemicals could enter the human food chain through the consumption of contaminated fresh or seawater creatures. Health impacts included endocrine disruptions, carcinogenic effects, and other health issues. The review highlighted the urgent need for monitoring and regulating plastic additives and improving waste management practices to mitigate human health risks. Most plastic waste in riverine contexts stems from land-based use.

Tajima et al. (2019) investigated factors influencing waste disposal behavior of the residents living near canals, especially the Lat Phrao Canal in Bangkok, Thailand. The study employed the Theory of Planned Behavior to hypothesize that perceived behavioral control, attitude, norms, environmental values, perceived quality of environment, and willingness to act influence waste disposal practices. Through field investigations, interviews, and face-to-face questionnaire surveys, the authors identified that while municipal waste collection points were accessible, some residents still disposed of waste directly into the canal. Factors influencing this behavior included low risk perception, weaker social norms against improper disposal, and a lack of environmental values. Those valuing their environment and perceiving higher risks were less likely to dispose of waste into canals. The study concluded that unintentional waste disposal, such as overflow from collection points, significantly contributed to canal pollution. Recommendations included improving waste management at collection points and enhancing overall environmental quality in areas prone to waste accumulation. Ishigaki et al. (2018) studied the source of canal waste, especially in the Lat Phrao Canal, Bangkok. They found that household waste and construction debris were the most abundant types present in the canal. The study analyzed the composition of debris, highlighting timbers, plastics, and organic materials as prominent components. The waste obstructed the waterway and drainage, leading to flood risks during heavy precipitation. Field surveys and interviews highlight that waste often falls into canals due to poorly constructed living spaces above them and insufficient waste collection infrastructure. The proposed solutions include resettling communities away from canal edges and enhancing land-based waste collection services. The study concludes that a combined approach of resettlement and improved waste collection infrastructure is essential to mitigate canal waste issues effectively, stressing the need for localized strategies and further research to validate implementation outcomes. The distance between residential areas and the canal is one of the important factors influencing waste leakage into the water body. Velis and Cook (2021) found that residents residing within 500 meters of water systems could increase the likelihood of waste ending up in these water bodies. This research result coincides with Tajima et al. (2019) and Ishigaki et al. (2018), who found that waste collection infrastructure and awareness of the communities could exacerbate waste leakage problems. Proper waste collection also helped improve the quality of the environment and the health of people in the area.

3.4 LIFE CYCLE ASSESSMENT APPROACH IN WASTE MANAGEMENT STRATEGY

Waste management is increasingly challenged by the complexity of waste entering the system and the demand for sustainable solutions that are affordable, protect the climate, and contribute to a circular economy. Transitioning to circular economic and production systems, which mimic the self-sustaining closed-loop systems found in nature, such as the water cycle, can significantly reduce the consumption of natural resources and the amount of waste generated. A circular economy aims to transform waste back into a resource by reversing the dominant linear trend of extracting, processing, consuming, or using and then disposing of raw materials. This approach preserves natural resources, maintains economic growth, and minimizes environmental impacts. By improving resource and energy efficiencies, a circular economy can reduce environmental impacts, such as global warming (Cobo et al. 2018, Christensen et al. 2020). The environmental aspects and possible environmental impacts of waste management systems could be assisted by Life Cycle Assessment (LCA) (Torkayesh et al. 2022).

LCA assesses the potential environmental impact and resource usage within a waste management system that includes all relevant waste types. This assessment considers an integrated system that begins with waste generation and includes the transport, treatment, and disposal of various fractions and residues, along with material and energy exchange with the surrounding society. The broad scope of a waste management LCA is valuable for establishing an overview of all system flows, avoiding residues and emissions for both upstream exchanges (such as material and energy usage) and downstream exchanges (such as the use of recovered materials, energy, and nutrients). Additionally, it highlights the long-term considerations associated with landfilling and the use of residue in construction materials (Christensen et al. 2020).

However, the current LCA models face significant limitations, including the incompleteness of accounting for the environmental impacts of microplastics, which are often omitted in Life Cycle Impact Assessment (LCIA) models. This gap results in an underestimation of the environmental footprint of plastic waste, particularly its pervasive presence in marine and freshwater ecosystems. Moreover, plastics' end-of-life options, such as waste-to-energy and landfilling, are problematic because they remove plastics from the circular loop, necessitating the extraction of raw materials for new production (Brander et al. 2024). To enhance the effectiveness of LCA in supporting circular economy goals, it is crucial to address these gaps by including the impacts of microplastics and by prioritizing waste management strategies that keep materials within the loop, such as recycling, rather than disposal options that lead to resource loss (Kamalakkannan et al. 2022).

3.5 CANAL-BANK COMMUNITY BEHAVIOR AND WASTE MANAGEMENT

The effective separation of waste at its source in riverine communities is crucial for SDG14.1.1.b. As stated by the United Nations, Goal 14 is about conserving and sustaining the oceans, seas, and marine resources, specifically addressing floating plastic debris. Nevertheless, the global promotion of waste separation at the source remains a significant challenge. According to Heydari et al. (2021) and Zambrano-Monserrate et al. (2020), the household waste separation rate remains notably low, particularly in developing countries. This section aims to elucidate the influencing factors related to waste separation and present the learning outcomes derived from a select group of countries, drawing insights from literature spanning from 2014 to 2024.

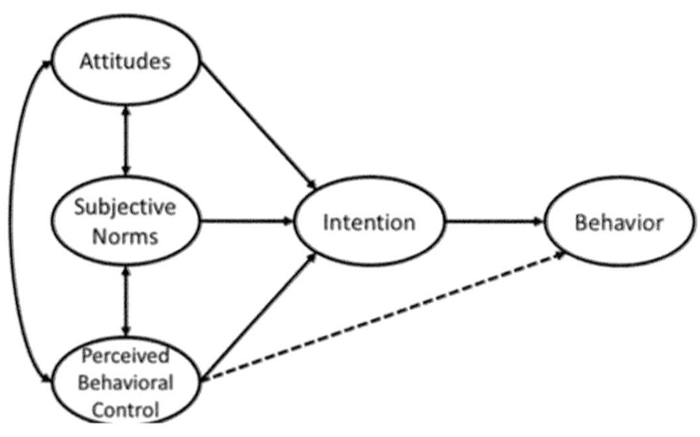

Figure 3.8 Theory of planned behavior (Ajzen and Fishbein 2005).

In terms of research on waste separation at its source, Shi et al. (2021) highlight the increasing focus on understanding the factors influencing recycling behavior. The widely employed theory of planned behavior (TPB), a psychological and economic framework, offers valuable insights (Figure 3.8). This approach is seen as more effective in fostering long-term commitment to waste separation than analyzing social and demographic attributes. However, previous research on social and demographic factors has highlighted the significant influence of gender, indicating a positive correlation with waste separation intention in females. Furthermore, non-linear relationships regarding waste separation intention have been observed between income and age. These findings are supported by data from Vietnam (Nguyen 2023), South Africa (Roos et al. 2022), and China (Zhang and Wen 2014; Xu et al. 2017a; Heydari et al. 2021).

The methodology, aligned with the Theory of Planned Behavior, emphasizes the influential factors of attitude, subjective norms, and perceived behavioral control on waste Separation Intention and behavior.

The research findings demonstrated the following outcomes. The intention to separate waste was influenced by attitude, as seen in observations from Thailand (Leknoi et al. 2024), Korea (Arachchilage and Chang 2024), Vietnam (Nguyen 2023), Ghana (Arkorful et al. 2023), China (Liao et al. 2018; Ma et al. 2018; Yu et al. 2018; Yang et al. 2021), Mongolia (Negash et al. 2021), and Malaysia (Ayob et al. 2017). Additionally, Arachchilage and Chang (2024) and Chuah et al. (2023) discussed the impact of environmental knowledge on attitudes and emphasized the influence of moral obligation on attitudes. The behavior of waste separation was influenced by attitudes, as demonstrated in the observations from Iran (Fallah-Nejad et al. 2023), Bolivia (Lozano Lazo and Gasparatos 2022), China (Liao et al. 2018; Yu et al. 2018), subjective norm, as evidenced in the observations from China (Xu et al. 2017b; Yu et al. 2018), and perceived behavioral control, as evident in the observations from China (Yuan et al. 2016; Xu et al. 2017a; Liao et al. 2018; Ma et al. 2018; Yu et al. 2018), Malaysia (Ayob et al. 2017). Furthermore, intention played a role, as evident in the observations from Malaysia (Yu et al. 2022) and China (Xu et al. 2017b; Wu et al. 2021). Several studies have investigated the gap between waste separation intentions and behavior. For instance, Cheng et al. (2024) used donations to measure and found that about half of the people donated less than they intended. Additionally, Li et al. (2022) suggested three main aspects for promotion to bridge this gap: ecological livability status, policy affect perception, and villagers' environmental protection behavior perception.

Furthermore, previous studies indicate that various critical factors influence waste separation behavior. These include the availability of recycling infrastructure, such as accessible recycling bins (Zhang and Wen 2014; Carchesio et al. 2015; Liao et al. 2018; Oyekale 2018; Qi and Liu 2021; Shabanova 2021; Yang et al. 2021; Pongpunpurt et al. 2022; Zhang et al. 2022; Cheng et al. 2023), trust in the recycling system (Shabanova 2021; Leknoi et al. 2024), knowledge (Carchesio et al. 2015; Briguglio et al. 2016; Challcharoenwattana and Pharino 2016; Ma et al. 2018; Zambrano-Monserrate et al. 2020; Wu et al. 2021; Cheng et al. 2023), individual emotions, moral obligations, social incentives and sanctions, perceptions (Lou et al. 2020; Shabanova 2021; Kim 2023) and policies (Zhang and Wen 2014; Challcharoenwattana and Pharino 2016; Xu et al. 2017a; Wu et al. 2021; Zhao et al. 2021; Roos et al. 2022).

Strategies to improve waste separation near the river for marine debris protection involve enhancing facilities, conducting educational campaigns, implementing government incentives, and promoting individual intentions and attitudes.

3.6 CONCLUSIONS

- Three key findings that stand out from the presented research include: (1) Canals can be major sources and transport pathways of riverine debris to the ocean. (2) Take-out consumer items greatly contribute to urban water plastic pollution. (3) Unintentional waste disposal, i.e., overflow from collection points and hotspots, significantly contributed to canal pollution. The distance between residential areas or hotspots and canals is one of the determining factors of waste leakage into water bodies. Socio-economic factors are significantly correlated with the abundance of plastic litter in waters.
- Based on the case study conducted over 20 months at the Lat Luang canal that flows through a densely populated community in Bangkok and discharges into the Chao Phraya River, the debris trapped at the sluice gates was mainly rigid single-use containers, e.g. bottles, boxes, and cups (PET, HDPE, PP); thin films (e.g. soft plastic sheets for food packaging); and multimaterial items. The urban lifestyle contributes significantly to the existence of food-contact plastic bags in riverine waste.
- The influencing factors for waste separation, as attained through learning outcomes from a select group of countries, include social and demographic factors, which are then reflected through attitude, perceived behavioral control, and subjective norms.

3.7 ACKNOWLEDGEMENTS

The authors thank Tito's Handmade Vodka and the donors of The Ocean Cleanup for partial funding of the research.

REFERENCES

Ajzen, I. and M. Fishbein. 2005. The influence of attitudes on behavior. pp. 173–221. *In*: D. Albarracín, B.T. Johnson and M.P. Zanna (eds). The Handbook of Attitudes. Lawrence Erlbaum Associates Publishers, Mahwah, NJ, US.

Arachchilage, C.S.M.W. and M. Chang. 2024. A study on factors influencing behavioral intentions for separating recyclable waste in Korea. J. Mar. Isl. Cult. 13.

Arkorful, V.E., Z. Shuliang and B.K. Lugu. 2023. Investigating household waste separation behavior: the salience of an integrated norm activation model and the theory of planned behavior. J. Environ. Plann. Manage. 66: 2195–2221.

Ayob, S.F., L. Sheau-Ting, R. Abdul Jalil and H.-C. Chin. 2017. Key determinants of waste separation intention: Empirical application of TPB. Facilities. 35: 696–708.

Blondel, E. and F.A. Buschman. 2022. Vertical and horizontal plastic litter distribution in a bend of a Tidal river. Front. Environ. Sci. 10.

Brander, S.M., K. Senathirajah, M.O. Fernandez, J.S. Weis, E. Kumar, A. Jahnke, et al. 2024. The time for ambitious action is now: Science-based recommendations for plastic chemicals to inform an effective global plastic treaty. Sci. Total Environ. 949: 174881.

Briguglio, M., L. Delaney and A. Wood. 2016. Voluntary recycling despite disincentives. J. Environ. Plann. Manage. 59: 1751–1774.

Carchesio, M., F. Tatàno, M. Goffi and M. Radi. 2015. Environmental and social sustainability of the proximity waste collection system: A case-study evaluation at an Italian local scale. Sustainability 7: 7492–7511.

Challcharoenwattana, A. and C. Pharino. 2016. Wishing to finance a recycling program? Willingness-to-pay study for enhancing municipal solid waste recycling in urban settlements in Thailand. Habitat Int. 51: 23–30.

Chen, Q., Q. Wang, C. Zhang, J. Zhang, Z. Dong and Q. Xu. 2021. Aging simulation of thin-film plastics in different environments to examine the formation of microplastic. Water Res. 202: 117462.

Cheng, X., W. Li, J. Yang and L. Zhang. 2023. How convenience and informational tools shape waste separation behavior: A social network approach. Resour. Policy. 86: 104152.

Cheng, X., L. Zhang, W. Li and R. Long. 2024. Intentions and actions in household waste separation: A machine learning approach on the gap and determinants. Environ. Impact Assess. Rev. 107: 107566.

Christensen, T.H., A. Damgaard, J. Levis, Y. Zhao, A. Björklund, U. Arena, et al. 2020. Application of LCA modelling in integrated waste management. Waste Manage. 118: 313–322.

Chuah, S.-C., L. Putit and J.S.K. Singh. 2023. Assessing residents' intention towards municipal solid waste source separation: A case study of Malaysia. Planning Malaysia. 21.

Cobo, S., A. Dominguez-Ramos and A. Irabien. 2018. From linear to circular integrated waste management systems: A review of methodological approaches. Resour. Conserv. Recycl. 135: 279–295.

Dihan, M.R., T. Naha and K. Kirtania. 2024. Comparative assessment of waste-to-energy scenarios to mitigate GHG emission from MSW in a developing mega city. Energy Sustainable Dev. 82: 101540.

DMCR. 2022. Thailand's Statistics of Marine Debris in 2022. Department of Marine and Coastal Resources.

Fallah-Nejad, M.J., M. Shamsi, R.D. Cheshmeh Soltani, M. Khorsandi and R. Moradzadeh. 2023. The effect of educational intervention based on the theory of planned behavior on the promotion of waste separation at source behavior in primary school students in Khorramabad, Iran. Health Serv. Res. 19: 192–199.

González-Fernández, D., A. Cózar, G. Hanke, J. Viejo, C. Morales-Caselles, R. Bakiu, et al. 2021. Floating macrolitter leaked from Europe into the ocean. Nat. Sustainability 4: 474–483.

Gundogdu, S., C. Cevik, B. Ayat, B. Aydogan and S. Karaca. 2018. How microplastics quantities increase with flood events? An example from Mersin Bay NE Levantine coast of Turkey. Environ Pollut. 239: 342–350.

He, Y., J. Lu, C. Li, X. Wang, C. Jiang, L. Zhu, et al. 2024. From pollution to solutions: Insights into the sources, transport and management of plastic debris in pristine and urban rivers. Environ Res. 245: 118024.

Heydari, E., M. Solhi, L. Janani and M. Farzadkia. 2021. Determinants of sustainability in recycling of municipal solid waste: Application of community-based social marketing (CBSM). Challenges in Sustainability. 9: 16–27.

Hoornweg, D., P. Bhada-Tata and C. Kennedy. 2013. Environment: Waste production must peak this century. Nature. 502: 615–617.

Ishigaki, T., R. Thaweesub, R. Kubota, R. Tajima, P.K. Lieu and C. Chiemchaisri. 2018. Investigation on solid debris in urban drainage system in tropical Asian cities. Proceedings of the Annual Conference of Japan Society of Material Cycles and Waste Management. 29: 541.

Jiao, M., L. Ren, Y. Wang, C. Ding, T. Li, S. Cao, et al. 2022. Mangrove forest: An important coastal ecosystem to intercept river microplastics. Environ. Res. 210: 112939.

Kamalakkannan, S., A. Abeynayaka, A.K. Kulatunga, R.K. Singh, M. Tatsuno and P.J.D. Gamaralalage. 2022. Life cycle assessment of selected single-use plastic products towards evidence-based policy recommendations in Sri Lanka. Sustainability. 14: 14170.

Kim, S. 2023. Antecedents of compliance intention and its impact on waste separation behavior: Based on rational choice theory and deterrence theory. Behav. Sci. 13: 424.

Lebreton, L., M. Kooi, T. Mani, S.M. Mintenig, M.B. Tekman, T. van Emmerik, et al. 2022. Plastics in Freshwater Bodies. 199–225. Plastics and the Ocean.

Leknoi, U., A. Yiengthaisong and S. Likitlersuang. 2024. Social factors influencing waste separation behaviour among the multi-class residents in a megacity: A Survey analysis from a community in Bangkok, Thailand. Sustainable Futures. 7: 100202.

Li, X.-Y., S.-W. Huang, Q. Lin, Q.-J. Lu and Y.-S. Zhang. 2022. A Study on the factors influencing the intention and behavior deviation of rural residents in waste separation—Based on LOGIT-ISM-MICMAC combination model. Sustainability. 14: 15481.

Liao, C., D. Zhao and S. Zhang. 2018. Psychological and conditional factors influencing staff's takeaway waste separation intention: An application of the extended theory of planned behavior. Sustainable Cities Soc. 41: 186–194.

Lithner, D., A. Larsson and G. Dave. 2011. Environmental and health hazard ranking and assessment of plastic polymers based on chemical composition. Sci. Total Environ. 409: 3309–3324.

Lou, T., D. Wang, H. Chen and D. Niu. 2020. Different perceptions of belief: Predicting household solid waste separation behavior of urban and rural residents in China. Sustainability. 12: 7778.

Lozano Lazo, D.P. and A. Gasparatos. 2022. Factors influencing household-level positive and negative solid waste management practices in rapidly urbanizing cities: Insights from Santa Cruz de la Sierra, Bolivia. Environ. Res.: Infrastruct. Sustainability. 2: 015002.

Ma, J., K.W. Hipel, M.L. Hanson, X. Cai and Y. Liu. 2018. An analysis of influencing factors on municipal solid waste source-separated collection behavior in Guilin, China by Using the Theory of Planned Behavior. Sustainable Cities Soc. 37: 336–343.

Mani, T., R. Ebner, L. Lebreton, P. Khamdahsag, D. Phihusut, Y. Hawangchu, et al. 2025. The tidal trap — Seasonal transport of floating macrodebris in the bi-directional Chao Phraya River network on the Gulf of Thailand. Mar. Pollut. Bull. 212: 117605.

Meijer, L.J.J., T. Van Emmerik, R. Van Der Ent, C. Schmidt and L. Lebreton. 2021. More than 1000 rivers account for 80% of global riverine plastic emissions into the ocean. Sci. Adv. **7**: eaaz5803.

Mellink, Y.A.M., T.H.M. van Emmerik and T. Mani. 2024. Wind- and rain-driven macroplastic mobilization and transport on land. Sci. Rep. 14: 3898.

Morales-Caselles, C., J. Viejo, E. Martí, D. González-Fernández, H. Pragnell-Raasch, J.I. González-Gordillo, et al. 2021. An inshore–offshore sorting system revealed from global classification of ocean litter. Nat. Sustainability. 4: 484–493.

Negash, Y.T., A.M. Hassan, B. Batbaatar and P.-K. Lin. 2021. Household waste separation intentions in Mongolia: Persuasive communication leads to perceived convenience and behavioral control. Sustainability. 13: 11346.

Nguyen, T.K.N. 2023. The determinants of individuals' waste separation intention in an urbanizing city: A case study of Hanoi, Vietnam. Habitat Int. 137: 102835.

Oyekale, A.S. 2018. Determinants of households' involvement in waste separation and collection for recycling in South Africa. Environ. Dev. Sustainability. 20: 2343–2371.

Poletti, S.A. and T. Landberg. 2021. Using nature preserve creek cleanups to quantify anthropogenic litter accumulation in an urban watershed. Freshw. Sci. 40: 537–550.

Pongpunpurt, P., P. Muensitthiroj, P. Pinitjitsamut, P. Chuenchum, P. Painmanakul, N. Chawaloesphonsiya, et al. 2022. Studying waste separation behaviors and environmental impacts toward sustainable solid waste management: A case study of Bang Chalong Housing, Samut Prakan, Thailand. Sustainability. 14: 5040.

Qi, X. and P.-P. Liu. 2021. Construal level and guilt promote pro-environmental behavioral intention: Based on analysis of variance. 2021 2nd International Conference on Information Science and Education (ICISE-IE). 743–748.

Roos, C., C. Baron, D. Cilliers, R. Alberts, F. Retief and J. Moolman. 2022. Investigating waste separation at source behaviour among South African households: the case of Abaqulusi Local Municipality. S. Afr. Geogr. J. 104: 467–483.

Ryan, P.G. 2015. A brief history of marine litter research. 1–25. Marine Anthropogenic Litter.

Shabanova, M.A. 2021. Separate waste collection as Russians' voluntary practice: The dynamics, factors and Potential. Sotsiologicheskie issledovaniya.

Shi, K., Y. Zhou and Z. Zhang. 2021. Mapping the research trends of household waste recycling: A bibliometric analysis. Sustainability. 13: 6029.

Tajima, R., R. Kubota and S. Janamporn. 2019. Generation of and solutions to canal waste: A case study in Bangkok. Proceedings of the Annual Conference of Japan Society of Material Cycles and Waste Management. 30: 511.

Torkayesh, A.E., M.A. Rajaeifar, M. Rostom, B. Malmir, M. Yazdani, S. Suh, et al. 2022. Integrating life cycle assessment and multi criteria decision making for sustainable waste management: Key issues and recommendations for future studies. Renew. Sustain. Energy Rev. 168: 112819.

Ubando, A.T., J.A. Gonzaga, I.H.V. Gue and J.B.M. Biona. 2024. Life cycle assessment of plastic and paper carrying bags in the Philippines. Cleaner Environ. Syst. 14: 100217.

van Emmerik, T. and A. Schwarz. 2020. Plastic debris in rivers. WIREs Water. **7**: e1398.

Velis, C.A. and E. Cook. 2021. Mismanagement of plastic waste through open burning with emphasis on the global South: A systematic review of risks to occupational and public health. Environ. Sci. Technol. 55: 7186–7207.

Weideman, E.A., V. Perold and P.G. Ryan. 2020. Limited long-distance transport of plastic pollution by the Orange-Vaal River system, South Africa. Sci. Total Environ. 727: 138653.

Wu, X., L. Zhang, J. Huang, W. Li, Y. Chen and W. Qiu. 2021. Evolutionary game analysis on behavioral strategies of government and residents in municipal household waste separation. Sustainability. 13: 11421.

Xu, L., M. Ling, Y. Lu and M. Shen. 2017a. External influences on forming residents' waste separation behaviour: Evidence from households in Hangzhou, China. Habitat Int. 63: 21–33.

Xu, L., M. Ling, Y. Lu and M. Shen. 2017b. Understanding household waste separation behaviour: Testing the roles of moral, past experience and perceived policy effectiveness within the theory of planned behaviour. Sustainability. 9: 625.

Yang, X., X. Chen, X. Xiao, H. Xi and S. Liu. 2021. College Students' Willingness to separate municipal waste and its influencing factors: A case study in Chongqing, China. Sustainability. 13: 12914.

Yin, L., X. Wen, C. Du, J. Jiang, L. Wu, Y. Zhang, et al. 2020. Comparison of the abundance of microplastics between rural and urban areas: A case study from East Dongting Lake. Chemosphere. 244: 125486.

Yu, S., T. Lu, X. Qian and W. Zhou. 2018. Behavioral intention analysis of waste separation in China—Case study of Hangzhou using theory of planned behavior. Int. Rev. Spat. Plan. Sustain. Dev. 6: 63–77.

Yu, P.L., N. Ab Ghafar, M. Adam and H.C. Goh. 2022. Understanding the human dimensions of recycling and source separation practices at the household level: An evidence in Perak, Malaysia. Sustainability. 14: 8023.

Yuan, Y., H. Nomura, Y. Takahashi and M. Yabe. 2016. Model of Chinese household kitchen waste separation behavior: A case study in Beijing City. Sustainability. 8: 1083.

Zambrano-Monserrate, M.A., M.A. Ruano and C. Yoong-Parraga. 2020. Households from developing countries do not sort their solid waste: truth or myth? J. Environ. Plann. Manage. 63: 2577–2592.

Zhang, H. and Z.-G. Wen. 2014. Residents' household solid waste (HSW) source separation activity: A case study of Suzhou, China. Sustainability. 6: 6446–6466.

Zhang, Y., G. Wang, Q. Zhang, Y. Ji and H. Xu. 2022. What determines urban household intention and behavior of solid waste separation? A case study in China. Environ. Impact Assess. Rev. 93: 106728.

Zhao, Y., R. Chen, M. Yabe, B. Han and P. Liu. 2021. I am better than others: Waste management policies and self-enhancement bias. Sustainability. 13: 13257.

Zhou, G., Q. Wang, J. Zhang, Q. Li, Y. Wang, M. Wang, et al. 2020. Distribution and characteristics of microplastics in urban waters of seven cities in the Tuojiang River basin, China. Environ. Res. 189: 109893.

Chapter 4

Current Status of Plastic Waste Pollution and Management in Vietnam: Challenges and Opportunities Toward a Circular Economy

Bui Quoc Lap*[1], Pham Thi Thom[2], Diep Thi Thu Thuy[3] and Huyen Thi Thanh Dang[4]

[1]Faculty of Chemistry and Environment, Thuyloi University, 175 Tay Son Street, Dong Da District, Hanoi, Vietnam

[2]Faculty of Natural Resources and Environment, Bac Giang Agriculture and Forestry University, Viet Yen Ward, Bac Ninh Province, Vietnam

[3]Faculty of Environment, Ha Long University, 258 Bach Dang Street, Uong Bi Ward, Quang Ninh Province, Vietnam

[4]Faculty of Environmental Engineering, Hanoi University of Civil Engineering, Hanoi, Vietnam

4.1 INTRODUCTION

According to the World Bank report in April 2024, Vietnam is one of the countries that has successfully reformed and undergone significant development, i.e., from

*For Correspondence: buiquoclap@tlu.edu.vn

DOI: 10.1201/9781003531463-4

one of the poorest countries in the world in the 1990s to a lower-middle-income country in recent years. Nevertheless, the rapid economic growth has created some environmental issues. Plastic waste (PW) is one of them. In 2021, the total collection of PW amounted to 6,581 tons/day, corresponding to an annual figure of 2.93 million tons/year (WWF 2023). Even though the plastic collection rate in some cities was high (i.e., Hanoi 99%, Ho Chi Minh City 100%), the rate of plastic recycling has been as low as 26.3% (WWF 2023). The rest has been dumped into the environment, causing both immediate and long-term influences.

The PW leakage has caused water, soil, and air pollution. Specifically, the plastic pieces decomposed in the soil environment can change soil composition and structure, porosity, enzymatic activities, nutrient transport, etc. It reduces the permeability of rainwater and irrigation water, negatively affecting the water-holding capacity, and can cause soil anaerobic conditions (Sajjad et al. 2022; WWF 2023). The microplastics, which originate from the deterioration of discarded plastic products through physical, chemical, and biological processes, cause a significant impact on the environment. They are claimed to play an important role in the transportation of toxic chemicals such as plasticizers, polycyclic aromatic hydrocarbons (PAHs), antibiotics, and potentially toxic elements (PTEs) into the soil (Sajjad et al. 2022). Most studies have been conducted on the abundance and characteristics of microplastics in seawater, followed by freshwater, wastewater, and groundwater (Kye et al. 2023).

Besides the significant impacts on the environment, the PWs have some negative impacts on socio-economic development. The clearest impact is on tourism. The floating PW on some famous beaches and bays makes them less aesthetic than they are supposed to be. For instance, Phu Quoc Island, one of the best beaches in Vietnam, has been suffering with about 155 tons of dry solid waste generated per day, of which only 91 tons were collected (WWF and USAID 2018).

Vietnam has taken strong actions to reduce PW pollution through comprehensive institutionalization in managing solid wastes (SW) in general and the PW in particular. This is expressed in both national strategies and legal frameworks. A number of measures have been introduced. These include an impending ban on the production and import of nylon bags for domestic use by 2026 and most single-use plastic products by 2031. A Policy Roadmap to reduce the use of the most common single-use plastics (SUP) was recommended by the World Bank to support the Government's plastic pollution reduction programs. In addition, the "Plastic Reduction Cities" program has been implemented since 2018 to promote the "Plastic Reduction Cities" initiative—a global initiative of the World Wide Fund for Nature (WWF). To date, 11 localities in Vietnam have signed commitments with the International Organization for Conservation of Nature to improve the management, collection, and classification of household solid waste, and reduce PW according to the model "Plastic-reducing urban areas" (WWF 2024).

Vietnam's Law on Environmental Protection 2020 stipulates that organizations and individuals are responsible for limiting the use, minimizing, classifying, and disposing of waste that is single-use plastic products and non-biodegradable plastic packaging according to regulations. It is not allowed to dispose of PW directly into drainage systems, ponds, lakes, canals, rivers, or oceans.

Considering technical aspects, a number of solutions can be implemented to increase the reuse of PWs and reduce the use of non-biodegradable plastics; however, one critical step is to raise community awareness and behavior toward plastic products. Limiting PW requires the consensus and joint contributions of individuals, organizations, unions, and social classes, starting from the smallest actions of each person to the macro policies of the country.

The following section will not only detail the current status of PW generation, treatment, and management in Vietnam but also reveal the activities from local to national levels to improve PW management, in particular, to address the challenges and opportunities to handle this issue toward a circular economy.

4.2 CURRENT STATUS OF PLASTIC PRODUCTION AND CONSUMPTION

Compared to long-established industries such as mechanical engineering, electronics, textiles, and chemicals, the plastic industry is relatively new in Vietnam but has experienced rapid growth in recent years. During the period from 2010 to 2020, the plastic industry was one of the fastest-growing industries with an annual growth rate ranging from 16 to 18%. Some product categories even experienced growth rates close to 100% per year. The growth stems from a broad market, significant potential, and notably because Vietnam's plastic industry is still in its early stages of development compared to the global scale. Plastic products are used in various aspects of life, such as packaging, construction materials, household items, and high-tech applications (Tu 2022).

Up to now, the industry has over 4,000 enterprises, with more than 99% being privately owned businesses, mainly concentrated in the southern region (accounting for over 80% of plastic enterprises nationwide). This concentration is due to the high presence of industrial processing zones, which serve as outlets for plastic packaging products. The majority of plastic industry enterprises are small and medium-sized, with over 90% of Vietnamese plastic enterprises engaged in processing for foreign markets without establishing their own brands. In recent years, many plastic enterprises have been investing in higher-tech solutions from South Korea and Japan for technical plastic products, or German technology for construction plastic products (Tu 2022). At present, Vietnamese plastic products are in place in 151 markets worldwide, including markets that demand high quality and optimal technical standards, such as Japan, the United States, and the EU.

Vietnam has a strength in producing plastic items, particularly in packaging, consumer goods, construction materials, and some advanced products like oil pipes, automotive components, and computer parts. However, the range of designs and types of Vietnamese plastic products remains limited, and there are few high-value-added products (Table 4.1).

Plastic packaging currently holds the largest market share in Vietnam's plastic industry. In the packaging sector, plastics are used in diverse areas such as healthcare, food and beverages, consumer goods, personal care products, home care, and gardening. Looking ahead, this industry is expected to become one of

Table 4.1 The main plastic product groups produced by Vietnam

Products	Proportion	Main Technology	Main Materials	Main Products	Main Output
Packaging plastic	39%	Blow molding technology	Polyethylene (PE), Polypropylene (PP), Polyethylene Terephthalate (PET)	Thin film packaging, plastic bags, and PET plastic bottles	Food and beverage processing enterprises, retail systems, and supermarkets
Household plastic	32%	Injection molding technology	PP, Polystyrene (PS), and Acrylonitrile Butadiene Styrene (ABS)	Household products	Consumers
Construction plastic	14%	Extrusion molding technology	PE and Polyvinyl Chloride (PVC)	Plastic pipes for construction, plastic building materials	Real estate and construction companies, consumers
Technical plastic	9%	Injection blow molding technology	PVC, PP, and Polyurethane (PU)	Types of components and accessories	Electronics, electrical, automotive, and motorcycle companies

Source: Le Anh Tu, "Overview of the Vietnamese Plastics Industry," 2022.

the fastest-growing in Vietnam, with growth rates projected to range from 15 to 20% (Carnovn 2023).

The consumption of plastics in Vietnam has been continuously increasing over the past several years, at a rate of about 10.6% annually. According to statistics from the Vietnam Plastics Association, the production and consumption of plastic in the country were approximately 5 million tons in 2015, of which 80% of the raw materials for plastic production were imported. In 2019, the average plastic consumption per capita was 41 kg per person, which is more than 10 times the 3.8 kg per person recorded in 1990. As plastic consumption continues to rise, Vietnam has become one of the countries with the highest levels of plastic waste generation globally (VPAS 2024).

4.3 CURRENT STATUS OF PLASTIC WASTE GENERATION

4.3.1 Categories and Sources of Plastic Waste

Plastic waste can originate from various sources, including domestic solid waste (DSW), agricultural and industrial solid waste, construction solid waste, and other waste from different activities. DSW is the primary source of PW leakage into the environment, as other sources are relatively well managed.

Plastic waste in DSW originates from various sources and can be categorized based on its composition. Here are the main categories and sources of PW:

- **Single-Use Plastics (SUPs):** These are disposable plastic items intended for one-time use before being discarded. SUPs include items like plastic bags, straws, disposable utensils, food wrappers, and beverage containers (Lindwall 2024).
- **Packaging Waste***:* Packaging materials contribute significantly to PW in DSW. This category encompasses food packaging, beverage bottles, plastic wraps, blister packs, and other packaging used for consumer goods.
- **Textile Waste:** Synthetic fibers from clothing and textiles contribute to plastic pollution when disposed of. Common synthetic fibers include polyester, nylon, and acrylic, which are found in clothing, linens, and household textiles.
- **Microplastics:** Microplastics are tiny plastic particles less than 5 mm in size. They can originate from the breakdown of larger plastic items, such as packaging or textiles, or from microbeads used in personal care products like exfoliating scrubs and toothpaste.
- **Construction and Building Materials***:* Plastics are commonly used in construction for items such as pipes, fittings, insulation, and siding. When buildings are demolished or renovated, these materials contribute to PW in DSW.

- **Electronics and Appliances:** Electronic devices and appliances contain plastic components, including casings, wiring insulation, and circuit boards. Discarded electronics contribute to electronic waste (e-waste), which contains a significant amount of plastic.
- **Fishing Gear:** Lost or discarded fishing gear, such as nets, lines, and traps, is a major source of plastic pollution in oceans and waterways. This category includes both commercial fishing gear and recreational fishing equipment.
- **Personal Care and Household Products:** Plastics are commonly used in personal care items like toothbrushes, shampoo bottles, and cleaning product containers. When these items are discarded, they contribute to PW in DSW.

The results of the analysis from surveys conducted in 34 provinces and cities of Vietnam indicated that the proportion of PW in DSW ranged widely from 2.6 to 26%, with an average of 12% (this ratio is related to the level of socioeconomic development and people's habits). The volume of PW generated in provinces and cities across the country has been increasing in recent years. From 2018 to 2021, both the Inland and Coastal regions demonstrated steady growth. During these years, the coastal region consistently posted higher figures compared to the Inland one. The disparity between the two was most significant in 2019 and has been diminished gradually by 2021 (Figure 4.1) (WWF 2023). The predicted values (using the TREND function in Excel 2016) indicate an increasing trend in plastic

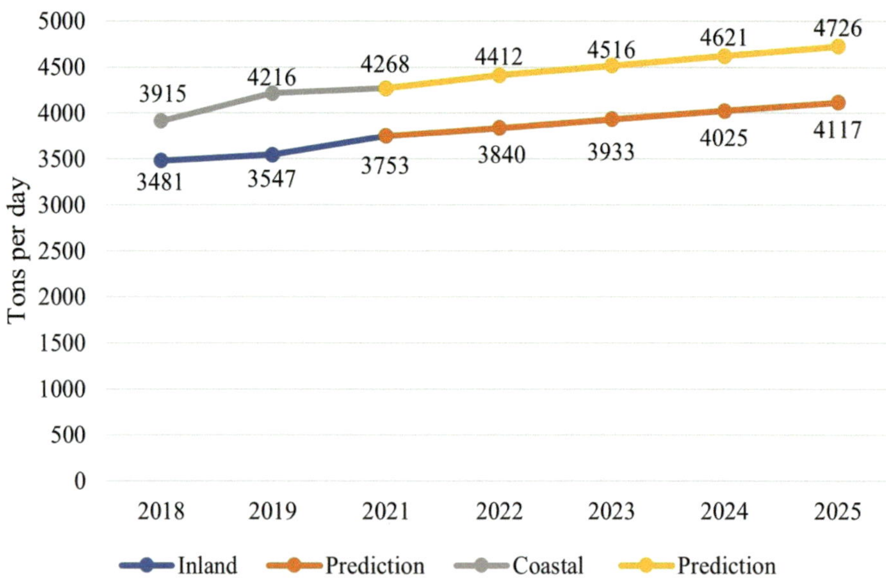

Data analyzed from: MONRE, WWF, Report on Plastic Waste generation in 2022. *Hanoi:* Thanh Nien Publishing House 2023; 6–120.

Figure 4.1 Plastic waste generated in Vietnam during the period from 2018 to 2021, and the forecast extends to 2025.

waste generation in both areas, in which the coastal area remains ahead of the inland region in terms of the generated volume across the entire time frame, with the gap expected to widen slightly by 2025. The predicted results also indicated that the generated volume in the coastal region will increase to 4726 tons/day by 2025 (over 10% increase compared to the volume of 2021). Similarly, by 2025, the generated volume in the inland areas will increase to 4,117 tons/day (nearly 10% increase compared to the volume generated in 2021). Notably, several tourism cities in Phu Quoc, Kien Giang Province, the Mui Ne tourist area, Phan Thiet City, Binh Thuan Province, Da Nang, Tuy Hoa, Quang Ninh, etc., all faced the overload of PW from boats and tourists (Minh et al. 2020). Specifically, in Ha Long Bay, Quang Ninh Province, an average of 4 tons of solid waste was generated per day, mainly from PW floating in the sea. In Da Nang City, 1,100 tons of waste were generated daily, of which PW accounts for 17%. In Tuy Hoa City, Phu Yen Province, 18.31% of PW from 524 tons of waste was discharged into the environment per day.

The generation of PW varied across different regions of Vietnam. It was related to the trends in socioeconomic development, demand, and people's habits regarding the consumption of plastic products. Figure 4.2 shows Vietnam's regions with their localities inside, and Table 4.2 indicates the amount of plastic waste generated and the generation rate among these regions in 2021.

In 2021, the total volume of PW generated in urban areas was approximately 4,460 tons/day, equivalent to 1.63 million tons/year. In rural areas, the total volume of PW generated was about 3,561 tons/day, corresponding to 1.3 million tons/year.

Table 4.2 The amount of plastic waste generated and the generation rate by economic region in 2021

Regions	Plastic Waste Generation (Tons Per Day)			Plastic Waste Per Capita (Kg Per Person Per Year)		
	Urban	Rural	All	Urban	Rural	All
Northern Midlands and Mountain Areas	300	462	762	41	16	22
Red River Delta	1,074	840	1,914	45	21	30
North Central and Central Coastal Areas	745	851	1,596	41	22	28
Central Highland	196	277	473	41	24	29
South East	1,612	226	1,838	48	13	37
Mekong River Delta	533	905	1,438	42	26	30
Total	4,460	3,561	8,021	43 (on average)	20.33 (on average)	29.33 (on average)

Source: Data synthesized from "National State of Environment Report 2019 and estimated results based on the reported environmental protection in 2021 of provinces and centrally-run cities" (MONRE 2020, 2022a).

The Red River Delta generated the highest total PW of 1,914 tons/day, with urban areas contributing 1,074 tons and rural areas generating 840 tons. Following

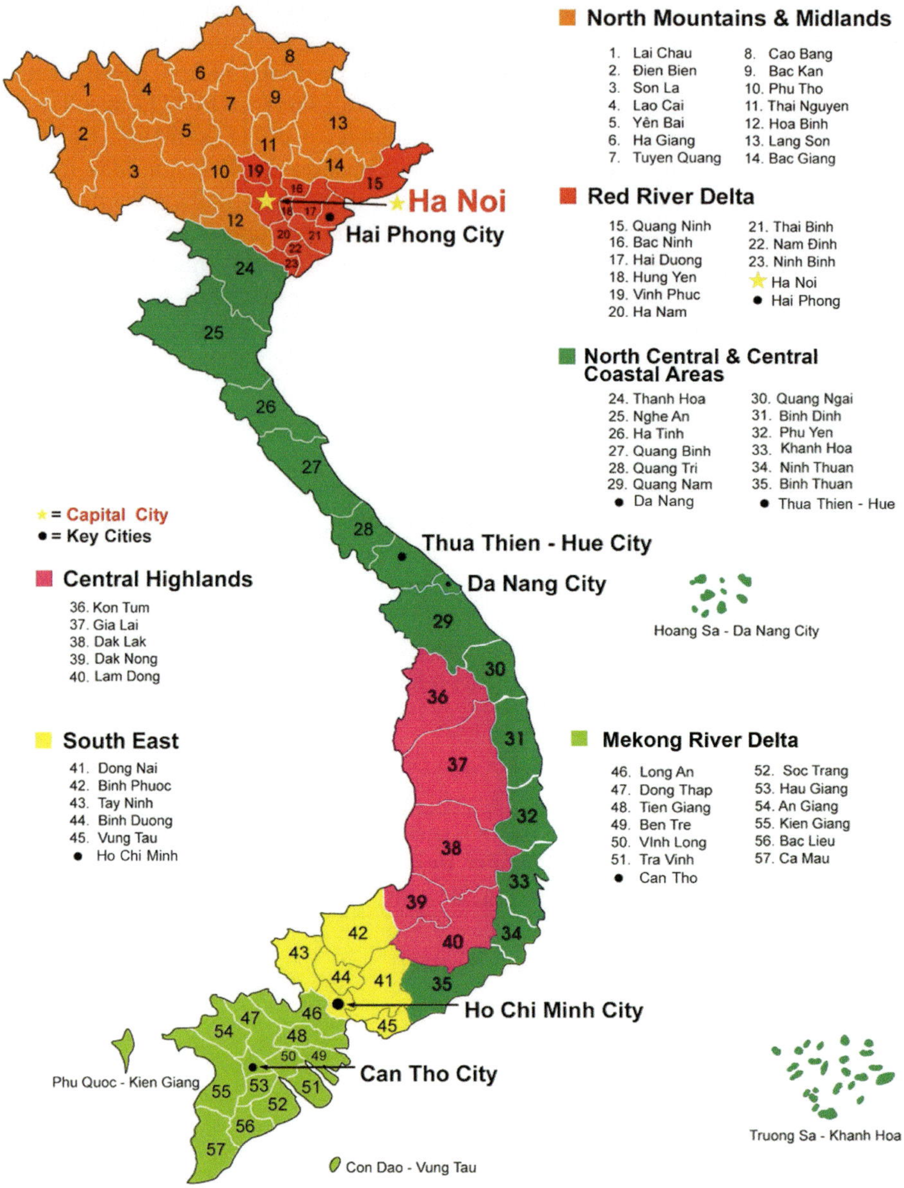

Figure 4.2 Map of Vietnam with its economic regions and the localities inside the regions.

closely, the Southeast Region produced 1,838 tons/day, with a significant difference between urban (1,612 tons) and rural (226 tons) contributions. The North Central and Central Coastal Areas, Mekong River Delta, and Northern Midlands and Mountain Areas all show a trend where rural areas generated more PW than urban areas. The lowest PW generation was in Central Highlands at 473 tons/day, with rural areas (277 tons) contributing more than urban areas (196 tons). Most urban areas had relatively consistent per capita PW generation rates ranging from 40 to 50 kg/person/year (43 kg/person/year on average). Areas with concentrated socioeconomic development activities, such as the Red River Delta, North Central and Central Coastal areas, and the South East region, had higher per capita PW generation rates compared to other regions. The Southeast Region's urban areas have the highest per capita PW at 48 kg/person/year, whereas the rural areas in the same region have the lowest volume of 13 kg/person/year. Table 4.2 also showed that per capita PW volume generated in urban areas (43 kg/person/year) was two times higher than that in rural areas (20.33 kg/person/year) on average in 2021.

4.3.2 Environmental Pollution by Plastic Waste and Microplastics

Plastic waste leakage into the environment can impact land, air, water, and sediment environments in various water bodies (lakes, rivers, seas, etc.). The transport and dispersion characteristics of PW in the environment depend on the size, weight, and properties of each type of plastic. High-density plastics like PVC tend to accumulate near the source of discharge, whereas lighter plastics like LDPE and PS can travel over longer distances. Additionally, hollow plastic particles and materials containing trapped air can float on water and move further away. Weather conditions, geographical features, and obstacles along transportation routes also influence the movement of plastics in the environment.

4.3.2.1 Soil Environment

The rapid increase in the generation of PW increasingly threatens the soil environment. Sources of plastic pollution in the soil environment include various types of plastics used directly in agricultural practices, such as plastic mulch films used for covering and greenhouse construction, packaging for fertilizers and pesticides, and other PW generated from household activities. PW on the soil surface will degrade into smaller pieces due to physical, chemical, and biological factors. These plastic fragments in the soil environment can be classified into two forms. Pieces smaller than 5 mm are typically identified as microplastics, while those smaller than 0.1 μm are identified as nanoplastics (Alabi et al. 2019; Zhang et al. 2022).

PW accumulation in soil can disrupt the balance of soil ecosystems by affecting the abundance and diversity of soil organisms, such as bacteria, fungi, earthworms, and insects. On the other hand, plastics are highly resistant to degradation, persisting in the soil environment for hundreds to thousands of years. Their long-term presence can pose a persistent threat to soil health, ecosystem integrity, and human health (Iqbal et al. 2020; Wan et al. 2019).

Despite the gravity of the issue, microplastic pollution in Vietnamese soil environments remains inadequately studied. A 2022 study analyzing microplastics in coal ash soil in Long An province found that the number of microplastic particles ranged from 0 to 360 particles/kg. The predominant components were microplastic fragments (70%), foam (17%), fibers (8.7%), and film forms (4.3%). The main chemical components were PP, PE, and PVC plastics (WWF 2023).

4.3.2.2 Atmospheric Environment

PW not only impacts terrestrial and aquatic environments but also has indirect effects on the atmospheric environment. The production, transportation, and disposal of PW contribute to greenhouse gas emissions. Manufacturing plastics often involves the combustion of fossil fuels, releasing carbon dioxide (CO_2) and other greenhouse gases into the atmosphere. Additionally, when PW is incinerated, it releases CO_2 and other pollutants, further contributing to climate change. Moreover, PW that has been landfilled eventually decomposes, releasing CO_2 and CH_4 into the air, and they can trap the emitted heat and hinder it from escaping from the Earth, causing global warming (Alabi et al. 2019). Improper disposal of PW, such as open burning, can release toxic chemicals and particulate matter into the air. Burning plastics emits harmful pollutants like dioxins, furans, and polycyclic aromatic hydrocarbons (PAHs), which can degrade air quality and pose health risks to humans and wildlife (Sarkingobir et al. 2021; Alabi et al. 2019). On the other hand, microplastics can become airborne through various processes such as weathering, fragmentation, and mechanical abrasion. Once airborne, microplastics can travel long distances and be dispersed by wind, leading to the formation of microplastic aerosols in the atmosphere. These aerosols may carry contaminants and contribute to air pollution.

In Vietnam, studies investigating and assessing microplastic pollution in the air are still relatively scarce, mainly focused on some big cities and urban areas. The results of surveys on microplastic density in the air in the Phuoc Hiep landfill area, Ho Chi Minh City, in 2022 showed a deposition rate of microplastics of up to 1,367 particles/m^2/day, which was 50 times higher than the results observed in Paris in the same year. Measurements in Ho Chi Minh City also detected plastic fragments ranging from 5 to 200 μm in size in the atmosphere. The plastic composition in the atmosphere was mainly in the form of fibers (accounting for 64%) and fragments (accounting for 36%). Plastic fragments were found to have higher densities during the dry season and lower densities during the rainy season (WWF 2023).

The survey results of 2022 on microplastics in road dust in Da Nang City showed an average density of 20 ± 14 particles/m^2, with some areas recorded at densities ranging from 22 to 40 particles/m^2 and an average size of 791 ± 530 μm. In areas with heavy traffic density, the size of microplastics was usually smaller. The predominant plastic components consisted of two main groups: packaging plastics (PET, PS, PP, and PE) and rubbers (PU, EPDM, SBR, and EPC). The shapes of microplastics mainly included fragments, films, fibers, and pellets, with fragments and films being the predominant forms (WWF 2023).

4.3.2.3 Water and Sediment Environment

PW has a profound impact on both water and sediment environments, affecting ecosystems and wildlife in various ways:

Plastic debris, ranging from large items like bottles and bags to microplastics, accumulates in water bodies such as rivers, lakes, and oceans. This accumulation alters the physical environment, affecting water flow patterns and sediment distribution. Large plastic items can entangle aquatic organisms, causing injury, suffocation, or death, while smaller plastics can be ingested, leading to internal blockages or toxicity. PW contains additives such as plasticizers, flame retardants, and colorants, which can leach into the water over time, posing risks to aquatic organisms, disrupting hormone systems, impairing reproduction, and weakening immune functions. PWs act as carriers for other pollutants, such as heavy metals, pesticides, and persistent organic pollutants (POPs). Plastic pollution alters aquatic habitats, reducing biodiversity and disrupting ecosystem functions (Yin et al. 2023).

Microplastics can accumulate in sediments over time. They disrupt the nutrient cycle, alter sediment properties, and affect the distribution and abundance of sediment-dwelling organisms.

The accumulation of PW in water bodies in Vietnam is increasing, causing pollution to the water environment and the sediment of these water bodies. In the year 2021, estimates revealed a staggering amount of PW leakage, reaching 1,152 tons/day, equivalent to an alarming 420,373 tons/year. Of particular concern is the significant portion of PW that finds its way into aquatic environments, with 66,900 tons/year, or 183 tons/day, posing a grave threat to marine ecosystems. The Mekong River Delta has emerged as a hotspot for plastic leakage, recording a daily emission of 56 tons, totalling 20,484 tons/year. In stark contrast, the Central Highlands exhibited the lowest PW leakage, with a modest rate of 19.8 tons/day, translating to 7,216 tons/year (MONRE 2022a). Moreover, the detrimental impact of PW extends beyond conventional sources, with aquaculture and fishing activities significantly contributing to environmental degradation. Reports from WWF-Vietnam underscore the scale of this issue, revealing that coastal fishing vessels in Vietnam alone generated a staggering 64,143 tons/year of PW, with 3,814 tons of it finding its way into marine environments. Similarly, lobster farming operations contributed 2,875 tons/year of PW, of which 139 tons infiltrate marine ecosystems. Shrimp farming, a vital economic activity, generated approximately 301,477 tons/year of PW, with an astonishing 164,644 tons attributed solely to pond liners (WWF 2020). A study conducted from July 2020 to April 2021 by the World Bank surveyed 24 locations in 5 main river basins in Vietnam, including the Red River, rivers in Central Vietnam, the Dong Nai—Saigon River, the Mekong River, and rivers in Phu Quoc. The research results identified the 10 most common types of plastics at riverbank sites, accounting for 82% (Mekong River) to 93% (Red River) of the total PW, including: plastic bags, food packaging boxes, soft plastic fragments, food packaging, plastic strapping, hard plastic fragments, other plastic packaging, fishing gear, and other plastics. Single-use plastic components are often more prevalent than other types of plastics, with the highest levels found in Can Tho City (25 pieces/0.3 m^3), Lao Cai Province (23 pieces/0.3 m^3), and Ho Chi Minh City (22 pieces/0.3 m^3) (WWF 2023).

The increasing amount of plastic debris on the surface of river water also leads to an increase in the volume of plastic settling at the riverbed. Observations in the Tien River and Hau River systems in 2022 showed that an average of

0.7–932 grams of PW per 100 square meters of riverbed was recorded. Large volumes of PW were observed at riverbed sections near the cities of Can Tho and Vinh Long. The volume of PW was higher during the dry season compared to the flood season (WWF 2023).

Normally, microplastics in freshwater were the first to be affected by land-based waste. At the same time, the microplastics in wastewater were the result of anthropogenic activity, and those in seawater were affected by the marine industry or received the final destination of inland waters (Kye et al. 2023). They found that major microplastic concentration (>10,000 particles/m^3) in the seawater was present near Europe and Asia regions, much less in Africa, America, and Australia regions (100–1,000 particles/m^3). In Vietnam, studies on microplastic pollution in water environments have also been conducted in several locations, including the Red River, the lower reaches of the Day River, the Nhue—To Lich River system, the Han River, the Saigon – Dong Nai River system, West Lake, Trị An Lake, and several lakes in Hanoi. The average microplastic density in river water in 2021, as determined by these studies, was 2.3 particles/m^3 in the Red River, 2.7 particles/m^3 in the Han River, and 3.9 particles/m^3 in the Dong Nai River. At the lower reaches of the Day River, microplastics in the water ranged from 269.96 ± 60.62 particles/m^3 to 863 ± 131.9 particles/m^3. Microplastic density in the river water was higher during the rainy season than during the dry season. The predominant forms of microplastics were fibers (accounting for 92–96%), while the remaining were fragments (only accounting for 3–7%). The main components of microplastics were PE and PP (WWF 2023). Microplastics in lake sediments in Ha Noi City ranged from 2,767 ± 240 to 2,833 ± 176 particles/kg with mostly PE and PP components (Le et al. 2022). Eventually, it affects the soil biota, plant growth, and crop harvesting (Wang et al. 2020).

4.3.2.4 Marine Environment

PW in the marine environment originates from activities both on land and at sea. Sources of land-based plastic emissions into the marine environment include rivers, coastal communities, and various coastal economic activities, such as tourism, port operations, and coastal economic zones. These sources account for over 80% of ocean PW. PW in the marine environment, due to its lightweight nature, is easily carried by currents, dispersed by waves and wind (Li et al. 2016).

Plastic debris can smother and suffocate marine habitats such as coral reefs, seagrass beds, and mangroves, disrupting ecosystems and reducing biodiversity. Floating PW can become entangled in coral reefs and cover their surfaces. Corals may ingest plastics, leading to growth impairment, bleaching phenomena, and disease outbreaks within coral communities. Pathogenic microorganisms can thrive on the surface of PW and cause diseases that contribute to coral bleaching. Seagrass beds can reduce water flow velocity, making it easier for PW to accumulate in this ecosystem. Large PW caught in seagrass beds can block sunlight, reducing the photosynthetic productivity of seagrass beds and promoting the growth of invasive plant species. Mangrove forests are also suffering significant damage from plastic pollution (Danh and Hoi 2019). Marine animals, including fish, seabirds, turtles, and marine mammals, often mistake plastic debris for food. Ingesting plastics can lead to internal injuries, blockages, and starvation. Plastic pollution affects marine-based industries such as tourism, fishing, aquaculture, etc.

Vietnam ranks fourth in Asia in terms of PW generation, with an average of at least 30 kg of plastic products used per person each year. Therefore, millions of tons of PW were dumped into the seas and oceans annually, causing serious impacts on marine resources, ecosystems, and the reproductive health of coastal communities (Danh and Hoi 2019). Major rivers in Vietnam, such as the Red River and the Mekong River, transport large volumes of freshwater, sediment, and, along with them, PW into the marine environment every year. Sources of PW generation at sea include fishing activities, aquaculture, maritime tourism, and transportation. Fishing activities and aquaculture are recognized as the primary sources of PW emissions at sea. The investigation conducted by the World Bank in 2022 has identified 10 common types of PW along the coast of Vietnam, including Type 1 fishing gear (ropes, fishing nets, bait and fishing lines, hard plastic buoys), Type 2 fishing gear (Polystyrenes—EPS, floating buoys, foam boxes), hard plastic fragments, plastic straws, plastic bags size 1 (size 0–5 kg), plastic bags size 2 (size > 5 kg), soft plastic fragments, food packaging, other plastic fragments, plastic packing straps, and single-use plastic items. The composition of different types of PW varies among the Northern, Central, and Southern coastal regions of Vietnam. Along the Northern coastline, the most prevalent types of PW included fishing gear such as ropes, fishing nets, bait and fishing lines, hard plastic buoys, hard plastic fragments, and plastic straws. In the Central region, there was a higher concentration of soft plastic fragments, various types of fishing gear including ropes, fishing nets, bait and fishing lines, hard plastic buoys, expanded polystyrenes (EPS), floating buoys, foam boxes, and plastic bags. Along the Southern coastline, the predominant types of PW were fishing gear and food packaging foam boxes (WWF 2023).

Plastic debris on Vietnamese beaches is associated with plastic products for personal use (comprising 43% by count and 36% by weight), those used in fishing, aquaculture, and seafood trade (comprising 35% by count and 33% by weight), household PW (containing 20% by count and 22% by weight), and medical and agricultural PW (comprising 1% by count and 7% by weight). Among the surveyed plastic debris, the proportion of single-use plastic items was notably high, accounting for 38 to 63% of the total plastic debris. Beaches in Soc Trang, Khanh Hoa, Quang Nam Provinces, and Phu Quoc Island have a high proportion of single-use plastic items (Figure 4.3).

Figure 4.3 compares the proportion of single-use plastic waste and other types of plastic waste in several coastal regions of Vietnam. In particular, Thua Thien Hue, Quang Nam, and Phu Quoc have nearly equal amounts of single-use plastic waste compared to other plastics, while other provinces show a significant disparity. This can be explained by the fact that Thua Thien Hue, Quang Nam, and Phu Quoc have made considerable efforts to reduce single-use plastics in various sectors, especially in tourism development. For example, plastic waste reduction practices have been implemented in the tourism industry in Hue and

Hoi An, Phu Quoc has set a target to ban single-use plastics completely by 2025, and Quang Nam has issued a directive to combat plastic waste (Hieu 2024; Tuan 2024; Loc 2024; Anh 2019). These efforts have helped raise awareness among the local communities and tourists, leading to behavioral changes and reduced use of single-use plastics.

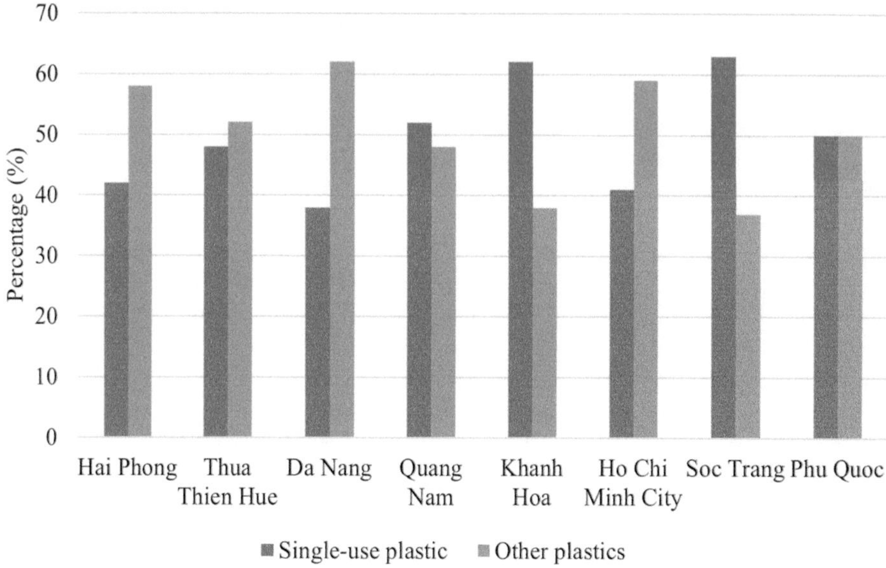

Figure 4.3 The percentage of single-use plastic items in total PW at beaches in some localities.

Source: Data synthesized from: Vietnam: Plastic pollution diagnostics. East Asia and Pacific Region, Marine Plastics Series (Washington, DC: World Bank, 2022).

4.4 CURRENT STATUS OF PLASTIC WASTE MANAGEMENT

4.4.1 Collection, Sorting, and Treatment

4.4.1.1 Current Status of Plastic Waste Collection in Vietnam

PW management in Vietnam is currently in a critical stage, demanding urgent attention due to escalating concerns over environmental degradation and sustainability. The effective collection and transportation of PW play a pivotal role in addressing these environmental challenges and ensuring a cleaner, healthier environment for both present and future generations. While commendable progress has been made in recent years, particularly in urban centers, significant challenges persist, particularly in rural and remote areas where infrastructure and resources are often scarce.

According to data from WWF-Vietnam, in 2021, the total volume of collected PW reached a substantial 6,581 tons/day, equivalent to around 2.4 million tons annually (WWF 2023). Urban areas accounted for 1.55 million tons/year, while rural areas contributed 0.85 million tons/year (MONRE 2022a). However, this apparent success in urban areas is juxtaposed with significant variances in collection rates across rural regions (Figures 4.4 and 4.5).

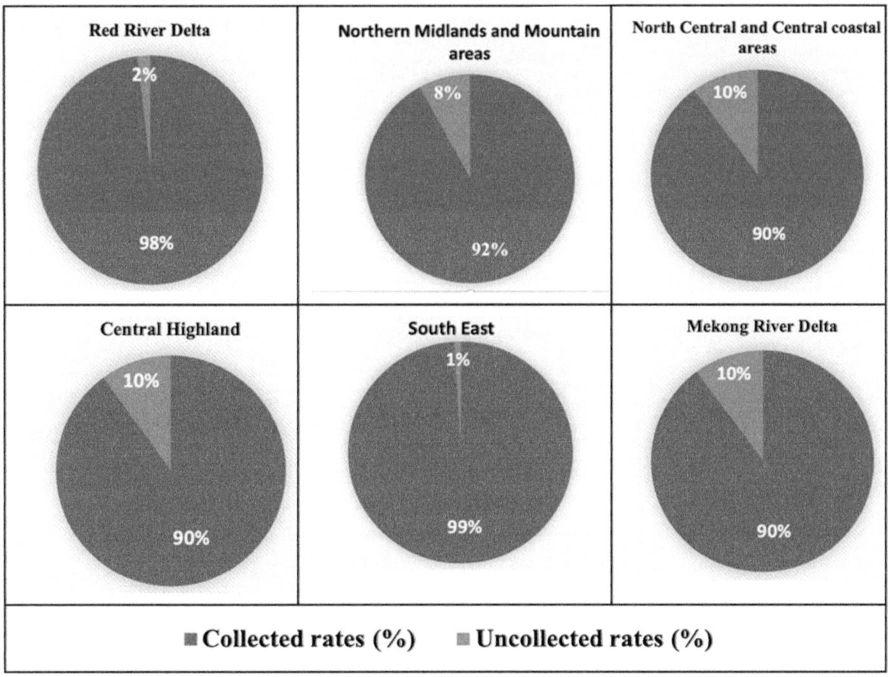

Figure 4.4 Collected and uncollected rates of PW in urban areas by economic regions in 2021.

(*Source:* Data synthesized from MONRE, "Summary of data from the environmental protection report in 2021 of localities," 2022a).

Urban areas typically boasted collection rates exceeding 85% (Figure 4.4), showcasing effective waste management infrastructure and initiatives. Conversely, rural areas experienced greater variability, with collection rates influenced by several factors, such as geographical remoteness, socioeconomic disparities, and limited access to resources. Regions such as the Northern Midlands and Mountains, Central Highlands, and Mekong River Delta struggled with lower collection rates compared to their urban counterparts, indicating the need for targeted interventions tailored to rural contexts (Figure 4.5).

Despite concerted efforts, a significant amount of PW remains uncollected, posing a significant environmental and public health risk. In 2021, an estimated 1,440 tons/day of PW went uncollected, totaling 0.53 million tons/year, constituting a substantial 18% of Vietnam's total PW output. Rural areas faced the most significant burden of uncollected waste, generating approximately 1,226 tons/day, or 0.45 million tons/year (MONRE 2022a). In contrast, urban centers produced around 213 tons/day, nearly 0.08 million tons/year (WWF 2023). Notably, regions such as the Central Highlands, Mekong River Delta, and Northern Midlands and Mountains contributed over 63% of the nation's uncollected PW (Figure 4.4). This highlighted the severe challenges faced by rural waste management systems.

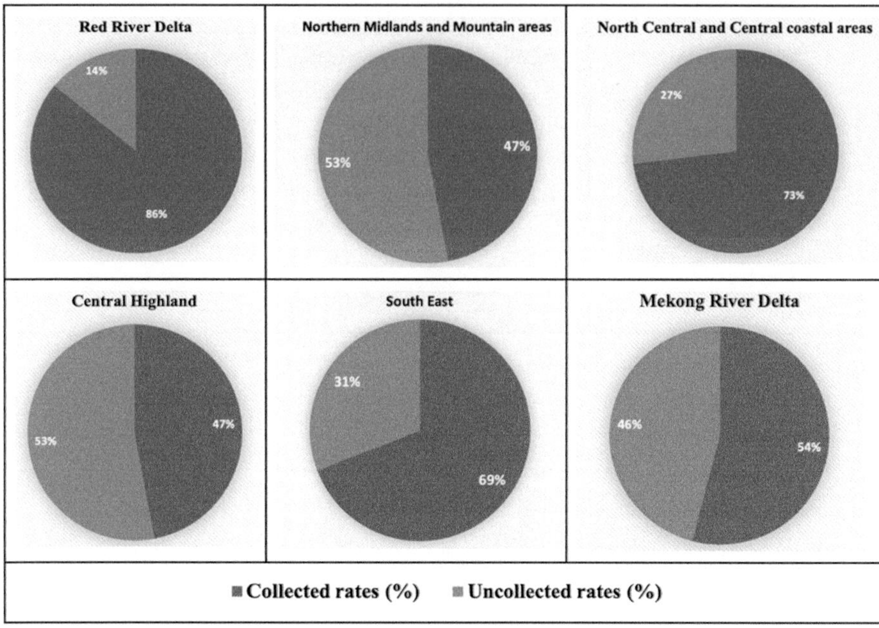

Figure 4.5 Collected and uncollected rates of PW in rural areas by economic regions in 2021.
(*Source:* Data synthesized from MONRE, "Summary of data from the environmental protection report in 2021 of localities," 2022a).

The current practices on the collection and transportation of PW within DSW in urban and rural settings encompass a range of vital activities. In urban and rural areas alike, collection at public locations serves as a cornerstone, utilizing expansive storage facilities to accommodate the influx of DSW. Meanwhile, in residential areas, residents deposit their waste into designated collection vehicles or facilities, strategically positioned at specified collection points. These collection vehicles and facilities then embark on a journey to a central collection point before ultimately being transported to either a transfer station or a treatment facility for further processing.

Moreover, home collection plays a crucial role in ensuring comprehensive waste management. Personnel are tasked with the responsibility of visiting each household, meticulously gathering and depositing DSW into specialized collection vehicles. In addition to these conventional methods, alternative avenues for waste collection are explored. Notably, PW within DSW is often collected through informal channels, with households sorting and selling their PW to informal sectors, who then redistribute it to scrap collection points.

In urban locales, the onus of collecting and transporting DSW primarily falls upon local Urban and Environmental services companies, as well as other city and district public service entities, alongside private enterprises. Conversely, rural areas typically rely on dedicated environmental sanitation teams to manage the collection and transport of DSW to designated landfills. However, despite their essential role, many rural sanitation teams grapple with limited investment in specialized equipment and vehicles, compounded by challenges stemming from temporary waste disposal sites and insufficient funding. These obstacles, in turn, exacerbate environmental pollution, underscoring the pressing need for sustainable waste management practices.

Furthermore, the crucial role of women workers in DSW collection cannot be overstated (NPAP-Vietnam 2023). Across both urban and rural landscapes, women assume significant responsibilities in waste collection efforts, contributing to the effective management of solid waste within communities. Their involvement underscores the importance of gender-inclusive approaches to waste management, highlighting the need for equitable participation and recognition within the sector. Ultimately, by addressing these multifaceted challenges and harnessing the collective efforts of all stakeholders, strides can be made towards fostering a more sustainable and environmentally conscious approach to PW collection and management.

4.4.1.2 Plastic Waste Sorting

The process of sorting PW in Vietnam is multifaceted, involving both formal and informal sectors, with significant implications for waste management practices nationwide.

PW sorting for recycling in Vietnam involves two main sources: collected and uncollected. Sorting activities for collected PW primarily occur at waste treatment plants, processing centers, or designated collection points for DSW. On the other hand, uncollected PW undergoes sorting by informal sectors such as waste pickers and garbage collectors. In 2021, around 0.79 million tons of PW were separated for recycling from collected sources, while 0.11 million tons were sorted from informal collection sources (Table 4.3). Consequently, the total sorted PW for recycling amounted to about 0.89 million tons, representing roughly 30% of the total volume of PW generated in Vietnam in 2021 (Table 4.3).

Article 75 of Vietnam's Law on Environmental Protection 2020 specifies that DSW from households and individuals is classified into three categories: waste that can be reused or recycled, food waste, and other types of DSW (NA-Vietnam 2020). PW, as part of the first group eligible for reuse and recycling, is suitable for sorting at the source. The sorting of DSW at its source has been tested in various areas in Vietnam, including centrally designated cities such as Hanoi, Ho Chi Minh, Da Nang, and Can Tho, as well as other provinces like Dong Nai (2016–2018), Bac Ninh (2014), Binh Duong (2017–2018), Hung Yen (2012–2014), and Ha Tinh (2019) (MONRE 2020; Dung 2022, 2023). This initiative aimed to promote efficient waste management practices and encouraged recycling efforts nationwide.

Table 4.3 The total amount of PW is classified for recycling by economic regions in 2021

Regions		SORTED PLASTIC WASTE FOR RECYCLING PROCESSES		Total (Tons/day)
		Sorted PW for recycling from formal collected sources (tons/day)	Sorted PW for recycling from informal collected sources (tons/day)	
Red River Delta		586.6	27.3	613.9
Northern Midlands and Mountain areas		162.7	54.0	216.7
North Central and Central coastal areas		425	61.6	486.6
Central Highland		100.7	33.5	134.2
Southeast		577.9	17.3	595.2
Mekong River Delta		318.9	94.2	413.1
All	Tons/day	**2171.8**	**287.9**	**2459.7**
	million tons/year	**0.79**	**0.10**	**0.89**

Source: Data synthesized from MONRE, WWF, "Report on Plastic Waste Generation in 2022," 2023.

However, the implementation of source separation of waste programs across various localities has largely been experimental, lacking uniformity, decisiveness, and formalization. For instance, 18 years ago, Hanoi initiated a source separation project (3R) funded by the Japan International Cooperation Agency (JICA). This $4 million project was tested in four central Hanoi wards: Phan Chu Trinh, Nguyen Du, Thanh Cong, and Lang Ha. Between 2006 and 2009, 18,000 households received training on separating waste at the source (Dap 2024; Dung 2023). Despite initial efforts, the practice of solid waste separation did not continue after the project finished. This was due to insufficient preparation, as well as unsuitable and inconsistent waste treatment processes and technologies. Waste was categorized into recyclable, inorganic, and organic categories. However, the recycling market for certain types of waste, including plastics, was unstable and challenging, with no organizations ensuring consistent processing and utilization.

Similarly, Ho Chi Minh City initiated a pilot program for sorting solid waste at the source in 2002, following a well-defined step-by-step plan. By 2017, each district was supposed to execute the program in at least one ward, expanding to 3–5 wards by 2018, and reaching citywide implementation by 2020. Despite these endeavors, the results have been limited (Dung 2022, 2023).

The main challenges include inadequate and inappropriate collection and transportation equipment for sorted solid waste, the application of a single treatment method for all collected waste, high costs associated with sorting waste at the source, and the lack of a comprehensive communication program on source separation of waste. Additionally, the COVID-19 pandemic has disrupted waste management practices, leading to temporary suspensions and delays in program implementation (MOH 2022; Nguyen et al. 2023).

While there have been efforts to improve the sorting of PW in Vietnam, the results have been mixed due to various systemic issues. Continued investment in technology, infrastructure, and public awareness is essential for achieving more significant progress in the future.

4.4.1.3 Plastic Waste Treatment Status in Vietnam

The treatment and recycling of PW in Vietnam face significant challenges and limitations. According to recent statistics, an overwhelming 90% of PW was managed through traditional methods such as landfilling and incineration, with only a mere 10% being recycled. This imbalance highlights the outdated and inadequate waste management infrastructure in Vietnam. Despite efforts to improve, the recycling rate remains low due to insufficient technology, limited capacity, and a lack of comprehensive waste segregation at the source.

- **Plastic Waste Recycling:** Recycling PW transcends mere environmental responsibility; it emerges as a cornerstone of sustainable resource management and economic advancement. The statistics from the Ministry of Natural Resources and Environment (MONRE) underscored a concerning reality: Vietnam generated approximately 3.27 million tons of PW annually as of 2022 (WWF 2023). Particularly concerning was the situation in urban centers like Hanoi and Ho Chi Minh, where a combined daily influx of 80 tons of PW and nylon bags inundated the environment (WWF 2023). Despite this staggering amount of waste, the rate of PW recycled remained alarmingly low at about 10% (WWF 2023). A report by the International Finance Corporation (IFC) and the World Bank estimated that Vietnam squanders nearly 3 billion USD due to its failure to fully recycle PW from daily activities (World Bank Group 2021).

  A significant portion of plastic recycling in Vietnam is conducted by the informal sector, which accounts for over 90% of recycling activities. These informal recycling operations are primarily located in craft villages. These villages help alleviate poverty and create jobs, but their recycling activities are typically small-scale and family-run, often leading to unregulated and unsafe working conditions (VNCPC 2018).

  Unutilized PW was either dumped near recycling areas or buried in local landfills, exacerbating environmental concerns. Outdated recycling technology and the handling of low-value and contaminated PW further hindered recycling efficiency and the profitability of recycling businesses. Despite these challenges, the informal recycling sector remains a vital part of Vietnam's recycling ecosystem, highlighting the need for better regulation and support to improve efficiency and sustainability.

- **Waste Incineration:** In Vietnam, waste incineration, including the burning of PW, is implemented through two primary methods: open burning and controlled burning.

  According to the Environmental Pollution Control Department, by the end of 2023, Vietnam had approximately 467 municipal solid waste incinerators (DEPC 2023). The incineration of waste has been increasing, with about 20% of the total waste incinerated. Of this, 9.3% was incinerated with energy recovery, and 10.7% was incinerated without energy recovery, representing about 7% increase compared to 2019 (DEPC 2023). Despite the widespread practice of waste incineration, including PW, in many Vietnamese localities, the number of incinerators that comply with environmental standards remains limited. Many facilities lacked adequate waste storage, prompting low-temperature burning that contributed to air pollution. Future efforts must focus on enhancing the capacity of existing facilities to handle PW effectively. Public awareness campaigns and educational programs are essential to inform communities about the risks of open burning and the benefits of controlled burning and waste-to-energy technologies. Investing in advanced technologies and infrastructure, such as high-efficiency incinerators and comprehensive waste segregation systems, will be crucial for Vietnam to achieve sustainable waste management.

In recent years, there has been a marked increase in the development of waste-to-energy projects across the country, with about 15 waste-to-energy plants currently under construction, reflecting Vietnam's commitment to integrating energy recovery into waste management (DEPC 2023). Among these, three plants have commenced power generation: the Can Tho Waste Treatment Plant with a capacity of 400 tons/day, demonstrating significant progress in waste-to-energy technology; the Soc Son Waste-to-Energy Plant with a capacity of 4,000 tons/day; and the Green Star Environmental Company's High-Tech Waste-to-Energy Plant in Bac Ninh, with a capacity of 180 tons/day (100 tons/day of municipal solid waste and 80 tons/day of industrial waste, including PW), currently in trial operation. Additional plants in Bac Ninh, Thua Thien Hue, and other provinces, such as Phu Tho, Da Nang, Ho Chi Minh City, and Thai Nguyen, are either in trial operation or under development.

Furthermore, in response to the increasing challenges of PW management, several cement plants in Vietnam have adopted innovative strategies to incorporate non-recyclable PW into their production processes. For instance, the Lam Thach Cement Plant, located in Quang Ninh Province, has successfully applied over 200 tons of PW from local plastic crafting villages, including Minh Khai (Hung Yen Province) and Trang Minh (Hai Phong City), as alternative energy for its cement production. This practice not only provides an environmentally sound solution for handling PW but also significantly reduces the plant's dependence on coal. By replacing 35–40% of coal consumption with PW, the plant has reported both economic and environmental advantages, such as lower energy costs and reduced carbon emissions (VTV 2024). Moreover, this approach maintains the high temperatures required for clinker production while enhancing overall energy efficiency in the cement manufacturing process.

- **Landfill:** PW treatment in Vietnam is a pressing environmental issue, with landfilling being the most common method used for disposal. As of the end of 2023, Vietnam has approximately 1,207 municipal solid waste landfills scattered across the country (DEPC 2023). However, only about 20% of these landfills were classified as sanitary landfills, which were designed to minimize environmental impacts through measures like proper lining and leachate management. The remaining 80% were unsanitary landfills or open dumping sites that lacked these critical features, leading to significant environmental hazards (MONRE 2020).

Sanitary landfills in major urban centers are facing significant challenges due to overcapacity. For instance, the Nam Son Landfill in Hanoi, despite its extensive area of 83 hectares, is struggling to manage the approximately 4,000 tons of waste it receives daily. The site frequently deals with issues such as inadequate leachate treatment and strong odors, leading to protests from residents. Similarly, the Da Phuoc Landfill in Ho Chi Minh City, which spans around 128 hectares and handles about 5,000 tons of waste per day, has been the center of controversy due to severe environmental impacts, including water pollution and health issues reported by nearby communities. Overcapacity in these landfills increases environmental risks, operational challenges, and health hazards, exacerbating the already critical situation (Nguyen and Le 2021).

Several challenges hinder effective PW management through landfilling in Vietnam. Many landfills lack essential infrastructure such as proper lining, leachate collection, and gas management systems. There is a need for stronger regulatory frameworks and enforcement to ensure that landfills meet international environmental standards. Rapid urbanization and economic development have led to increased waste generation, often outpacing the development of adequate waste management facilities. This results in overcrowded and improperly managed landfills. Additionally, low levels of public awareness and participation in waste segregation and recycling practices exacerbate the problem of PW management.

4.4.2 Institutional and Legal Framework and Policies

Vietnam has shown significant commitment to addressing PW pollution through a comprehensive set of strategies, policies, and legal frameworks that foster sustainable production and consumption, reduce plastic pollution, and promote a circular economy. These efforts involve multiple stakeholders, including government ministries, local authorities, and the private sector, to ensure a coordinated and effective response to the PW challenge.

4.4.2.1 Strategies, Policies, and Legal System Related to Plastic Waste Management

To address environmental pollution caused by PW, government strategies and action plans have been developed. They emphasized the urgent need to tackle plastic pollution. Clear targets have been established to tackle waste management, plastic pollution, and marine litter, encompassing waste generated by sectors such as retail and tourism. These strategies also highlighted the necessity for additional plastic-related policies to foster reduction and promote better and eco-friendly designs of plastic products. These include:

- **The Environmental Protection Law of 2020:** Introduces measures for extended producer responsibility (EPR) targeting manufacturers of plastic packaging and mandates a plan for reducing the production and importation of SUPs, along with other non-biodegradable plastic packaging and products. Moreover, it also stipulates issues related to the circular economy, including PW management (NA-Vietnam 2020).
- Law on Environmental Protection Tax No. 57/2010/QH12, dated 15 November 2010, levies taxes on manufacturers of non-biodegradable plastic bags. (NA-Vietnam 2010).
- **Vietnam's Law of the Sea No. 18/2012/QH13:** Regulates activities related to marine and coastal environments, including measures to prevent and control marine pollution. It plays a crucial role in managing marine PW by setting standards for waste discharge, promoting sustainable use of marine resources, and enforcing regulations to minimize the impact of PW on marine ecosystems (NA-Vietnam 2012).
- The Law on Natural Resources and Environment of Sea and Islands No. 82/2015/QH13 outlines a comprehensive framework for managing marine and island resources, emphasizing sustainable development. It includes specific provisions for the management and reduction of marine plastic pollution, emphasizing the importance of protecting marine biodiversity and promoting eco-friendly practices to support a circular economy. This law encourages the use of renewable resources and the recycling of materials to reduce PW in marine and coastal areas (NA-Vietnam 2015).
- **Decree No. 08/2022/ND-CP, dated 10 January 2022:** Giving detailed guidelines on the enforcement of the Law on Environmental Protection 2020, with a particular focus on EPR implementation and restrictions on SUPs (GOV-Vietnam 2022). Key measures include:
    - ***EPR Implementation:*** Producers are mandated to take responsibility for the entire lifecycle of their plastic products, including take-back schemes, recycling, and disposal. This encourages manufacturers to design eco-friendly products and reduce PW generation.
    - ***SUP Restrictions:*** The decree outlines a roadmap to phase out the production, importation, and consumption of SUPs. It includes targets and timelines for reducing the use of non-biodegradable plastic bags and other disposable plastic items, aiming to shift toward sustainable alternatives and support the development of a circular economy.

To execute the strategies, policies, and laws on solid waste management, including PW, various ministries and agencies have issued numerous legal documents for guidance on the enforcement. These documents comprise:

- **Circular No. 02/2022/TT-BTNMT, dated January 10, 2022:** Issued by MONRE, this circular details specific articles of the Law on Environmental Protection, providing guidelines for effective implementation (MONRE 2022b).
- **Circular No. 08/2018/TT-BTNMT, dated September 14, 2018:** This circular promulgates national technical regulations on the environment for imported scrap used as raw production materials, ensuring that imports meet environmental standards (MONRE 2018).
- **Decision No. 2346/QD-BTNMT, dated December 13, 2021:** This decision by MONRE implements Decision No. 1316/QD-TTg, approved by the Prime Minister on July 22, 2021, to strengthen PW management in Vietnam. It outlines specific actions and plans to reduce PW and promote sustainable practices (MONRE 2021b).

The implementation of key legal documents, notably the Law on Environmental Protection 2020 and Decree No. 08/2022/ND-CP, signifies a pivotal step forward in addressing pressing environmental concerns in Vietnam. These legislative instruments introduce comprehensive regulations targeting the pervasive issues of microplastics, disposable plastic products, and non-biodegradable packaging.

As indicated in the National Action Plan for Marine PW management issued in Decision No.1746/QD-TTg dated December 4, 2019, by Vietnam's Prime Minister, the Government has set ambitious targets for addressing plastic pollution. These targets include reducing marine litter by 50% by 2025 and 75% by 2030, collecting 100% of fishing gear waste by 2030, preventing the use of SUPs and non-biodegradable plastic bags, conducting clean-up campaigns, and monitoring the country's marine plastic debris, especially in tourist areas, to make 100% of tourism locations plastic-free by 2030 (GOV-Vietnam 2019). Furthermore, starting on January 1, 2026, Vietnam will not produce or import non-biodegradable plastic bags with dimensions smaller than 50 cm × 50 cm. After December 31, 2030, the Government requires stopping the production and import of single-use plastic products, plastic packaging that is difficult to biodegrade, and products and goods containing microplastics under Decree No.08/2022/ND-CP (GOV-Vietnam 2022). This Decree also directs the Provincial People's Committees to restrict the distribution and use of SUPs in commercial centers, supermarkets, hotels, and tourism areas after 2025, and it prescribes a gradual reduction of the production and importing of other SUPs until their eventual ban starts in 2031 (GOV-Vietnam 2022).

Additionally, the Law on Environmental Protection 2020 has introduced regulations on the circular economy. Article 5, Clause 11 of the Law states that "the State's policy on environmental protection integrates and promotes circular and green economy models in the formulation and implementation of strategies, planning, plans, programs, schemes, and socio-economic development projects." Article 142 further specifies the concept of the circular economy (NA-Vietnam 2020).

These regulations are not merely about enacting laws but about instigating a paradigm shift in social attitudes and behaviors toward plastic consumption and waste management. They were designed to evoke a collective sense of responsibility among producers and the general public alike, emphasizing the urgent need for concerted action in curbing plastic pollution. Furthermore, these legal frameworks are instrumental in fostering the transition towards a circular economy model, wherein resources are utilized efficiently, waste is minimized, and environmental impacts are mitigated. Timelines for the development of the laws and policies that Vietnam has adopted for the management of PW are shown in Figure 4.6.

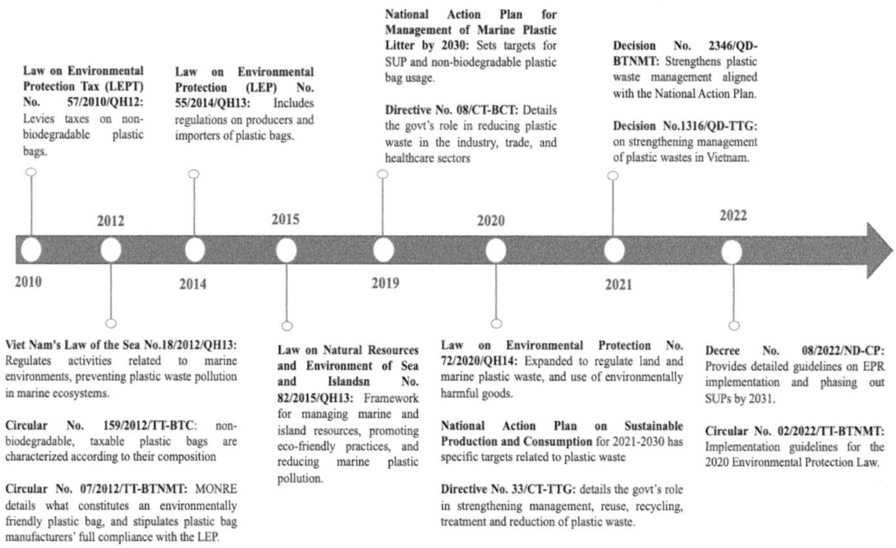

Figure 4.6 Policy and regulatory timelines on PW management in Vietnam.

4.4.2.2 Responsibilities for Waste Management

Effective waste management, especially for PW, is essential for environmental sustainability. Successful PW management requires a comprehensive approach with clearly defined roles and responsibilities for various stakeholders. Effective waste management in Vietnam is a collective effort involving multiple stakeholders, including government agencies, local authorities, private enterprises, and the general public. Table 4.4 outlines the roles and responsibilities of these stakeholders in managing PW.

By clearly defining these roles and responsibilities, it is possible to create a cohesive and effective system for managing PW. Coordination and cooperation among all stakeholders are essential to achieving significant improvements in waste reduction, recycling rates, and overall environmental sustainability.

4.4.2.3 Financial Resource

As per Prime Ministerial Decision 1316/QD-TTg, dated July 22, 2021, endorsing a proposal to enhance the management of PW in Vietnam, financial support for the execution of PW management strategies is sourced from state budgets and

Table 4.4 Stakeholders and responsibilities for PW management

Responsible Institutions	Main Roles and Responsibilities in PW Management
	Government Agencies
Ministry of Natural Resources and Environment (MONRE)	MONRE serves as the authoritative body entrusted with overseeing environmental functions, encompassing waste management and plastic. Directive No. 33/CT-TTg (PM-Vietnam 2020), issued on August 20, 2020, designates MONRE as the central hub for coordinating policies and initiatives related to plastic and waste management. MONRE assumes a pivotal role in the formulation, implementation, and oversight of these policies, emphasizing rigorous research and monitoring to gauge their efficacy and environmental outcomes.
Ministry of Finance (MOF)	MOF is tasked with researching, amending, and supplementing the Law on Environmental Protection Tax, aiming to increase tax rates on plastic bags, packaging, and other plastic products, while providing incentives and support for environmentally friendly plastic bags and recyclable plastic products.
Ministry of Industry and Trade (MOIT)	MOIT is responsible for formulating and enforcing standards and technical regulations governing the quality and design of plastic products, ensuring compliance with recycling, reuse, and durability requirements, as well as sustainable production and consumption practices.
Ministry of Agriculture and Rural Development (MARD)	MARD is the focal point for implementing PW management activities in rural areas, particularly within the agriculture and fisheries sectors.
Ministry of Health (MOH)	MOH is the central authority responsible for overseeing activities related to PW management in the healthcare sector. This includes assessing the health implications of waste, regulating the quality of recycled plastics, and implementing protocols to minimize waste generation, promote reuse, and enhance recycling practices within medical and pharmaceutical facilities.
Ministry of Education and Training (MOET)	MOET develops and implements plans for waste segregation and mitigation of PW within schools, educational institutions, and training centers.
Ministry of Science and Technology (MOST)	MOST promotes innovation, creativity, research, and technology transfer aimed at developing environmentally sustainable materials as alternatives to plastics in industrial and commercial sectors.
Ministry of Culture, Sports and Tourism (MCST)	MCST advocates for standards aimed at minimizing the use of SUPs and non-degradable plastic bags in businesses, tourist accommodations, service establishments, as well as during cultural, sports, and tourism events.

(*Contd.*)

Responsible Institutions	Main Roles and Responsibilities in PW Management
Ministry of Information and Communications (MOIC)	MOIC, Vietnam Television, Voice of Vietnam Radio, and the Vietnam News Agency are primarily responsible for coordinating with agencies, organizations, and individuals to develop information campaigns that raise awareness about the impacts of PW and relevant regulations. Their responsibilities also include disseminating information on strategies for reducing PW, as well as methods for sorting, collection, recycling, and treatment of PW.
Local Governments	
Provincial and Municipal People's Committees	These local authorities are crucial in executing national policies at the grassroots level. They implement waste management programs, conduct awareness campaigns, and monitor compliance with local waste management regulations.
Private Sector	
Manufacturers and Producers	Under the EPR scheme, producers are required to manage the entire lifecycle of their products. This includes designing eco-friendly products, reducing packaging waste, and implementing take-back and recycling programs.
Retailers and Distributors	Retailers are encouraged to reduce the use of SUPs and promote reusable and biodegradable alternatives. They are also responsible for educating consumers about the environmental impact of PW and encouraging sustainable consumption habits.
Waste Management Companies	These companies are responsible for collecting, transporting, and processing PW. They play a crucial role in the recycling process and in ensuring that waste is disposed of in an environmentally friendly manner. Collaboration with local authorities is essential to enhance waste collection efficiency and recycling rates.
Civil Society and Public Participation	
Non-Governmental Organizations (NGOs)	NGOs play a vital role in advocating for environmental protection and raising public awareness about plastic pollution. They collaborate with government agencies and the private sector to promote sustainable practices and conduct clean-up campaigns.
Community Groups and Volunteers	Local community groups and volunteers are instrumental in organizing and participating in clean-up activities, recycling programs, and educational workshops. Their grassroots efforts help to instill a sense of environmental stewardship in the broader community.
General Public	Individual actions collectively contribute to waste reduction. The public is encouraged to minimize plastic use, participate in recycling programs, and support initiatives aimed at reducing plastic pollution.

diversified through various channels, including equity capital from enterprises, credit institution loans, and international sponsorships (PM-Vietnam 2021). These funds are directed toward a spectrum of activities aimed at enhancing state management, encompassing the review and amendment of mechanisms, policies, and legal regulations, as well as the establishment of standards, technical regulations, and capacity-building initiatives. Market development endeavors, propaganda campaigns, and awareness-raising initiatives are also included within this framework.

Allocation of funds from state budgets is facilitated through the annual plans and programs of ministries, agencies, and localities, ensuring a systematic approach to resource distribution. Investment activities aimed at developing technologies for producing environmentally friendly products and equipment for PW treatment are financed through a combination of sources. These include investments, domestic and international loans, sponsorships, as well as partial contributions from the Vietnam Environmental Protection Fund, local development investment funds, public-private partnerships, and other legal capital sources.

However, Vietnam is currently facing challenges due to the lack of a transparent financial mechanism for managing PW. The MOF is actively engaged in reviewing and amending financial mechanisms and policies to address this gap.

- **Funding for Environmental Protection:** In 2016–2022, the Government of Vietnam underscored its commitment to sustainable development through policies emphasizing that investments in environmental protection are investments in long-term sustainability.

  A minimum allocation of 1% of the total state budget (for environmental protection); Vietnam has consistently increased its budgetary provisions in this area, with a notable 55.8% rise observed in 2021 compared to 2015. The allocation distribution between central and local levels stands at 15% and 85%, respectively. Notably, funding for environmental initiatives, including household solid waste management, has seen steady growth over the years, with expenditures rising from 6,230 billion VND in 2010 to 21,424 billion VND in 2021 (MONRE 2021a).

  Between 2016 and 2022, significant progress was achieved in securing financial support and loans from various countries and international organizations to enhance environmental conservation efforts in Vietnam (WWF 2023). In particular, in 2020, the Ministry of Finance secured non-refundable aid exceeding 4.76 million USD (approximately 110.5 billion VND), specifically allocated for environmental protection initiatives. A notable initiative in addressing PW is the United Nations Environmental Program's implementation of a PW treatment project in the Mekong region, with Japan committing to support the project with 5.7 million USD.

- **Fund for the Collection and Transportation of PW:** Variations exist in the maximum fees imposed for household solid waste collection and transportation across different localities. For instance, in Ha Noi Capital, urban residents are charged 6,000 VND/person/month (equal to 0.24 USD), whereas individuals in rural communes and towns pay only 3,000 VND/person/month (equal to 0.12 USD), as stipulated in Decision No.54/2016/QD-UBND of Ha Noi (People's Committee of Ha Noi 2016). Conversely, in Tuyen Quang Province, fees for waste collection and transportation are categorized into multiple subjects, including households, businesses, schools, and healthcare facilities, with the lowest fee for individual households set at 10,000 VND/household/month (equal to 0.40 USD), as per Decision 432-QD/QD-UBND of Tuyen Quang province (People's Committee of Tuyen Quang 2017).

Despite fee collections from households and various establishments, these funds often fall short in covering expenses related to worker wages, labor protection, and equipment maintenance. Consequently, local budgets usually bridge the financial gap to ensure the effective collection and transportation of DSW.

- **Fund for Investment and Treatment of PW:** In the realm of PW investment and treatment, financial resources are earmarked for the development and construction of production technologies geared toward environmentally friendly products, as well as the establishment of equipment and facilities for waste treatment. These resources encompass various funding channels, including state budget allocations such as development investment capital, both at the central and local levels, Official Development Assistance (ODA) capital, and other pertinent sources. Additionally, partial support is extended from entities like the Vietnam Environmental Protection Fund, the Local Environmental Protection Fund, the Local Development Investment Fund, and other capital sources.

  Despite the critical need for investment in PW management, comprehensive statistics delineating the financial resources allocated explicitly for this purpose are currently unavailable. Nevertheless, the predominant source of funding for these endeavors continues to originate from state budget allocations.

4.5 CHALLENGES AND OPPORTUNITIES TOWARD A CIRCULAR ECONOMY

4.5.1 Overview of Circular Economy

The circular economy (CE) is an economic system in which materials are designed to be used, not used up. The maximum use is achieved from every process, material, and component. If we promote and apply properly, the circular economy benefits society, the environment, and the economy (Babaremu et al. 2022). Currently, a widely accepted definition of the CE by many countries and organizations is "the CE is a framework for systems solutions and transformation that tackles global challenges like climate change, biodiversity loss, waste, and pollution. It has three principles, all driven by design: (1) eliminate waste and pollution; (2) circulate products and materials; (3) regenerate nature (MacArthur 2021)."

The transition to a CE is a significant opportunity for sustainable development, not only achieving economic, social, and environmental goals but also addressing climate change (Thu 2020). Transitioning to a CE requires a system change that includes institutional changes, cultural shifts, technological innovation, and the creation of a stable regulatory environment. The CE also necessitates closer cooperation between the government and businesses (Salmela 2016).

As mentioned above, it is possible and doable to initiate the CE in the field of plastics management. This is due to the fact that the recycling and reuse of PW, as well as the mitigation of non-biodegradable plastic products, have been promoted and applied in Vietnam for decades.

CE does not mean we ban plastic usage; instead, we continue to use plastic products but make sure that these plastic products are reusable, recyclable, or compostable. This means no plastics end up in the environment.

The world is increasingly determined to reduce carbon emissions globally. Therefore, the plastic collection and recycling process is highly essential. A booming circular economy for plastics requires:

(i) Elimination of Waste
(ii) Reuse Models
(iii) Material Circulation

Examples for Elimination

In Vietnam, the elimination may involve the refilling of plastic containers at home, at stores, and at factories. Additionally, it requires the designing or redesigning of plastic products to make them recyclable. Consumers who use and return plastic products (e.g., containers) shall earn some credits.

Example for Reuse Models

This comes up with effective systems/solutions to recover the used plastic products. In Vietnam, it is quite common for people to reuse plastic packaging for packing other items and especially, make products out of PWs: light covers, stuff holders, floating belts for plants in the lakes, plant holders, etc. Many plastic reuse projects have been launched by young people locally.

Example for Material Circulation

It is to reuse plastics by recycling them, turning them into valuable materials for different kinds of products used in construction, fuel production, wastewater treatment, and other high-value-added products (Maitlo et al. 2022). In a new plastics economy, plastic never becomes waste or pollution, but raw materials for different purposes. Research has shown that PW can be reused to make fuels (after mixing with dolomite), produce construction materials, produce plastic beads for making recycled products, or can be used directly in wastewater treatment.

4.5.2 Challenges

PW collection and recycling activities take place across the country, but are only concentrated in the "waste pickers force." Currently, there are about 500 small and medium-sized companies nationwide that organize plastic recycling. Therefore, plastic recycling activities in Vietnam over the past 40 years have not been "really developed." Vietnam is the country with the 8th largest amount of ocean PW with 28,221 tons (VPA 2023). The collection rate is not high enough (only 30%), so it does not meet the requirement for industrial-scale application. Vietnam still must import PW from outside (China, Japan, etc.). It has been due to several reasons (ISPONRE 2022):

- There is a lack of specific regulations on circular economic development, including ecological design; a system of standards and technical regulations on recycled products; regulations on the mandatory proportion of recycled plastic in products; and regulations on financial contributions based on the product's recyclability (modulation fee).
- Manufacturing and importing enterprises continue to face numerous challenges in organizing, establishing, and operating recovery networks, as well as in implementing responsibilities for recycling and handling plastic packaging after disposal. Accessing preferential and supportive mechanisms and policies for businesses still face many barriers and obstacles.
- There is a lack of detailed instructions on incentives and support for land, capital, and taxes in general, particularly incentives and support for PW recycling activities. The processes and procedures for requesting incentives and support are still complicated and confusing.
- Criteria for determining which green projects are eligible for green credit and green bonds have not been established. Green procurement policies have not been strongly implemented, and as a result, they have not been a driving force to promote the development of ecological products in general and recycled plastic products in particular. Most of the raw materials for recycling activities are imported plastic scrap.
- The Extended Producer Responsibility (EPR) mechanism still lacks specific and detailed instructions on how to calculate financial contributions to the Vietnam Environmental Protection Fund; on financial management and selection of recycling projects; on establishing the Office, National Council on EPR, and operating regulations to ensure the effectiveness and efficiency of these institutions.
- Recycling activities still rely heavily on imported plastic scrap. The source of imported plastic scrap is stable, meets the cleanliness requirements, is classified according to the correct type, is more convenient for recycling activities, and ensures the quality of recycled products.
- Domestic scrap plastic sources have many impurities, which increase costs, affect the quality of recycled plastic products, and reduce the competitiveness of recycled products. The lack of regulations on sorting

at source and collection, and unstable scrap supply, leads to difficulties in planning and expanding production scale. The passivity in supply is also the reason why the plastic recycling industry finds it difficult to implement large-scale recycling projects.

- Informal plastic recycling activities are carried out by small-scale establishments, mostly in craft villages with outdated technology and equipment, consuming a lot of energy and causing environmental pollution. The area of facilities is small, making it difficult to expand production and adapt to changing technology. The quality of recycled plastic products from informal recycling activities is low, so they can only meet the needs of the low-quality market.
- Due to their specificity and flexibility in organizing production and business activities, informal recycling facilities in craft villages have more diverse access to scrap plastic sources, competing with official businesses in the recovery and recycling of plastic packaging after disposal. The capacity of informal recycling activities is limited, and the database is inadequate, which causes barriers and obstacles in developing PW recycling activities.
- The current workforce of the plastic recycling industry is mostly low-skilled, making it challenging to improve recycling efficiency. The capacity of the management team in informal plastic recycling facilities is very low compared to other manufacturing industries, which is a major obstacle to recycling technology innovation and production expansion.
- The lack of a database system and public and transparent information on materials and products from plastic recycling activities hinders the ability of recycling businesses and related parties to access information.

4.5.3 OPPORTUNITIES

4.5.3.1 National Level

With a commitment to net-zero emissions by 2050, the green and low-carbon economy has become a development trend in Vietnam. Recycling waste in general and PW in particular is an indispensable need to realize green growth and sustainable development. In 2025, for the first time, Vietnam's Environmental Protection Law stipulates that households must classify waste at source into three trash bins: recycling bins, organic bins, and other trash bins.

As the definition and criteria for the CE model have been institutionalized, CE will be integrated into development strategies, plans, and programs. Businesses will have to implement measures to effectively manage resources, minimize waste, and reduce their environmental impacts. The developed national action plan on CE will set out orientations and roadmaps as a basis for ministries, branches, and localities to implement.

As mentioned above, the National Action Plan for marine plastic waste management to 2030 sets a target of reducing PW in the sea and ocean by 75%

by 2030; by 2025, not to use non-degradable plastic bags and single-use plastic products in shopping centers, supermarkets, hotels and tourist areas; after 2030, not to produce or import non-degradable plastic bags and disposable plastic products.

In Vietnam, several non-profit organizations, such as the Vietnam Waste Recycling Association, the Vietnam Plastics Association, and the Packaging Recycling Organization Vietnam, have been established over the past ten years to promote awareness and recycling activities in the plastic industry.

4.5.3.2 Enterprise Level

A roadmap to reduce PW and eliminate single-use plastic products has been issued; The EPR mechanism for plastic packaging is applied from 2024. In which manufacturers and importers are responsible for recovering and recycling discarded plastic packaging through one of the following forms:

- Organize their own recovery, recycling, or;
- Entrust a third party to recover, recycle, or;
- Contribute recovery and recycling costs to the Vietnam Environmental Protection Fund.

Efforts to achieve zero-waste in plastic management shall initiate exciting innovations such as biodegradable packing/bags or even reusable bags, and give incentives for the reuse of bags/containers. An increasing number of recycled décor pieces are made from PWs. The enterprises shall design the production process to minimize waste generation and recycle waste to the maximum.

4.5.4 Roadmaps for Plastic CE in Vietnam

It will probably take 5 to 10 years for Vietnam to make the plastic industry as green as possible. The roadmap for promoting plastic CE at an industrial scale in Vietnam is illustrated in Figure 4.7 below.

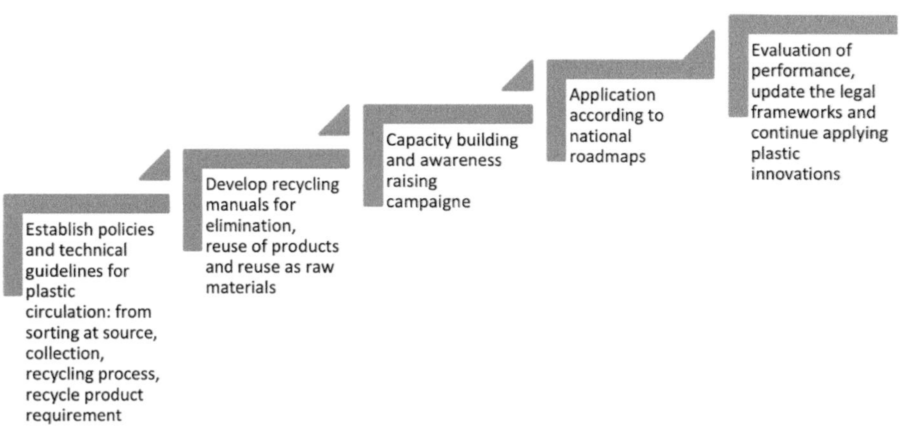

Figure 4.7 Roadmaps for the implementation of CE on plastic waste management.

Success shall depend greatly on the first step, which establishes the proper legal documents to instruct enterprises and communities to comply with. This should be based on best practices in similar conditions in countries in Asia, plus adapted to the conditions of Vietnam.

The support from the national level has a tremendous impact on the compliance and application of CE in PW management.

4.6 CONCLUSIONS

Plastic production and consumption in Vietnam are increasing in line with the trend of socio-economic development. However, the increasing levels of production and consumption have not increased in parallel with the management and mitigation to prevent PW from contaminating the environment. As a result, PW remains a major component of environmental pollution in the country. The studied results indicated that the proportion of PW in DSW in Vietnam is about 12% on average. The volume of PW generated in the country has been increasing in recent years. Consequently, the Red River Delta generated the highest total PW with 1,914 tons/day, while the Central Highlands recorded the lowest PW generation of 473 tons/day in 2021. On average, the generated volume of plastic waste per capita in urban areas was about 43 kg/person/year, which was over two times higher than that generated in rural areas (20.33 kg/person/year) in 2021. The single-use PW accounted for 38 to 63% of the total PW. In general, coastal areas have higher PW generated than the inland ones. Similarly, it is generated at a higher rate in urban areas than in rural areas. The collection rate of PW in urban areas was above 90%, which was much higher than that in rural areas (from 47% in Central Highlands and Northern Midlands and Mountain areas, to 86% in Red River Delta). PW has polluted the soil, air, water, and sediment of Vietnam through the decomposition of plastics into fragments, microplastics, and nanoplastics. However, studies on PW pollution in Vietnam, especially on microplastics, have been relatively few.

Regarding PW management, although Vietnam has made some advancements, the collection, sorting, and treatment of PW are facing many challenges, ranging from outdated technologies and a lack of budget to public awareness, etc. To further improve PW management, Vietnam has issued legal frameworks, strategies, and policies emphasizing the plastic production and consumption, the collection, sorting, and disposal of PW. Notably, the concept of CE, EPR, green credits, etc., for waste management, including PW, was officially mentioned in the legal documents from 2020 for the first time. During the implementation of these legal frameworks, strategies, and policies, there are also some opportunities and the associated challenges for PW management toward CE as presented above. These results will further support the formulation of suitable strategies and policies for PW management towards CE in Vietnam, as well as promote future research in this field.

REFERENCES

Alabi, O.A., K.I. Ologbonjaye, O. Awosolu and O.E. Alalade. 2019. Public and environmental health effects of plastic waste disposal: A review. J. Toxicol. Risk Assess. 5: 021. https://doi.org/10.23937/2572-4061.1510021.

Anh, L. 2019. Quang Nam Issues Directive on Combating Plastic Waste. Retrieved 09 18, 2024, from https://monre.gov.vn/Pages/quang-nam-ban-hanh-chi-thi-chong-rac-thai-nhua.aspx?cm=M%C3%B4i+tr%C6%B0%E1%BB%9Dng (In Vietnamese).

Babaremu, K.O., S.A. Okoya, E. Hughes, B. Tijani, D. Teidi, A. Akpan, et al. 2022. Sustainable plastic waste management in a circular economy. Heliyon 8(7): e09984. https://doi.org/10.1016/j.heliyon.2022.e09984.

Carnovn. 2023. Plastic Industry Report 2023–2028: Trends and Growth Forecast. Retrieved 04 19, 2024, from https://carnovn.com/bao-cao-nganh-nhua-xu-huong-va-du-bao/ (In Vietnamese).

Danh, N.T. and H.T. Hoi. 2019. Effects of plastic waste to sea environment in Vietnam. IOP Conf. Ser.: Earth Environ. Sci. 351(1): 012023. https://doi.org/10.1088/1755-1315/351/1/012023

Dap, N.D. 2024. Waste classification at the source according to the environmental protection law 2020. Vietnam Academy of Social Sciences. Retrievable at https://vass.gov.vn/nghien-cuu-khoa-hoc-xa-hoi-va-nhan-van/Phan-loai-rac-tai-nguon-theo-Luat-bao-ve-moi-truong-2020-215 (in Vietnamese).

Department of Environmental Pollution Control (DEPC). 2023. Report on collection and treatment of DSW in urban and rural areas in 2023 (in Vietnamese).

Dung, N.H. 2022. Current status of household waste sorting at source and waste recycling technology in Vietnam. J. Environ. Urban Stud. Retrievable at https://www.moitruongvadothi.vn/thuc-trang-phan-loai-rac-tai-nguon-va-cong-nghe-tai-che-rac-thai-tai-viet-nam-a114927.html (in Vietnamese).

Dung, N.H. 2023. Current status of household waste sorting at source and waste recycling in Vietnam. J. Environ. Urban Studies. Retrievable at https://www.moitruongvadothi.vn/thuc-trang-phan-loai-rac-thai-sinh-hoat-tai-nguon-va-tai-che-rac-thai-tai-viet-nam-a138677.html (in Vietnamese).

Government of Vietnam (GOV-Vietnam). 2019. Decision on issuing the national action plan on ocean plastic waste management until 2030, No.1746/QD-TTg. Hanoi (in Vietnamese). Retrievable at https://thuvienphapluat.vn/van-ban/Tai-nguyen-Moi-truong/Quyet-dinh-1746-QD-TTg-2019-Ke-hoach-hanh-dong-quoc-gia-ve-quan-ly-rac-thai-nhua-dai-duong-430378.aspx

GOV-Vietnam. 2022. Decree elaboration of several articles of the law on environmental protection No. 08/2022/ND-CP. Hanoi. Retrievable at https://faolex.fao.org/docs/pdf/vie212268.pdf

Hieu, D.N. 2024. The tourism industry in Thua Thien Hue is making efforts to reduce plastic waste. Retrieved 09 18, 2024, from https://thuathienhue.gov.vn/vi-vn/Chinh-sach-moi/tid/Nganh-du-lich-Thua-Thien-Hue-no-luc-giam-thieu-rac-thai-nhua/newsid/5E1E31DE-CCF5-45E4-9F2C-B1C50083D66E/cid/2BEA0540-FCA4-4F81-99F2-6E8848DC5F2F (In Vietnamese).

Institute of Strategy and Policy on Natural Resources and Environment (ISPONRE). 2022. Report on the Current status and solutions to promote plastic waste recycling in Vietnam (In Vietnamese).

Iqbal, S., J. Xu, S.D. Allen, S. Khan, S. Nadir, M.S. Arif, et al. 2020. Unraveling consequences of soil micro- and nano-plastic pollution on soil-plant system: Implications for nitrogen (N) cycling and soil microbial activity. Chemosphere. 260: 127578. https://doi.org/10.1016/j.chemosphere.2020.127578.

Kye, H., J. Kim, S. Ju, J. Lee, C. Lim and Y. Yoon. 2023. Microplastics in water systems: A review of their impacts on the environment and their potential hazards. Heliyon. 9(3): e14359. https://doi.org/10.1016/j.heliyon.2023.e14359.

Le, N.D., T.T.H. Hoang, T.T. Duong, X. Lu, T.M.H. Pham, T.X.B. Phung, et al. 2022. First observation of microplastics in surface sediment of some aquaculture ponds in Ha Noi, Vietnam. J. Hazard. Mater. Adv. 6: 100061. https://doi.org/10.1016/j.hazadv.2022.100061.

Li, W., C. H.F. Tse and L. Fok. 2016. Plastic waste in the marine environment: A review of sources, occurrence and effects. Sci. Total Environ. 566–567: 333–349. https://doi.org/10.1016/j.scitotenv.2016.05.084.

Lindwall, C. 2024. Single-Use Plastics 101. Retrieved 05 03, 2024, from NRDC: https://www.nrdc.org/stories/single-use-plastics-101#what.

Loc, N.X. 2024. Phu Quoc is a pioneer in reducing Plastic Waste. Retrieved 09 18, 2024, from https://dangcongsan.vn/xa-hoi/phu-quoc-tien-phong-giam-rac-thai-nhua-667399.html (In Vietnamese).

MacArthur, E. 2021. How the circular economy tackles climate change. Completing the Picture. Retrieved 06 13, 2024. https://materialeconomics.com/node/36#

Maitlo, G., I. Ali, H.A. Maitlo, S. Ali, I.N. Unar, M.B. Ahmad, et al. 2022. Plastic waste recycling, applications, and future prospects for a sustainable environment. Sustainability. 14: 11637. https://doi.org/10.3390/su141811637.

Minh, T., P. Trung and N. Xuan. 2020. Reducing plastic waste for the competitiveness of Vietnamese tourist attractions. E3S Web Conf. 210: 01011. https://doi.org/10.1051/e3sconf/202021001011

Ministry of Health of Vietnam (MOH). 2022. Re-examination of treatment of COVID-19 related waste for pandemic prevention and control. Retrievable at https://moh.gov.vn/tin-noi-bat/-/asset_publisher/3Yst7YhbkA5j/content/ra-soat-anh-gia-xu-ly-chat-thai-y-te-trong-phong-chong-dich-covid-19. Accessed April 24, 2024.

Ministry of Natural Resources and Environment (MONRE). 2018. Circular promulgation of national technical regulations on environment, No. 08/2018/TT-BTNMT. Hanoi. Retrievable at https://lawnet.vn/en/vb/Circular-08-2018-TT-BTNMT-promulgation-of-national-technical-regulations-on-environment-61305.html

MONRE. 2020. National Environmental Status Report of Vietnam 2019 - Topic: DSW management (in Vietnamese). Retrievable at http://dwrm.gov.vn/uploads/news/2020_11/bao-cao-hien-trang-moi-truong-2019.pdf

MONRE. 2021a. National Environmental Status Report of Vietnam 2016-2020 (in Vietnamese). Retrievable at https://moit.gov.vn/upload/2005517/fck/files/20211108_Bao_cao_HTMT_2016-2020_F_a4980.pdf

MONRE. 2021b. Decision No. 2346/QD-BTNMT. Hanoi (in Vietnamese). Retrievable at https://thuvienphapluat.vn/van-ban/Tai-nguyen-Moi-truong/Quyet-dinh-2436-QD-BTNMT-2021-thuc-hien-Quyet-dinh-1316-QD-TTg-quan-ly-chat-thai-nhua-497260.aspx

MONRE. 2022a. Summary of data from the environmental protection report in 2021 of localities (in Vietnamese)

MONRE. 2022b. Circular detailing a number of articles of law on environmental protection, No. 02/2022/TT-BTNMT. Hanoi. Retrievable at https://lawnet.vn/en/vb/Circular-02-2022-TT-BTNMT-Detailing-articles-of-law-on-environmental-protection-7BF03.html

National Assembly of Vietnam (NA-Vietnam). 2010. Law on Environmental Protection Tax No. 57/2010/QH12. Hanoi. Retrievable at https://thuvienphapluat.vn/van-ban/Thue-Phi-Le-Phi/Law-No-57-2010-QH12-Environmental-protection-Tax-123975.aspx

NA-Vietnam. 2012. Vietnam's Law of the Sea No. 18/2012/QH13. Hanoi. Retrievable at https://vietnamlawmagazine.vn/the-2012-law-of-the-sea-of-vietnam- 4904.html

NA-Vietnam. 2015. Law on Natural Resources and Environment of Sea and Islands 2015 No. 82/2015/QH13. Hanoi. Retrievable at https://www.economica.vn/Content/files/LAW%20%26%20REG/82_2015_QH13%20Law%20on%20natural%20resources%20and%20environment%20of%20sea%20and%20island.pdf

NA-Vietnam. 2020. Law on Environmental Protection No. 72/2020/QH14. Hanoi. Retrievable at https://winrock.org/wp-content/uploads/2023/02/LuatBVMT-Eng_20230203.pdf

Nguyen, T.H. and M.T. Le. 2021. Challenges in municipal solid waste management in urban areas of Vietnam. J. Environ. Manag. 270: 111499.

Nguyen, T.D.T., T. Nakakubo and K. Kawai. 2023. Analysis of COVID-19 waste management in Vietnam and recommendations to adapt to the 'new normal' period. J. Mater. Cycles Waste Manag. 25: 835–850. https://doi.org/10.1007/s10163-022-01563-x

NPAP-Vietnam. 2023. Gender context assessment of the plastic value Chain in Vietnam. Retrievable at https://weforum.ent.box.com/s/8y10a2qcpko0qnoc8lprgff52yg8gw11

People's Committee of Ha Noi. 2016. Decision 54/2016/QD-UBND, on fees for waste collection and transportation in Ha Noi. Hanoi, (in Vietnamese). Retrievable at https://thuvienphapluat.vn/van-ban/Tai-chinh-nha-nuoc/Quyet-dinh-54-2016-QD-UBND-gia-dich-vu-thu-gom-van-chuyen-rac-thai-sinh-hoat-ve-sinh-moi-truong-Ha-Noi-336776.aspx

People's Committee of Tuyen Quang. 2017. Decision 432/QD-UBND, on fees for waste collection and transportation in Tuyen Quang. Tuyen Quang (in Vietnamese). Retrievable at https://thuvienphapluat.vn/van-ban/Tai-chinh-nha-nuoc/Quyet-dinh-432-QD-UBND-2017-quy-dinh-tam-thoi-gia-toi-da-dich-vu-thu-gom-rac-thai-Tuyen-Quang-369095.aspx

Prime Minister of Vietnam (PM-Vietnam). 2020. Directive regarding strengthening of management, reuse, recycling, disposal, and reduction of plastic waste, No. 33/CT-TTg. Hanoi. Retrievable at https://thuvienphapluat.vn/van-ban/EN/Tai-nguyen-Moi-truong/Directive-33-CT-TTg-2020-regarding-strengthening-of-management-reduction-of-plastic-waste/475128/tieng-anh.aspx

PM-Vietnam. 2021. Decision Approving the scheme for strengthening management of plastic wastes in Vietnam, No. 1316/QD-TTg. Hanoi. Retrievable at https://english.luatvietnam.vn/decision-no-1316-qd-ttg-approving-the-scheme-for-strengthening-management-of-plastic-wastes-in-vietnam-206004-doc1.html

Salmela, M. 2016. Circular economy business models—Case: UPM plywood. Lahti Univ. Appl. Sci. Ltd. https://core.ac.uk/download/pdf/80989476.pdf

Sajjad, M., Q. Huang, S. Khan, M.A. Khan, Y. Liu, J. Wang, et al. 2022. Microplastics in the soil environment: A critical review. Environ. Technol. Innov. 27: 102408. https://doi.org/10.1016/j.eti.2022.102408.

Sarkingobir, Y., M. Bello and H. Yabo. 2021. Harmful effects of plastics on air quality. Academia Letters. https://doi.org/10.20935/AL2967

Thu, N.T. 2020. Mof.gov.vn. Retrieved 06 14, 2024, from https://mof.gov.vn/webcenter/ShowProperty?nodeId=/UCM12/MOFUCM187894//idcPrimaryFileandrevision=latest releasedandrid=1.

Tu, L.A. 2022. Overview of the Vietnamese Plastics Industry. Retrieved 04 19, 2024, from https://vioit.org.vn/vn/chien-luoc-chinh-sach/tong-quan-nganh-nhua-viet-nam---phan-1--4666.4050.html (In Vietnamese).

Tuan, Q. 2024. Hoi An practices the reduction of single-use plastic waste. Retrieved 09 18, 2024, from https://baoquangnam.vn/hoi-an-thuc-hanh-giam-thieu-rac-thai-nhua-dung-mot-lan-3136145.html (In Vietnamese).

Vietnam Plastics Association (VPA). 2023. Workshop report on Plastic recycling—Towards circular economic development. April 11, 2024, Hanoi, Vietnam (In Vietnamese).

Vietnam Television (VTV). 2024. Using plastic waste as fuel for cement kilns. VTV online. Retrievable at https://vtv.vn/xa-hoi/su-dung-rac-thai-nhua-lam-nhien-lieu-cho-lo-dot-20240422091238222.htm. Accessed April 12, 2024.

VNCPC. 2018. Report on Current Situation of Phan Boi and Minh Khai Villages; VNCPC: Hanoi, Vietnam, 2018.

VPAS. 2024. Reducing Plastic Waste Pollution in Vietnam Through Digital Data. Rechieved from vpas.vn: https://vpas.vn/thong-tin/tin-tuc/giam-thieu-o-nhiem-rac-thai-nhua-o-viet-nam-thong-qua-du-lieu-so.html?utm'

Wang, W., J. Ge, X. Yu and H. Li. 2020. Environmental fate and impacts of microplastics in soil ecosystems: Progress and perspective. Sci. Total Environ. 708: 134841. https://doi.org/10.1016/j.scitotenv.2019.134841.

Wan, Y., C. Wu, Q. Xue and X. Hui. 2019. Effects of plastic contamination on water evaporation and desiccation cracking in soil. Sci. Total Environ. 654: 576–582. https://doi.org/10.1016/j.scitotenv.2018.11.123.

World Bank Group. 2021. Market Study for Vietnam: Plastics Circularity Opportunities and Barriers. Marine Plastics Series; © World Bank, Washington, DC. http://hdl.handle.net/10986/36313 License: CC BY 3.0 IGO. Retrievable at https://openknowledge.worldbank.org/entities/publication/f94e9a28-d95e-5c6b-b060-ef25e0fef974. Accessed June 12, 2024.

World Bank. 2022. Vietnam: Plastic pollution diagnostics. East Asia and Pacific Region, Marine Plastics Series. Washington, DC: World Bank. © World Bank. https://openknowledge.worldbank.org/handle/10986/37693 License: CC BY 3.0 IGO

WWF. 2020. National survey on the contribution of plastic waste from fisheries and aquaculture to ocean plastic debris (in Vietnamese).

WWF. 2023. Report on Plastic Waste Generation in 2022. Retrievable at https://wwfasia.awsassets.panda.org/downloads/eng_wwf_a4_bao-cao-chat-thai-nhua_260124.pdf

WWF and USAID. 2018. Solid waste survey report focusing on reducing plastic waste in Phu Quoc (in Vietnamese).

WWF. 2024. Impact Report 2019–2024: No Plastic in Nature Initiative – Driving Systems Change in the Global Fight Against Plastic Pollution. Retrievable at https://wwfint.awsassets.panda.org/downloads/npin-report_25.pdf'

Yin, M., B. Yan, H. Wang, Y. Wu, X. Wang, J. Wang, et al. 2023. Effects of microplastics on nitrogen and phosphorus cycles and microbial communities in sediments. Environ. Pollut. 318: 120852. https://doi.org/10.1016/j.envpol.2022.120852

Zhang, J., S. Ren, W. Xu, C. Liang, J. Li, H. Zhang, et al. 2022. Effects of plastic residues and microplastics on soil ecosystems: A global meta-analysis. J. Hazardous Mater. 435: 129065. https://doi.org/10.1016/j.jhazmat.2022.129065

Chapter **5**

Microplastic Pollution in the Water Column and Sediment of Cochin Estuary, India: Abundance, Characteristics, and Potential Ecological Risk Assessment

N.V. Viji and E.V. Ramasamy*
School of Environmental Sciences, Mahatma Gandhi University, Kottayam, Kerala, India-686560

5.1 INTRODUCTION

Marine ecosystem services support the livelihoods of over 40% (three billion people) of the global human population, and they contribute over 60% of the economic value of all life on earth (Andersen 2021). However, the oceans have become a sink for almost all kinds of pollutants, from plastics to toxic chemicals. Plastics represent the largest, harmful, and most persistent proportion of marine litter. About 4.8 to 12.7 million tons of plastic waste reach global oceans annually (Jambeck et al. 2015), making oceans the largest waste bins of plastic discards. The plastic debris that reaches oceans (or during their journey) is fragmented

*For Correspondence: evramasamy@mgu.ac.in

DOI: 10.1201/9781003531463-5

into smaller pieces known as "Microplastics" (MPs) when their size is less than 5 mm; if the size is less than 1 μm, they are known as nano plastics (Lang et al. 2020). The major drivers of this fragmentation of plastics include: ultraviolet (UV) radiation, wind and wave action, physical abrasion, and, to some extent, microbial degradation (Ramasamy and Harit 2023).

Microplastics (MPs) can originate from various sources such as discharge from wastewater treatment plants, untreated sewage, atmospheric deposition, surface run-off, and landfill leachate (Saikia and Handique 2024). MPs pass through the filtration units of sewage or wastewater treatment plants and end up in nearby water bodies and ultimately in oceans, posing a severe threat to aquatic life. Due to their small size and ubiquitous presence across various ecosystems, MPs are available for ingestion by almost all trophic level organisms; thus, the potential for detrimental effects through the food chain and food web is substantial. The MPs, with their inbuilt additives such as Bisphenols, Phthalates, and Colorants, cause damage to the biota that ingest them. Besides the additives, the MPs also sorb many pollutants from their surrounding environment; these contaminants vary from persistent organic pollutants (POPs), persistent bioaccumulative and toxic (PBT) substances, heavy metals, pesticides, etc. (Ashton et al. 2010; Cole et al. 2011; Wagner et al. 2014; Brennecke et al. 2016). The microbial colonization (as biofilms) on the surface of MPs also takes place, thus MPs in aquatic systems act as vectors transporting the cocktail of chemicals and microbes to distant places and across the food chain and food web. Considering the long-term impacts these MPs could cause on the aquatic and terrestrial biota, they have been identified as contaminants of emerging concern - CEC (Huang et al. 2021).

Literature on aquatic microplastics has plenty of reports on marine microplastics, while studies on MPs in freshwater bodies and estuaries are relatively less. The studies in India also reflect a similar trend. In the case of wetlands, there are only limited studies reported on MPs pollution from India, especially with Ramsar sites (Saikia and Handique 2024). Focusing on this research gap, the current study was conducted in Cochin Estuary (CE), which is a part of Vembanad Lake (VL)—a Ramsar site—in Kerala, India.

Vembanad Lake (Lat. 9° 30′–10° 10′ N and Long. 76° 10′–76° 25′ E) with an area of 151,250 ha, receives water from six rivers: Meenachil, Manimala, Achancovil, Pamba, Muvattupuzha, and Periyar (Chandran and Ajaykumar 2018) and annual rainfall of 300 cm. The lake has an artificial barrier, Thannermukkom Bund (TM bund), which divides VL into a saltwater-dominant northern part and a freshwater-dominant southern part (Sruthy and Ramasamy 2017). Cochin Estuary is in the northern part of VL.

CE is located near Kochi, the most populous and industrial city of Kerala. The discharges from industries, municipal solid waste, domestic sewage, and run-off water from the city are released into the estuary. All these may also cause MPs pollution in the CE. The current study aims to assess the MPs' abundance, types, and characteristics in the water column and in sediment samples of CE, also to study the potential risks posed by the MPs in the study area.

5.2 METHODS

5.2.1 Study Area

Cochin Estuary (CE), which is a part of Vembanad Lake (VL)—a Ramsar site—in Kerala, India, is the study area selected for this work (Figure 5.1). Cochin Estuary (Latitude: 9° 40′ and 10° 12′ N and Longitude: 76° 10′ & 76° 30′ E) is in the northern part of Vembanad Lake (VL). Cochin Estuary, the third largest coastal wetland system in India, has been established as one of the most productive estuarine systems (Thasneem et al. 2018). This estuary consists of a network of shallow brackish water with two permanent openings to the Arabian Sea, with a tidal amplitude of 1.0 m (Salas et al. 2017). The study area receives an annual rainfall of 300 cm, which may vary from year to year. (Gireeshkumar et al. 2013). The following anthropogenic activities are commonly noticed in and around CE: transportation of goods, tourism, fishing, and industrial activities (Srinivas et al. 2003). All these activities, in turn, contribute a massive quantity of waste, including municipal solid waste, industrial waste, sewage, and other domestic wastes into CE (Naidu et al. 2018).

5.2.2 Sampling

The sampling was done during the Monsoon of 2021. The Sub Surface Water (SSW) and Bottom Water (BW) samples (in duplicates) were collected from the selected eight locations of CE (Figure 5.1) using the depth water sampler (Hydro-bios, Germany) with a capacity of 2 litres. The collected samples (2 L) were screened on board using a metal sieve having a pore size of 53 μm to reduce the volume of the sample. The particles retained on the surface of the sieve were washed into a glass bottle of 250 ml capacity with a glass lid, then transported to the laboratory for further analysis. In order to avoid contamination, pre-filtered, ultra-pure water was used in all steps of the sample processing. Sediment samples were collected from the eight locations using the grab sampler (Hydro-bios, Germany); about one kg of samples (in duplicate) from each site were collected. The sediment samples were transferred into a metal tray and homogenized using a stainless-steel scoop; after homogenization, the samples were packed in stainless steel boxes closed with a lid, then transported to the laboratory for further analysis.

5.2.3 Extraction of Microplastics from the Samples

5.2.3.1 Water Samples

The extraction of microplastics from the samples was done as per the procedure of the National Oceanic and Atmospheric Administration (NOAA) protocol (Masura et al. 2015). The processing of the samples for the extraction of MPs and subsequent analysis was carried out in a special glass hood (6 × 2 × 3.5’ size), specially fabricated for this purpose, in order to avoid airborne MPs contamination

(Figure 5.2). The screened water samples were subjected to Wet Peroxide Oxidation (WPO) using 30 percent hydrogen peroxide and ferrous sulphate (20 ml each) in order to digest the organic matter present in the samples (Rodrigues et al. 2018). This step was repeated two to three times for a few samples, depending on the organic matter content. After digestion, the samples were subjected to density separation in a glass beaker containing sodium chloride (NaCl) solution (Alam et al. 2019). After density separation, the supernatant along with MPs was decanted into the flask of the vacuum filter apparatus, and the same was filtered through a Whatman glass fiber filter (GF/A) paper having a pore size of 1 μm.

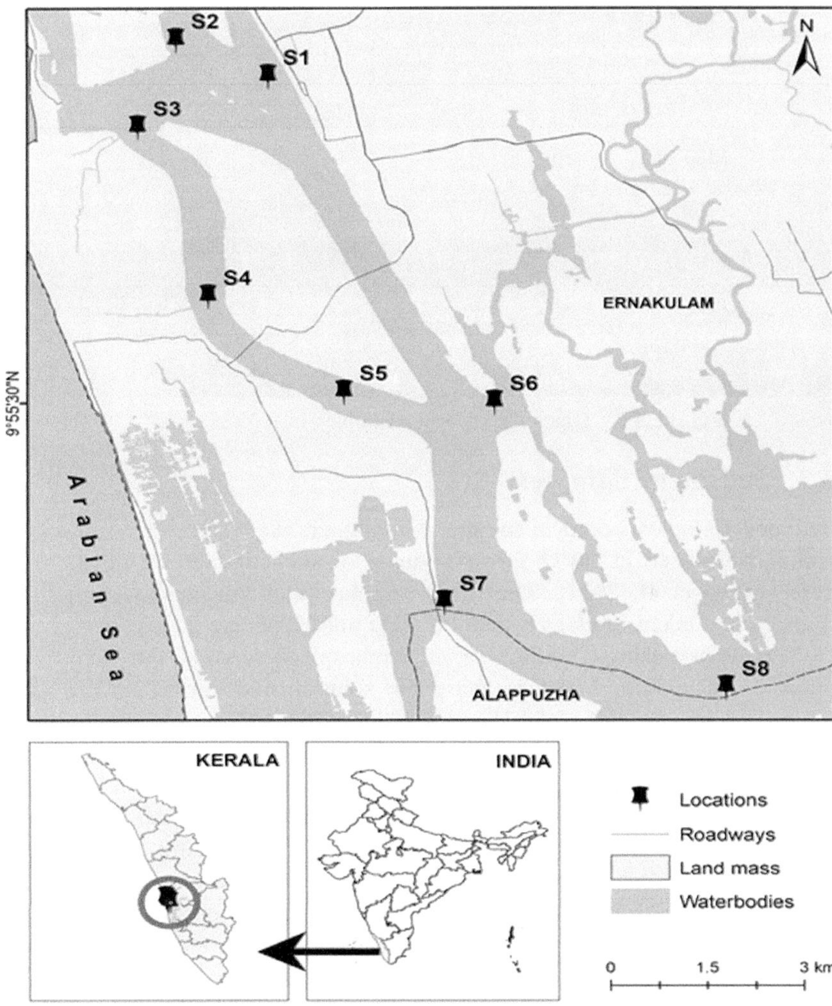

Figure 5.1 Sampling locations in Cochin Estuary.

After filtration, the filter paper was carefully transferred to a glass petri plate and dried in an oven at 40 degrees Celsius until the paper was completely dry. The dried filter papers were subjected to further microscopic observations.

Figure 5.2 Digestion of water samples in a glass hood.

5.2.3.2 Sediment Samples

The sediment samples stored in the metal containers were transferred into metal trays and oven-dried at 50°C for 3–4 days for the complete removal of the moisture (Liang et al. 2022). The dried samples (in duplicates) were weighed (100 g each) and taken in glass beakers (Saikia and Handique 2024). The samples were disaggregated using a sodium hexametaphosphate solution and were treated with 300 ml of sodium hexametaphosphate solution and stirred on the rotary shaker for 1 hour (Amrutha and Warrier 2020). Then the sample was subjected to density separation using a zinc chloride solution. The mixture was stirred manually with a metal spoon, and the sample was left overnight, allowing the particles to settle down. The supernatant was decanted into a tall glass beaker and subjected to WPO for the digestion of organic matter using 30% hydrogen peroxide (WPO). After digestion, the sample was filtered through a vacuum filtration unit using a Whatman glass fiber filter (GF/A) paper having a pore size of 1 μm. The filter papers were dried in the same manner as explained in the previous section.

5.2.4 Observation and Identification of Microplastics

The particles over the filter papers were observed under a stereo microscope (Carl Zeiss Stemi 508, Germany). Physical features like size, shape, and colour were recorded during microscopic observation. Based on shape, MPs were classified into five types: Fiber, Fragment, Film, Foam, and Beads. Likewise, the MPs based on their size were categorized into six size classes: <100 μm, 100–500 μm, 500–1,000 μm, 1,000–2,000 μm, 2,000–3,000 μm, and 3,000–5,000 μm. Six different colours were noted with the MPs: blue, red, transparent, green, black, and yellow. The suspected MPs were subjected to micro-Raman spectroscopy (WITec Alpha 300 RA, Germany) coupled with a confocal microscopic observation in order to confirm that these particles are plastics and to identify the polymer profile of the MPs. Raman spectroscope has a single-mode 532 nm DPSS (Diode Pumped Solid State) laser as the excitation source with 42 mV power. The spectral resolution is enhanced by a grating of 600 grooves/mm and 20 X objective. With an integration time of 10 s, averaged over 10 scans, a CCD detector detected signals to get the Raman spectra. The resulting spectra were analyzed using the *KnowItAll* software (spectral library) provided by Wiley for polymer confirmation.

5.2.5 Risk Assessment

The microplastic pollution in the study area was analyzed using the application of pollution risk assessment indices. Pollution load index (PLI) was used for pollution risk assessment in this study. Tomlinson et al. (1980) proposed a pollution load index (PLI), and the same was used to evaluate the overall pollution load of MPs present in the subsurface water, bottom water, and sediment samples of the estuary (Pan et al. 2021). The PLI was calculated using the following equation:

$$CF_i = \frac{C_i}{Cn_i}$$

$$\mathrm{PLI} = \sqrt{CF_i}$$

where, CF_i is the concentration factor; C_i is the concentration of MPs in each site; and Cn_i is the background value.

PLI can be calculated from the concentration factor (CF_i). For obtaining the CF_i value, the concentration of each microplastic (C_i) has to be divided by the baseline value or background value; the lowest value obtained among the sites is considered as the background value (Cn_i) (Yin et al. 2021).

5.3 RESULTS AND DISCUSSION

5.3.1 Abundance of Microplastics

Microplastics were recorded from all sampling sites, indicating the ubiquitous distribution of MPs in the estuary. In sub-surface water (SSW) samples, the abundance of MPs was in the range of 22 to 32 particles per L with an average concentration of 35.34 ± 7.68 particles per L (Figure 5.3a). Likewise, in the bottom water (BW) samples, the range of MPs varied from 25 to 51 particles per L with an average abundance of 26.87 ± 3.14 particles per L (Table 5.1) was noted (Figure 5.3b). The mean abundance of MPs was higher in SSW than in BW samples, likely due to the salinity of the water, which may be the reason for this difference. Additionally, the influence of the monsoon also causes more plastics to be discharged into the estuary and float on the surface (Anagha et al. 2023).

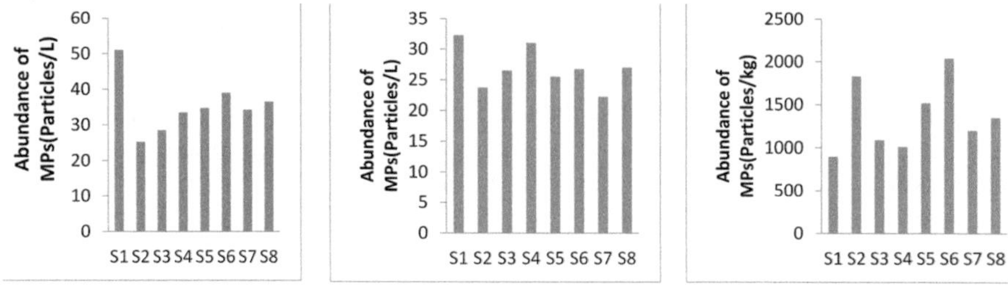

(a) SSW (b) BW (c) Sediment

Figure 5.3 Abundance of microplastics (a) sub-surface water, (b) bottom water, and (c) sediment.

The present study's findings are comparable with the MPs abundance reported from Kodaikanal Lake (Tamil Nadu, India) with 24.42 ± 3.22 items per L; however, the values are for surface water of the Kodaikanal Lake and the sampling was done using Manta net with a mesh size of 333 μm and a mouth diameter of 30 cm (Laju et al. 2022). Previous studies conducted in the southern part of Vembanad Lake also recorded an average MPs abundance of 26.79 ± 3.74 items per L in subsurface water (Anagha et al. 2023), which is less compared with the current study (Table 5.1). According to Yuan et al. (2019), the abundance of MPs in the waters of Poyang Lake was reported as 5–34 items per L, which is similar to the results of the CE study. In the case of bottom water, no study is available in the literature other than Anagha et al (2023), in which the BW samples were reported with an average abundance of 52.7 ± 5.43 items per L, which is double the concentration of MPs reported in the current study. The depth of the estuary is negatively correlated with the abundance of the MPs present in the water column. As depth increases, a decline in the abundance of MPs in the water column has been observed (r = –0.57231).

Table 5.1 Abundance (Mean) of MPs from each site of water and sediments

Sampling Station Code	Subsurface Water	Bottom Water	Sediment
S1	25.5 ± 3.89	16.12 ± 4.06	450 ± 127.28
S2	12.63 ± 1.24	11.87 ± 3.35	915 ± 77.78
S3	14.25 ± 1.77	13.25 ± 3.18	545 ± 205.06
S4	16.75 ± 6.01	15.5 ± 1.06	505 ± 21.21
S5	17.38 ± 5.83	12.75 ± 0.35	760 ± 70.71
S6	19.50 ± 6.36	13.37 ± 3.35	1,020 ± 509.12
S7	17.13 ± 4.07	11.12 ± 3.35	600 ± 28.28
S8	18.25 ± 2.12	13.5 ± 4.94	675 ± 219.20

The highest abundance was found in both subsurface and bottom water at Site 1, which is located near Marine Drive, one of the major tourist spots in

Kochi. This spot is popular for boating and sightseeing. Tourist activities and improper waste management practices may be the main reasons for the higher MP abundance in S1 (Suresh et al. 2020). In sediment samples, the mean abundance of all sites is 1367.5 ± 403.75 particles per kg with a range of 900–2,040 particles per kg. (Figure 5.3c). The MPs abundance found in CE sediment was higher compared to the abundance (15 to 632 particles per kg) observed in Renuka Lake, Himachal Pradesh (Ajay et al. 2021). The highest abundance was recorded from location S6, situated near Thevara, a place with a high human population density. This high population density may contribute to the increased MP pollution observed in the S6 sampling location. On the basis of the average abundance of MPs in the sediment samples, the following trend (Descending order) has been noted: S6 > S2 > S5 > S8 > S7 > S3 > S4 > S1.

5.3.2 Morphological Characteristics of Microplastics

5.3.2.1 Shape

Microplastics (MPs) were observed as five different morphotypes in the water and sediment samples of CE. Fibers, fragments, film, foam, and beads are the shapes observed under a stereomicroscope (Figure 5.4). Fibers are linear or filamentous, while fragments are thick and of irregular shape, originating from the fragmentation of larger plastic Items. Films have a flat surface, flexible with soft texture; foams are foamy plastics or spongy foams, and beads are granular or spherical structures (Zhao et al. 2024). Fragment is the predominant shape

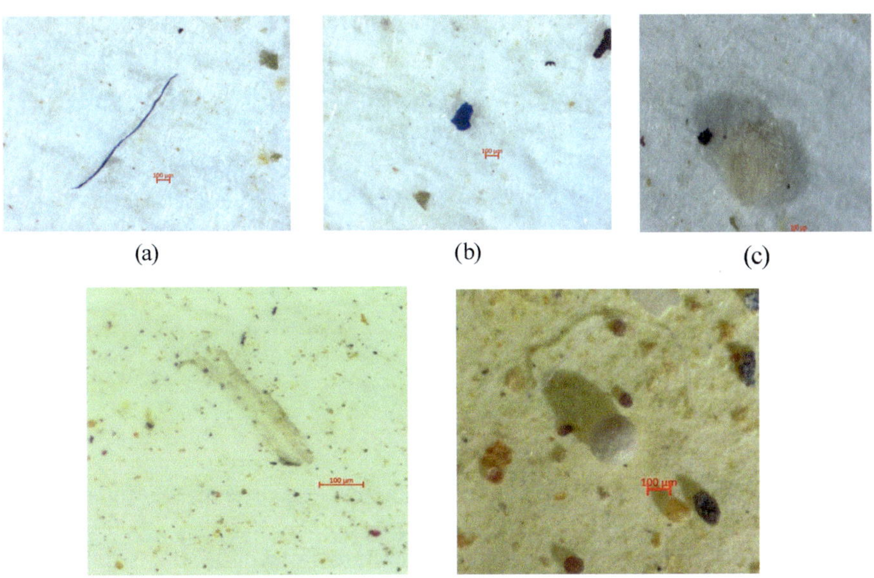

(a) (b) (c) (d) (e)

Figure 5.4 Different morphotypes of microplastics (a) Fiber, (b) Fragment, (c) Film, (d) Foam, and (e) Beads observed in both SSW and BW (Figure 5.5). In sediments, fibers were noted as the predominant shape of MPs, followed by fragments (Figure 5.5). These fiber shape MPs may originate from the weathering of synthetic clothes while washing and reach CE through laundry wastewater discharges, as well as from the peeling or disintegration of fishing nets and ropes (Browne et al. 2011; Cole et al. 2011). Fibers were also reported as the dominant shape of MPs reported from the Yangtze Estuary (Zhao et al. 2014). The shape of MPs is a crucial characteristic feature, as the further impacts of these MPs on the aquatic systems, especially on the aquatic organisms, are based on the shape and size of the MPs as well. The ingestion of MPs by aquatic fauna is more feasible when the MPs are in fiber shape, while the fragments with better surface area are prone to biofouling and sorption of other contaminants, which may ultimately result in the sinking or rising of microplastics in the water column.

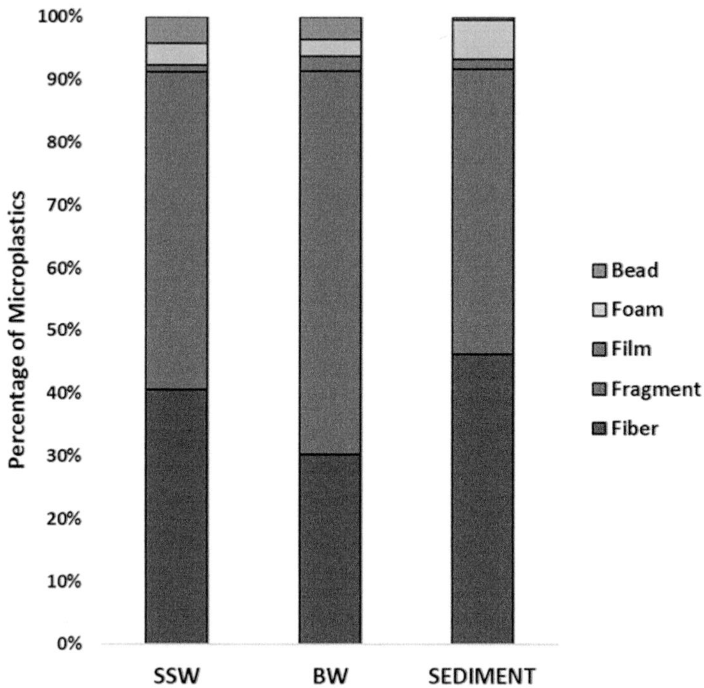

Figure 5.5 Proportion of different morphotypes of MPs present in the water column and sediment samples of Cochin Estuary.

5.3.2.2 Size

The microplastics identified from the samples are categorized into six size classes. <100 µm, 100–500 µm, 500–1,000 µm, 1,000–2,000 µm, 2,000–3,000 µm and 3,000–5,000 µm. Most of the particles are present in the size range of 100–500 µm (Figure 5.6). The particles greater than 5 mm, which are mesoplastics, are also present in small amounts in all samples. Physical and biochemical forces fragment the larger-sized particles in the aquatic environment to form smaller-sized MPs. Most of the small-sized microplastics are from the weathering of parent plastic products (Huang et al. 2021). The subsequent decrease in the size of MPs becomes more pronounced with a prolonged time period of the presence of plastic debris in the aquatic environment (Saad et al. 2022). The size is considered an important parameter because it affects the transport of microplastics in the water column (Kılıc et al. 2024). The size of MPs affects not only transportation but also distribution, bioavailability, pathways, fate, buoyancy, and the drag force. Accordingly, it determines the distribution of MPs in water and the accumulation of MPs in sediment. Additionally, the size of MP particles affects their mobility and long-distance transportation (Zhang et al. 2022). Moreover, MP size is correlated with hydrodynamic parameters, including rise velocity, settling time, and relaxation time; in the end, this correlation defines MP mobility and transport. For example, it has been found that small MPs have a lower settling velocity and relaxation time than bigger MPs, yet they sink more quickly due to biofouling. This suggests that small MPs are more mobile and have a higher transport rate than larger MPs (Kumar et al. 2021; Shamskhany et al. 2021). The small-sized MPs are more accessible to organisms and are also prone to forming microbial biofilm over their surface. The aquatic organisms easily feed on them by mistake; the small-sized particles have a large surface area, which results in the sorption of the contaminants present in the water column and sediments (Saad et al. 2024).

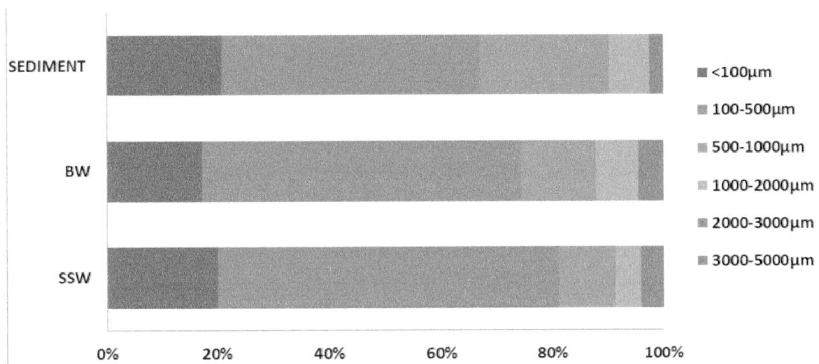

Figure 5.6 Size distribution of microplastics in the water column and sediment samples of Cochin Estuary.

5.3.2.3 Colour

The stereomicroscope was effectively used to identify the colour of microplastics. Six different colours of MPs were identified from both the water and sediment samples of Cochin Estuary. These colours include blue, red, transparent, green, black, and yellow (Figure 5.7). Plastic products are produced with different colours to attract consumers, which leads to the formation of coloured microplastics. (Kılıc et al. 2024). The dominance of blue-coloured MPs present in water samples may be because they arise from the weathering of fishing nets, while transparent plastics are the raw materials for most of the plastic products (Zhao et al. 2024). In sediment samples, transparent is the dominant colour, followed by blue and red. A similar observation of transparent as predominant colour was reported from

CE by Suresh et al. (2020). More coloured MPs indicate the influence of human activities, and colour can also be considered a handy parameter to identify the source of microplastics. The high abundance of coloured MPs implies that the colour exposure may pose significant health risks to the biota that ingest the MPs. Several studies have reported that the aquatic organisms feed on the coloured MPs, mistaking them as their prey (Saad et al. 2024). The ingestion of coloured MPs is of great concern because the coloured MPs consist of colorant agents like pigments and dyes. These colorants are toxic. The plastic and chemicals added to plastic during the production process are the main sources of coloured MP's toxicity (Onoja et al. 2022).

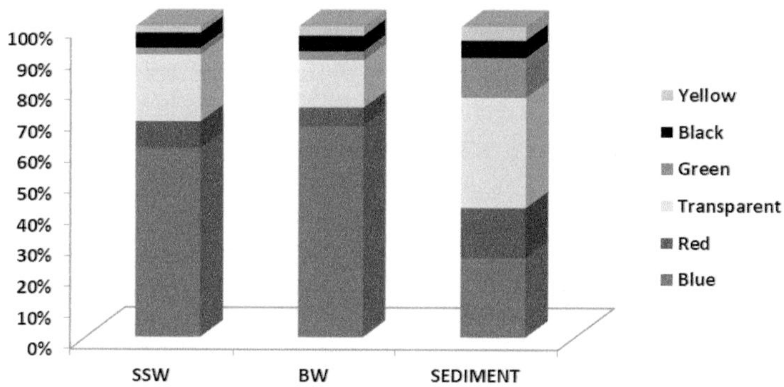

Figure 5.7 Microplastics with different colours observed in the water column and sediment samples of Cochin Estuary.

5.3.3 Polymer Identification of Microplastics

Raman spectra were used to identify the polymer content of the MPs (Figure 5.8). The spectral library software "*Knowitall*" from *Wiley* was used for this purpose. Only those spectra that matched over 70%with the reference spectrum of the spectral library were confirmed as the corresponding polymers (Peng et al. 2017). Different types of polymers were identified from the water column and sediment samples (Figure 5.8). In SSW, the polymers found were Polyethylene (PE), Polyamide (PA), Polypropylene (PP), and Low-Density Polyethylene (LDPE). Among these, Polyethylene (PE) and Polypropylene (PP) were dominant. In a similar observation, Polyester, Polypropylene, and Polyethylene were identified as the most common polymer types detected from the Meghna Estuary in Bangladesh (Alam et al. 2023). Excessive use of plastic carry bags, plastic wrap, and milk cartons, and discarding them into the environment without recycling, leads to the high abundance of PE in the environment. Furthermore, expanded polystyrene (EPS) and extruded polystyrene (XPS) are employed as buoyants in fishing nets. Polystyrene (PS) primarily originates from riverine fishing activities (Hossain et al. 2023). PVC and HDPE were dominant in BW because of their high density. Polymers like PET and PVC are getting into the bottom of the water body due to their density, which is about 1.4 per cm^3 (Huang et al. 2021).

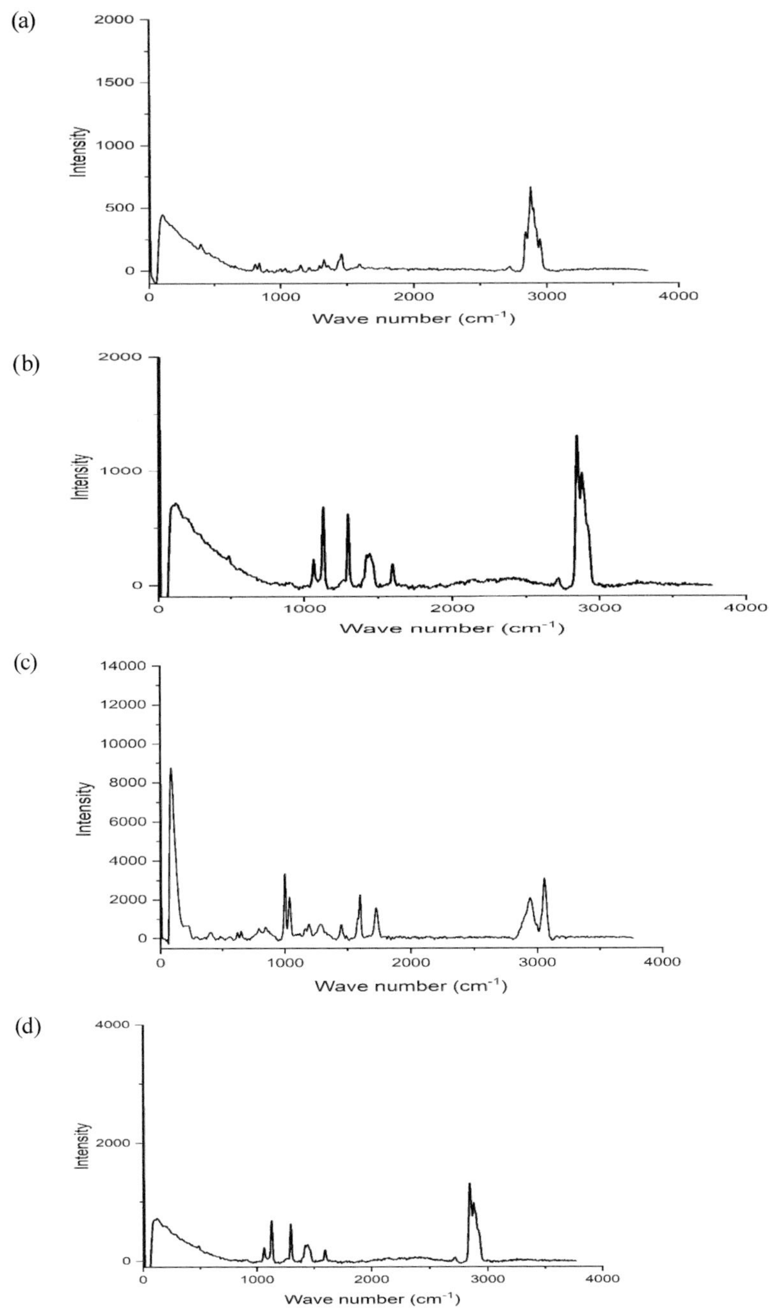

Figure 5.8 Raman spectral plots for four representative microplastic polymers. Panel (a) shows characteristic peaks of polypropylene (PP); panel (b) displays the spectral features of high-density polyethylene (HDPE); panel (c) presents the distinct peaks of polystyrene (PS); and panel (d) shows the polyethylene (PE) spectrum. Each plot highlights the unique vibrational signatures used to identify polymer types.

In addition to polymer density, other factors influencing the distribution of MPs in bottom waters include shape and surface-to-volume ratio, biofouling, age, and physical forces of the lake that can shift the position of particles in the water (Yang et al. 2023). In sediment samples, HDPE and PE were the dominant polymers. HDPE, a form of PE used in manufacturing Bottles, drums, gasoline tanks, pipes, wrapping films, and crates (Wang et al. 2018). PE is a low-density polymer, but its dominance was noted in BW and in sediments, which may be due to the biofouling. PE is mostly inert, yet it may progressively degrade in the environment due to structural defects or the action of natural substances (Chamas et al. 2020).

5.3.4 Risk Assessment

5.3.4.1 Pollution Load Index

Pollution load index (PLI) is used to quantify the MPs contamination (Ranjani et al. 2021) rather than to determine the level of pollution in each sampling site and also for the entire sampling area (Pan et al. 2021). When the calculated value for PLI falls below 10 (Ten), the hazard level of the contaminant is considered Hazard Level I (Level One), which belongs to a minor or low contaminant category (Nithin et al. 2023). In the current study, the calculated value of all sampling sites for the water samples (both SSW and BW) is in the range of 1.00 to 1.42; for the sediment samples, the values range from 1.00 to 1.51 (Table 5.2). Accordingly, for both water and sediment samples, as the PLI falls less than 10 (Figure 5.9), MPs contamination in this study area could be categorized as Hazard Level I (Level One), which indicates MPs contamination belongs to "minor" or "low" level. In PLI calculations, the selection of the background value is crucial; in the current study, the lowest value obtained among the sites has been considered as the background value index (Huang et al. 2023).

A study carried out by Ephsy and Raja (2023) in Kumaraswamy Lake, India, has also detected the hazard level I as the same as in the current study. The

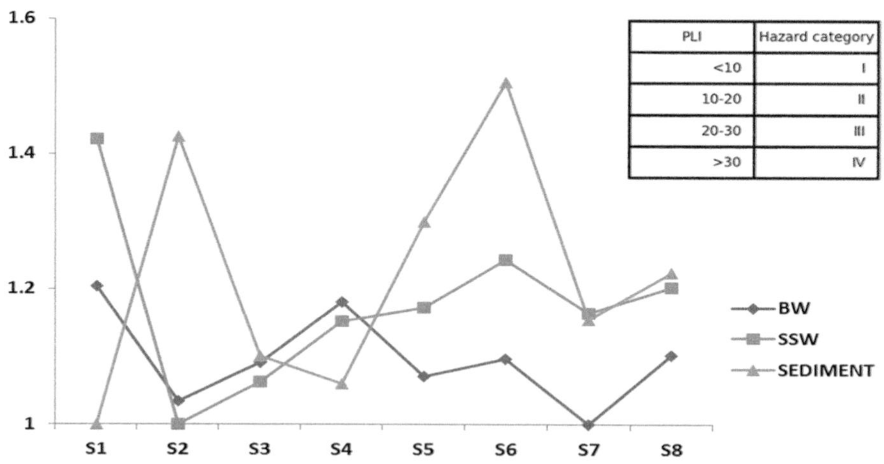

Figure 5.9 Pollution load index of water and sediment samples.

pollution load index of Deepor Beel lake in the Brahmaputra Plain, India, and the PLI for Karnaphuli River Estuary, Bangladesh, also showed similar hazard levels as those of the present study. (Saikia and Handique 2024; Rakib et al. 2022).

Table 5.2 Pollution load index of subsurface water, bottom water, and sediments

Sample Matrix	Sample ID	Ci	Cni*	Cfi	PLIi	PLI Indicator	Hazard Level
SSW	S1	51.00	25.25	2.02	1.42	Low	Level I
	S2	25.25	25.25	1.00	1.00	Low	Level I
	S3	28.50	25.25	1.13	1.06	Low	Level I
	S4	33.50	25.25	1.33	1.15	Low	Level I
	S5	34.75	25.25	1.38	1.17	Low	Level I
	S6	39.00	25.25	1.54	1.24	Low	Level I
	S7	34.25	25.25	1.36	1.16	Low	Level I
	S8	36.50	25.25	1.45	1.20	Low	Level I
BW	S1	32.25	22.25	1.45	1.20	Low	Level I
	S2	23.75	22.25	1.07	1.03	Low	Level I
	S3	26.50	22.25	1.19	1.09	Low	Level I
	S4	31.00	22.25	1.39	1.18	Low	Level I
	S5	25.50	22.25	1.15	1.07	Low	Level I
	S6	26.75	22.25	1.20	1.10	Low	Level I
	S7	22.25	22.25	1.00	1.00	Low	Level I
	S8	27.00	22.25	1.21	1.10	Low	Level I
Sediment	S1	900	900	1.00	1.00	Low	Level I
	S2	1,830	900	2.03	1.43	Low	Level I
	S3	1,090	900	1.21	1.10	Low	Level I
	S4	1,010	900	1.12	1.06	Low	Level I
	S5	1,520	900	1.69	1.30	Low	Level I
	S6	2,040	900	2.27	1.51	Low	Level I
	S7	1,200	900	1.33	1.15	Low	Level I
	S8	1,350	900	1.50	1.22	Low	Level I

* Cni: The lowest value obtained among the sites has been considered as the background value.

5.4 CONCLUSION

Microplastic (MPs) contamination and its impact on aquatic systems and aquatic organisms are attracting more attention in recent times. However, there exists a knowledge gap in the case of MPs contamination in estuarian systems. In this context, the present study assessed the MPs contamination in the water column and sediment samples of Cochin Estuary (CE) in Kerala, India. The findings of this study indicated that the abundance of MPs was higher in the subsurface water (SSW) than in the bottom water (BW). This may be attributed to the slightly saline water of the CE, due to which most of the MPs with polymers of lower densities may remain on the surface or subsurface region of the water column. Polyamide (PA), Polypropylene (PP), Low-Density Polyethylene (LDPE), Polyethylene Terephthalate (PET), Polyvinyl Chloride (PVC), Polystyrene (PS), and High-Density Polyethylene (HDPE) were the polymers found in the MPs extracted from this study area. Fragments and fibers were the predominant morphotypes in the water column and sediments, respectively. Blue, red, transparent, green, black, and yellow were the colours that composed the MPs. The majority of the microplastics were in the size range of 100–500 μm. Smaller-sized MPs are more dangerous, which means a decrease in the size of microplastics increases the possibility of intake of microplastics by the aquatic organisms. The study area is renowned for its tourism, boating, fishing, and industrial activities, all of which are considered major sources of microplastic pollution in this area. To mitigate this issue at its source, strict regulations are necessary to limit the entry of plastic debris into CE. Proper waste management practices are also to be followed. Proper awareness of the local people is also required.

5.5 ACKNOWLEDGEMENT

The authors acknowledge Kerala State Council for Science, Technology and Environment (KSCSTE), Thiruvananthapuram, for the financial support through a research project on Microplastics of Vembanad Lake [KSCSTE/295/2018-E and E dated 09-11-2018]. Ms. Viji gratefully thanks the E-grants from the Government of Kerala for the stipend. The authors extend their gratitude to the School of Environmental Sciences, M G University, for providing the laboratory facilities. The authors also express their special thanks to Ms. Anu Mathew of DST – SAIF Instrumentation Centre, M. G. University, for the technical support extended in Raman spectroscopic analysis of the MPs. The authors also express their gratitude to Ms. Devika Das and Mrs. Anagha P L for their support in sampling. Mr. Naveen Babu and Mr. Vishnu N G for their help during fieldwork.

REFERENCES

Ajay, K., D. Behera, S. Bhattacharya, P.K. Mishra, Y. Ankit and A. Anoop. 2021. Distribution and characteristics of microplastics and phthalate esters from a freshwater lake system in the Lesser Himalayas. Chemosphere. 283: 131132.

Alam, F.C., E. Sembiring, B.S. Muntalif and V. Suendo. 2019. Microplastic distribution in surface water and sediment river around slum and industrial area (case study: Ciwalengke River, Majalaya district, Indonesia). Chemosphere. 224: 637–645.

Alam, M.J., M. Shammi and S.M. Tareq. 2023. Distribution of microplastics in shoreline water and sediment of the Ganges River Basin to Meghna Estuary in Bangladesh. Ecotoxicology and Environmental Safety. 266: 115537.

Amrutha, K. and A.K. Warrier. 2020. The first report on the source-to-sink characterization of microplastic pollution from a riverine environment in tropical India. Sci. Total Environ. 739: 140377.

Anagha, P.L., N.V. Viji, D. Devika and E.V. Ramasamy. 2023. Distribution and abundance of microplastics in the water column of Vembanad Lake–A Ramsar site in Kerala, India. Mar. Pollut. Bull. 194: 115433.

Andersen, I. 2021. Tackling marine plastic pollution and protecting our oceans, speech delivered at the international symposium on plastics in the Arctic and sub-Arctic region hosted by the government of Iceland.

Ashton, K., L. Holmes and A. Turner. 2010. Association of metals with plastic production pellets in the marine environment. Mar. Pollut. Bull. 60(11): 2050–2055.

Brennecke, D., B. Duarte, F. Paiva and I. Cacador. 2016. Microplastics as a vector for heavy metal contamination from marine environments. Estuar. Coast. Shelf Sci. 178: 189–195.

Browne, M.A., P. Crump, S.J. Niven, E. Teuten, A. Tonkin, T. Galloway, et al. 2011. Accumulation of microplastic on shorelines woldwide: Sources and sinks. Environ. Sci. Technol. 45: 9175–9179.

Chamas, A., H. Moon, J. Zheng, Y. Qiu, T. Tabassum, J.H. Jang, et al. 2020. Degradation rates of plastics in the environment. ACS Sustainable Chem. Eng. 8(9): 3494–3511.

Chandran, M.S. and B. Ajaykumar. 2018. Quantitative geomorphological analysis to infer the hydrological behaviour of streams draining into the Vembanad Lake–A Ramsar site along the southwest coast of India. J. Geol. Soc. India. 92(1): 45–53.

Cole, M., P. Lindeque, C. Halsband and T.S. Galloway. 2011. Microplastics as contaminants in the marine environment: A review. Mar. Pollut. Bull. 62(12): 2588–2597.

Ephsy, D. and S. Raja. 2023. Characterization of microplastics and its pollution load index in freshwater Kumaraswamy Lake of Coimbatore, India. Environ. Toxicol. Pharmacol. 101: 104207.

Gireeshkumar, T.R., P.M. Deepulal and N. Chandramohanakumar. 2013. Distribution and sources of sedimentary organic matter in a tropical estuary, south west coast of India (Cochin estuary): A baseline study. Mar. Pollut. Bull. 66(1–2): 239–245.

Hossain, M.B., J. Yu, A.A.U. Nur, P. Banik, Y.N. Jolly, M. Al-Mamun, et al. 2023. Microplastics in surface water from a mighty subtropical estuary: first observations on occurrence, characterization, and contamination assessment. Environ. Res. 226: 115594.

Huang, D., X. Li, Z. Ouyang, X. Zhao, R. Wu, C. Zhang, et al. 2021. The occurrence and abundance of microplastics in surface water and sediment of the West River downstream, in the south of China. Sci. Total Environ. 756: 143857.

Huang, F., J. Hu, L. Chen, Z. Wang, S. Sun, W. Zhang, et al. 2023. Microplastics may increase the environmental risks of Cd via promoting Cd uptake by plants: A meta-analysis. J. Hazard. Mater. 448: 130887.

Jambeck, J.R., R. Geyer, C. Wilcox, T.R. Siegler, M. Perryman, A. Andrady, et al. 2015. Plastic waste inputs from land into the ocean. Science. 347(6223): 768–771.

Kılıç, E., N. Yücel, F. Bengil, E.G.T. Bengil and S.M. Şahutoğlu. 2024. Microplastic pollution levels in the surface water and sediment of Orontes basin: Urgent risk for endangered species. Mar. Pollut. Bull. 208: 116945.

Kumar, R., P. Sharma, C. Manna and M. Jain. 2021. Abundance, interaction, ingestion, ecological concerns and mitigation policies of microplastic pollution in riverine ecosystem: A review. Sci. Total Environ. 782: 146695.

Laju, R.L., M. Jayanthi, K.I. Jeyasanta, J. Patterson, N.G.G. Asir, M.N. Sathish, et al. 2022. Spatial and vertical distribution of microplastics and their ecological risk in an Indian freshwater lake ecosystem. Sci. Total Environ. 820: 153337.

Lang, M., X. Yu, J. Liu, T. Xia, T. Wang, H. Jia, et al. 2020. Fenton aging significantly affects the heavy metal adsorption capacity of polystyrene microplastics. Sci. Total Environ. 722: 137762.

Liang, T., Z. Lei, M.T.I. Fuad, Q. Wang, S. Sun, J.K.H. Fang, et al. 2022. Distribution and potential sources of microplastics in sediments in remote lakes of Tibet, China. Sci. Total Environ. 806: 150526.

Masura, J., J. Baker, G. Foster and C. Arthur. 2015. Laboratory methods for the analysis of microplastics in the marine environment: Recommendations for quantifying synthetic particles in waters and sediments.

Naidu, S.A., V. Ranga Rao and K.J.E.G. Ramu. 2018. Microplastics in the benthic invertebrates from the coastal waters of Kochi, Southeastern Arabian Sea. Environ. Geochem. Health. 40: 1377–1383.

Nithin, A., A. Sundaramanickam, M. Saha, M. Hassanshahian, M. Thangaraj and C. Rathore. 2023. Risk assessments of microplastics accumulated in estuarine sediments at Cuddalore, Tamil Nadu, southeast coast of India. Environ. Monit. Assess. 195(7): 890.

Onoja, S., H.A. Nel, M.A.E. Abdallah and S. Harrad. 2022. Microplastics in freshwater sediments: Analytical methods, temporal trends and risk of associated organophosphate esters as exemplar plastics additives. Environ. Res. 203: 111830.

Pan, Z., Q. Liu, R. Jiang, W. Li, X. Sun, H. Lin, et al. 2021. Microplastic pollution and ecological risk assessment in an estuarine environment: The Dongshan Bay of China. Chemosphere. 262: 127876.

Peng, G., B. Zhu, D. Yang, L. Su, H. Shi and D. Li. 2017. Microplastics in sediments of the Changjiang Estuary, China. Environ. Pollut. 225: 283–290.

Rakib, M.R.J., M.B. Hossain, R. Kumar, M.A. Ullah, S. Al Nahian, N.N. Rima, et al. 2022. Spatial distribution and risk assessments due to the microplastics pollution in sediments of Karnaphuli River Estuary, Bangladesh. Sci. Rep. 12(1): 8581.

Ramasamy, E.V. and A.K. Harit. 2023. Microplastics in Human Consumption. CRC Press.

Ranjani, M., S. Veerasingam, R. Venkatachalapathy, M. Mugilarasan, A. Bagaev, V. Mukhanov, et al. 2021. Assessment of potential ecological risk of microplastics in the coastal sediments of India: a meta-analysis. Mar. Pollut. Bull. 163: 111969.

Rodrigues, M.O., N. Abrantes, F.J.M. Gonçalves, H. Nogueira, J.C. Marques and A.M.M. Gonçalves. 2018. Spatial and temporal distribution of microplastics in water and sediments of a freshwater system (Antuã River, Portugal). Sci. Total Environ. 633: 1549–1559.

Saad, D., M., Ndlovu, G. Ramaremisa and H. Tutu. 2022. Microplastics in freshwater environment: The first evaluation in sediment of the Vaal River, South Africa. Heliyon. 8(10): e11118.

Saad, D., G. Ramaremisa, M. Ndlovu and L. Chimuka. 2024. Morphological and chemical characteristics of microplastics in surface water of the Vaal River, South Africa. Environ. Process. 11(1): 16.

Saikia, K.K. and S. Handique. 2024. Microplastics abundance and potential ecological risk assessment in sediment, water and fish of Deepor Beel—A Ramsar Wetland of the Brahmaputra plain, India. Environ. Monit. Assess. 196(10): 977.

Salas, P.M., C.H. Sujatha, C.R. Kumar and E. Cheriyan. 2017. Heavy metal distribution and contamination status in the sedimentary environment of Cochin estuary. Mar. Pollut. Bull. 119(2): 191–203.

Shamskhany, A., Z. Li, P. Patel and S. Karimpour. 2021. Evidence of microplastic size impact on mobility and transport in the marine environment: a review and synthesis of recent research. Front. Mar. Sci. 8: 760649.

Srinivas, K., C. Revichandran, P.A. Maheswaran, T.M. Asharaf and N. Murukesh. 2003. Propagation of tides in the Cochin estuarine system, southwest coast of India.

Sruthy, S. and E.V. Ramasamy. 2017. Microplastic pollution in Vembanad Lake, Kerala, India: The first report of microplastics in lake and estuarine sediments in India. Environ.Pollut. 222: 315–322.

Suresh, A., G. Vijayaraghavan, K.S. Saranya, K.V. Neethu and B. Aneesh. 2020. Microplastics distribution and contamination from the Cochin coastal zone, India. Reg. Stud. Mar. Sci. 40: 101533.

Thasneem, T.A., S.B. Nandan and P.N. Geetha. 2018. Water quality status of Cochin estuary, India.

Tomlinson, D.L., J.G. Wilson, C.R. Harris and D.W. Jeffrey. 1980. Problems in the assessment of heavy-metal levels in estuaries and the formation of a pollution index. Helgoländer Meeresuntersuchungen. 33: 566–575.

Wagner, M., C. Scherer, D. Alvarez-Muñoz, N. Brennholt, X. Bourrain, S. Buchinger. 2014. Microplastics in freshwater ecosystems: what we know and what we need to know. Environ. Sci. Eur. 26: 1–9.

Wang, W., W. Yuan, Y. Chen and J. Wang. 2018. Microplastics in surface waters of Dongting Lake and Hong Lake, China. Sci. Total Environ. 633: 539–545.

Yang, J., K. Song, C. Tu, L. Li, Y. Feng, R. Li, et al. 2023. Distribution and weathering characteristics of microplastics in paddy soils following long-term mulching: A field study in Southwest China. Sci. Total Environ. 858: 159774.

Yin, K., Y. Wang, H. Zhao, D. Wang, M. Guo, M. Mu, et al. 2021. A comparative review of microplastics and nanoplastics: Toxicity hazards on digestive, reproductive and nervous system. Sci. Total Environ. 774: 145758.

Yuan, W., X. Liu, W. Wang, M. Di and J. Wang. 2019. Microplastic abundance, distribution and composition in water, sediments, and wild fish from Poyang Lake, China. Ecotoxicol. Environ. Saf. 170: 180–187.

Zhang, Y., H. Wu, L. Xu, H. Liu and L. An. 2022. Promising indicators for monitoring microplastic pollution. Mar. Pollut. Bull. 182: 113952.

Zhao, S., L. Zhu, T. Wang and D. Li. 2014. Suspended microplastics in the surface water of the Yangtze Estuary System, China: first observations on occurrence, distribution. Mar. Pollut. Bull. 86(1–2): 562–568.

Zhao, W., J. Li, M. Liu, R. Wang, B. Zhang, X.Z. Meng, et al. 2024. Seasonal variations of microplastics in surface water and sediment in an inland river drinking water source in southern China. Sci. Total Environ. 908: 168241.

Chapter 6

Global Microplastics Contamination in Bottled and Tap Water: Prevalence and Sources

Karuna Gautam[1†], Rashika Pandit[1†], Yubraj Dahal*[2], Bikash Adhikari[1] and Sandhya Babel[3]

[1]Kathmandu University, Dhulikhel, Kavre, Nepal

[2]Nepal Academy of Science and Technology, Khumaltar, Lalitpur, Nepal

[3]Sirindhorn International Institute of Technology, Thammasat University, Bangkok, Thailand

6.1 INTRODUCTION

Plastic production surged from 1.5 million metric tons in 1950 to 400.3 million in 2022 (Garside 2024). This increase has led to significant plastic pollution, with 29.1 million tons of plastic waste recorded in 2020 (Silva 2021). Of all plastic waste, only 9% is recycled, 12% is incinerated, and the rest is discarded or landfilled (Geyer et al. 2017). In 2015, 60–99 million tons of plastic waste were improperly disposed of globally (Lebreton and Andrady 2019), harming ecosystems and resources (Jambeck et al. 2015; Geyer et al. 2017). Plastics degrade into tiny particles called Microplastics (MPs), sized <5 millimeters (mm) to 1 micrometer (μm) (Arthur et al. 2009; Gigault et al. 2018), and even smaller

*For Correspondence: yubraj.dahal369@gmail.com

†These authors contributed equally to this work.

DOI: 10.1201/9781003531463-6

nanoplastics (Koelmans et al. 2019), through biodegradation and photodegradation (Andrady 2011, Magalhães et al. 2020).

MPs are categorized into primary and secondary based on their sources. Primary MPs are intentionally manufactured for use in products, while secondary MPs result from the fragmentation of larger plastic pieces (Galafassi et al. 2019). MPs infiltrate soil, water, and the atmosphere, posing ecosystem risks, contaminating food sources, and impacting human health. They enter the environment through leaked plastics and products containing plastics, such as biosolids, wastewater, fertilizers, pesticides, agricultural mulch, polyamide (PA) fabrics, and cosmetics (Roy et al. 2022). Bioaccumulation occurs at all trophic levels (Miller et al. 2020). Primary MPs constitute 15–31% of marine MPs, while secondary MPs comprise 69–91% (European Parliament 2018). Sources of MPs release include textiles (Periyasamy and Tehrani-Bagha 2022), cosmetics (Praveena et al. 2018), and packaging (Jadhav et al. 2021). Figure 6.1(a) below shows the sources and fate of MPs and their circulation in the environment, and Figure 6.1(b) depicts various sources of MPs generation.

MPs are categorized by shape into beads/pellets, fibers, fragments, and films. Beads have smooth surfaces and are manufactured, not fragmented; fragments are irregular with angular edges from larger plastic pieces; fibers are elongated particles with much greater length than width; and films are thin, transparent plastic sheets (Rodríguez-Seijo and Pereira 2017; Sruthy and Ramasamy 2017; Lusher et al. 2020a).

MPs can also be classified by chemical composition, including Low-Density Polyethylene (LDPE), High-Density Polyethylene (HDPE), Polypropylene (PP), Polystyrene (PS), Polyethylene Terephthalate (PET), Polyethylene (PE), Polycarbonate (PC), Polyurethane (PU), and Acrylonitrile Butadiene Styrene (ABS) (Sun et al. 2019). PE is common in shopping bags, cosmetic beads, packaging, toys, containers, bottles, and pipes due to its affordability and recyclability. PS, created through vinyl polymerization of styrene, is used for food packaging and other products due to its useful properties. Polyvinyl Chloride (PVC), widely used in construction and pipes for its chemical resistance, is often not recycled and ends up in landfills, increasing environmental micro-particle risk. PP is used in food packaging, pipes, medicine bottles, and more for its resistance to chemicals and heat (Gausepohl and Nießner 2001; Ronca 2017; Sin and Tueen 2022; Amobonye et al. 2023; Wei et al. 2024).

MPs were first documented as environmental pollutants in the marine ecosystem in 1972 (Carpenter and Smith 1972). Since then, MPs have been detected globally, from Mount Everest to the Hadal zone (Napper et al. 2020; Gibbens 2018), and in various organisms from tiny phytoplankton to humpback whales (Lusher et al. 2020b). MPs are transported by wind and water via precipitation, surface runoff, infiltration, and riverine transport (Law 2017; Van Emmerik et al. 2019; Rech et al. 2015), becoming ubiquitous in aquatic (Ivleva et al. 2017; Pauna et al. 2019) and terrestrial ecosystems (de Souza Machado et al. 2018; Rillig and Lehmann 2020). Due to their widespread prevalence and potential health implications, the European Drinking Water Directive (DWD) plans to include MPs on the watch list of emerging compounds in 2024 (European Union (EU) 2020). MPs' pollution is

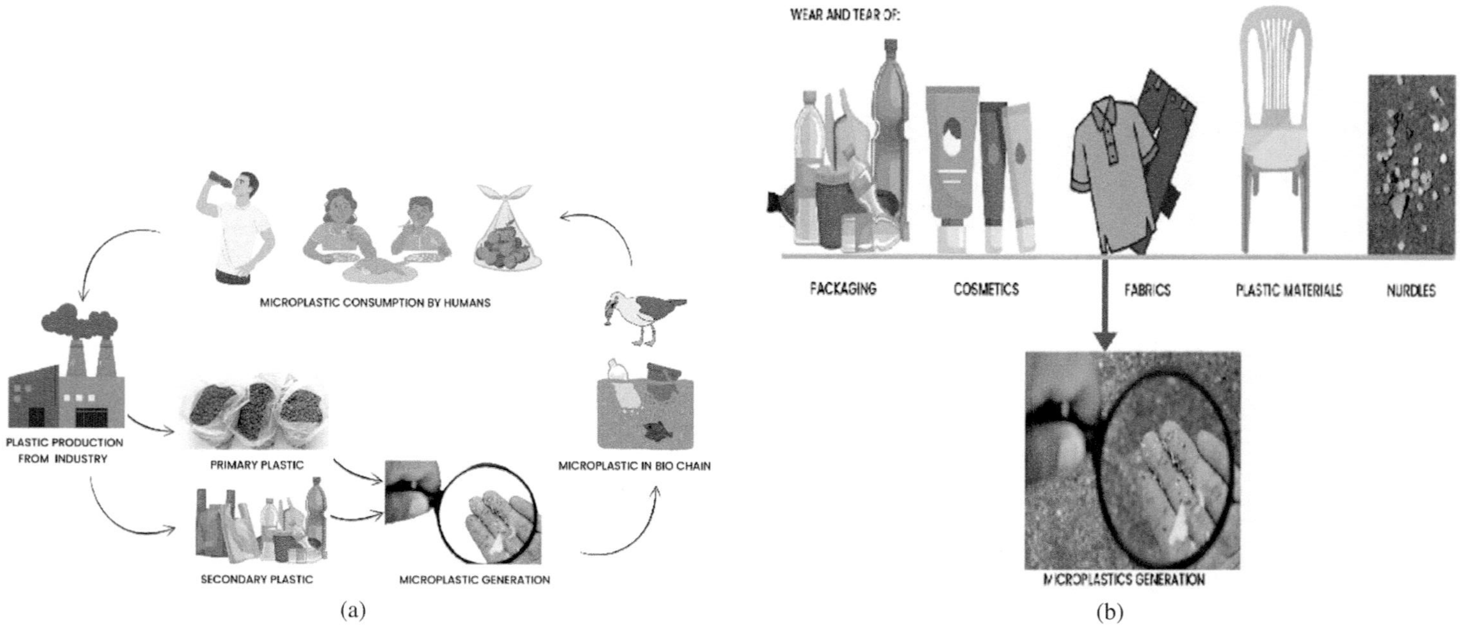

Figure 6.1 (a) Fates of plastic particles in the environment and their circulation. (b) Various sources of MPs generation.

a significant challenge recognized in the United Nations Sustainable Development Goals (UN-SDGs) under Goal 14, Target 14.1 (Walker and McKay 2021).

Ingestion through contaminated foods is a major pathway for MPs to enter the human body (Van Raamsdonk et al. 2020). MPs have been increasingly detected in various food products, including fish (Smith et al. 2018; Bessa et al. 2018), table salt (Iñiguez et al. 2017; Karami et al. 2017), fruits and vegetables (Conti et al. 2020), mussels (Van Cauwenberghe and Janssen 2014), beer (Liebezeit and Liebezeit 2014), honey (Liebezeit and Liebezeit 2013), and commercially packaged foods (Diaz-Basantes et al. 2020; Kedzierski et al. 2020).

Recent studies have detected MPs in human blood (Leslie et al. 2022), semen (Montano et al. 2023), heart (Yang et al. 2023), placenta (Ragusa et al. 2021), breast milk (Ragusa et al. 2022), and lungs (Amato-Lourenço et al. 2021). It is reported that 90% of ingested MPs are excreted (Smith et al. 2018), with particles >150 μm being excreted and particles <20 μm absorbed by organs (World Health Organization 2019). MPs are also found in human feces (Schwabl et al. 2019). Smaller particles (<2.5 μm) can reach the lungs, while larger particles pose a significant risk to the Gastrointestinal Tract (GIT). In the GIT, MPs can be engulfed by microfold cells in Peyer's Patches and enter the bloodstream (Wright and Kelly 2017). Figure 6.2 shows the pathways and fate of MPs in the human body.

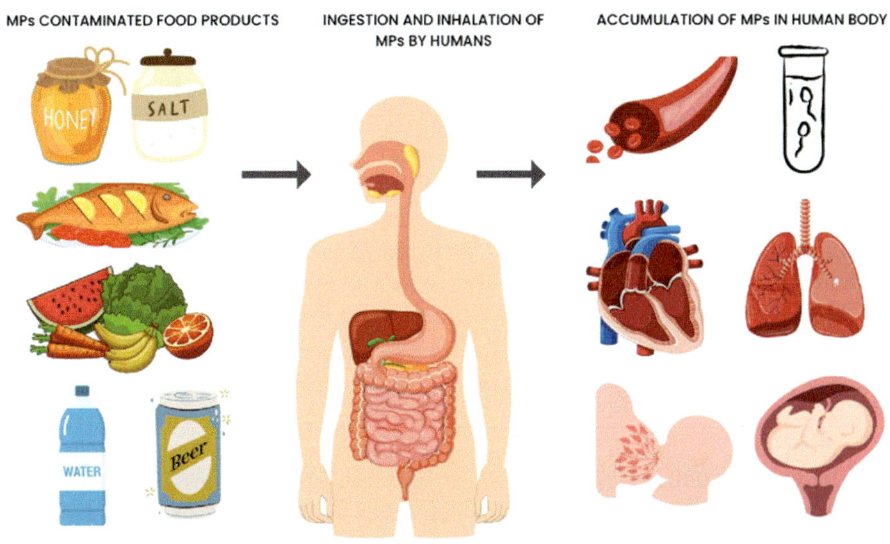

Figure 6.2 Pathways and fate of MPs in the human body.

MPs have been detected globally in bottled and tap water, with higher concentrations found in bottled water, indicating effective removal processes in water treatment facilities (Gambino et al. 2022). Humans consume 11,845 to 193,200 MPs annually, primarily from drinking water (Senathirajah et al. 2021). Humans are exposed to MPs through contaminated drinking water, posing health concerns based on the shape and type of polymer (Wright and Kelly 2017), with toxicity increasing as particle size decreases (Wu et al. 2019).

Additives in plastic polymers can alter their properties, making them prone to leaching into food, water, air, and bodies (Luo et al. 2019; Browne et al. 2013; Campanale et al. 2020). Depending on their physical and chemical properties, MPs can also absorb organic chemicals and trace metals (Rochman 2015). The United Nations Development Program (UNDP) in Kosovo (2023) linked chemical leaching from plastic bottles to endocrine disruption, weight gain, insulin resistance, decreased reproductive health, and cancer.

6.2 MPs CONTAMINATION IN BOTTLED WATER

Bottled water is a popular drinking source in many countries due to its purity, natural taste, and portability (Santana et al. 2014; Salazar-Beltran et al. 2018; Luo et al. 2018). According to the United States Food and Drug Administration, bottled water is defined as water packaged in bottles or other containers for human consumption without any added ingredients except for permissible microbial agents. Bottled water can be in plastic or glass containers (Salazar-Beltran et al. 2018). However, glass containers hold a tiny portion of the market. As reported by the IBWA in 2024, glass containers for bottled water hold only 2.3% of the market, while 97.3% of all bottled water containers are made of some polymer. Among these polymer containers, 78.8% are made of PET, 12% are made of PC, and 9.2% are made of HDPE. Primary packaging utilizes materials like HDPE, PET, and PC, while caps are made from HDPE, LDPE, and often PS as septa (World Packaging Organization 2008).

The concentration of MPs in bottled water is significantly higher than in tap water and natural water sources. Tap water, in particular, is reported to have the lowest concentration, indicating effective removal processes in water treatment facilities (Gambino et al. 2022). It is also reported that if water intake relies solely on bottled sources, the number of MPs ingested can increase to approximately 90,000 compared to 4,000 particles from tap water consumption (Cox et al. 2019).

6.2.1 Characteristics of MPs in Bottled Water

MPs abundance in bottled water (single-use, returnable, glass) was found to range from 0 to over 10,000 MPs/Liter (MPs/L), with sizes ranging from ≤ 1.5 to 5,000 μm. Fragments, fibers, beads, and films are present in bottled water, exhibiting various colors, usually transparent, blue, black, or green. The most dominant polymer found is PET, followed by PP, PE, PA, PS, and PVC, which might have occurred because MPs leach from the bottle due to mechanical stress, squeezing, and Ultraviolet (UV) exposure. Cellulose has also been detected in bottled water. Table 6.1 shows the overall scenario of bottled water, detailing the abundance, shape, size range, color, polymer type, and the analytical tools applied. MPs counting is mostly done by visual observation, and polymeric analysis is done by Fourier Transform Infrared (FTIR), followed by Raman spectroscopy.

Table 6.1 Analytical method, abundance, shape, size range, color, and polymer in bottled water

Abundance (MPs/L)	Analytical Tools	Shape	Polymers	Particle Size Range (μm)	Colors	Country	Reference
14 ± 14 (single-use PET), 118 ± 18 (returnable PET), 50 ± 52 (glass)	Single Particle Electron (SPE), μ-Raman	Fragment	Polyester + Polyethylene Terephthalate (PEST)*, PE, PP, PA, and others	Range: 5–1,359 5–20*	NG	Germany	Schymanski et al. (2018)
325	Nile Red (NR) tagging, Optical microscope, FTIR spectroscopy	Fragment*, Fiber, Pellet, and Film	PP*, Nylon (NY), PS, PE, and PEST	Range: >6.5–5,000 6.5–100*	NG	Global	Mason et al. (2018)
2,649 ± 2,857 (single-use PET), 6,292 ± 10,521 (glass), 4,889 ± 5,432 (reusable PET)	μ-Raman	NG	PET*, PP, and PE	Range: ≤1.5–>10 ≤1.5*	Blue and Violet	Germany	Oßmann et al. (2018)
148 ± 253	Scanning Electron Microscopy Energy Dispersive X-ray Spectroscopy (SEM-EDS)	NG	PET and HDPE	Range ≥0.5–<40 1–<5*	NG	Italy	Winkler et al. (2019)
5.42×10^7 p/L	SEM-EDS	NG	NG	Range 1.28–4.2 2.44 ± 0.66*	NG	Italy	Zuccarello et al. (2019)
140 ± 19 (bottled water), 52 ± 4 (glass water)	NR tagging, optical microscope, Attenuated Total Reflectance (ATR)—FTIR, Raman spectroscopy	Fiber* and Fragment	PET*, PE, PP, PA, and PVC	Range: 6.5–>50 6.5–20*	Blue, Bluish-Green, Red, Reddish Brown, Black, and Translucent	Thailand	Kankanige and Babel (2020b)

(Contd.)

Abundance (MPs/L)	Analytical Tools	Shape	Polymers	Particle Size Range (µm)	Colors	Country	Reference
2–23 p/bottle	SEM µ-FTIR	Fiber* and Fragment	Cellulose*, PS, PET, PE, PU, PVC, PP, Polyacrylic Acid (PAA), PA, Polyacrylamide (PAM), and Polyethylene Vinyl Acetate (PEVA)	Range: 25–5,000 100–300*	NG	China	Zhou et al. (2021)
8.5 ± 10.2	Rose Bengal (RB) staining, Stereomicroscope, ATR-FTIR, and Raman stereoscopy	Fragment* and Fiber	PET*, PS, and PP	1,000*	Colorless* and Black	Iran	Makhdoumi et al. (2021)
13 ± 19	LDIR	Fragment* and Fiber	PP*, PET, PA, PE, PS, PC, and PVC	Range: 6–480	NG	Australia	Samandra et al. (2022)
8,955 ± 5,205	Fluorescence Staining and Flow Cytometry	Fragment* and Fiber	NG	1–10*	NG	China	Tse et al. (2022)
11.7 ± 4.6	Visual Observation and FTIR	Fragment* and Fiber	PET, PP	Range: 25–5,000 100–300*	Transparent*, Blue, and Green	Malaysia	Praveena et al. (2022)
72.32 ± 44.64 items/L	LDIR	Film*, Pellet, and Fiber	Cellulose*, PVC and others	Range: 0–301 10–50*	NG	China	Huan et al. (2023)
391 ± 125	NR staining and Fluorescence Microscopy	Fiber	NG	Range: 5–>50 5–20*	NG	Chile	Nacaratte et al. (2023)

Abundance (MPs/L)	Analytical Tools	Shape	Polymers	Particle Size Range (µm)	Colors	Country	Reference
4.60 ± 3.85 (natural water), 12.60 ± 8.72 (mineral water)	Visual analysis and ATR-FTIR	Fiber*, Fragment, and Film	PE*, PP, PET, PA	Range: 10–>100	Blue*, Transparent, Grey, Purple, and Black	Turkey	Altunışık (2023)
35 ± 19	Visual observation and FTIR	Fiber*, Fragment, Film, and Bead	HDPE*, LDPE*, PET, Ethylene Vinyl Acetate (EVA)	Range: <500–1,000 <500*	Transparent*, Violet, Pink, Red, Blue, Green, Black, and Yellow	Bangladesh	Hossain et al. (2023)
212 ± 100 (Local), 72 ± 36 (National)	Visual Analysis, NR Staining, ATR-FTIR, and SEM	Fragment*, Fiber, and Other	PE* and PET	Range: 10–5,000	Black*, Transparent, Red, Orange, Green, Blue, and Yellow	India	Patil et al. (2024)
2.89 ± 0.48	Stereo zoom microscope, and ATR-FTIR	Fiber, Fragment	PP*, PET, NY, PVC, PA, and Cellulose	Range: 20–5,000 500–1,000*	Transparent*, Blue, Pink, Yellow, Black, Green, and Violet	India	Sekar et al. (2024)

Note: *Refers to the dominant type in the category. NG refers to Not Given.

6.2.2 Sources of MPs in Bottled Water

Plastic packaging is a significant source of MPs contamination in bottled water, but in the case of glass bottles, the hard glass material might have caused increased abrasion on the soft plastic of the bottle cap and seal, resulting in the release of particles (Schymanski et al. 2018). The factors contributing to the generation of MPs include mechanical stress from opening and closing bottles, as well as the physical stress subjected to the materials during the transport, storage, and bottling process, as well as contaminated raw water sources (Winkler et al. 2019; Makhdoumi et al. 2021; Kankanige and Babel 2020b; Mason et al. 2018; Zhou et al. 2021).

Studies have reported an average concentration of 708 ± 1,024 MPs/L of the additive Tris(2,4-di-tert-butylphenyl) phosphite in reusable PET bottles, which may leach from the bottle material. Studies have reported the presence of PET particles in bottles made of PET, indicating contamination from the packaging material, and the glass water sample was reported to contain PE, PP, or styrene-butadiene copolymer, likely coming from bottle caps, washing machinery, or other steps during the filling process (Oßmann et al. 2018). Additionally, frequently reused PET bottles had higher levels of MPs compared to the newer ones, suggesting the ageing of the material can lead to more leaching and fragmentation; a few antioxidant additives were also reported to leach out of the packaging (Oßmann et al. 2018).

It is now established that PET bottlenecks and HDPE caps significantly contribute to the MPs contamination in bottles, especially after a prolonged subjection to mechanical stress (Winkler et al. 2019). Besides mechanical stress, manufacturing and bottling processes, chemical leaching, and UV exposure are also shown to influence MPs contamination in bottled water. Chemical leaching from the bottle material may carry MP particles with it, while UV radiation further degrades the material, releasing MPs into the bottled water (Sekar et al. 2024). Figure 6.3 depicts various sources of MPs contamination in bottled water.

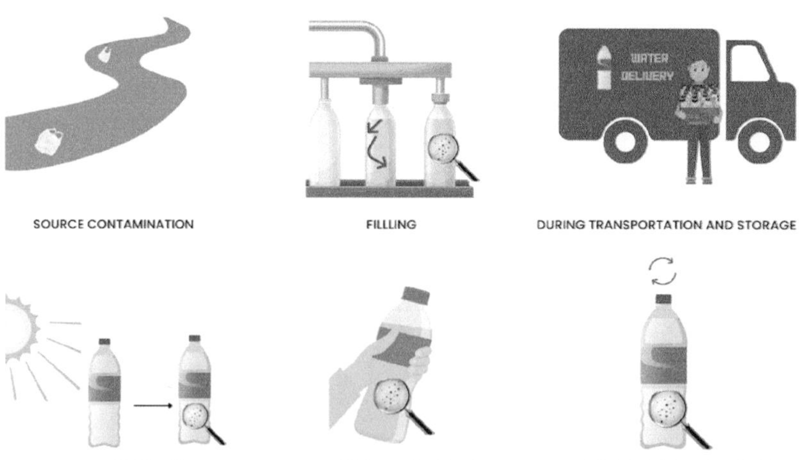

Figure 6.3 Sources of MPs contamination in bottled water.

Thus, bottled water, favored for its purity and convenience, is predominantly made of polymer containers such as PET, PC, and HDPE for packaging material, with glass bottles holding a minimal market share. Despite its popularity, bottled water exhibits significantly higher concentrations of MPs than tap and natural water sources, highlighting concerns over MP contamination. Plastic packaging, particularly from PET bottles and HDPE caps, is a major source of MPs, intensified by mechanical stresses during handling and storage. Research indicates that bottled water MPs can originate from the leaching of additives and abrasion within packaging materials. As environmental and health concerns grow, there is a critical need for stringent regulatory measures and sustainable practices to mitigate MP pollution in bottled water.

6.3 MPs CONTAMINATION IN TAP WATER

Tap water is a common source of drinking water that originates from surface or underground sources and is supplied to the community (Koelmans et al. 2019). The quality of tap water is crucial to human health as it is a primary route for exposure to various contaminants, and thus, strict quality and safety measures should be followed (Liu et al. 2022). Interestingly, a significantly large percentage of bottled water, estimated between 40% and 60% globally, is simply packaged tap water that is occasionally reprocessed (Doria 2006). Tap water is more strictly monitored and subjected to stricter standards than bottled water (Doria 2006). Consequently, communities with poor tap water quality will likely depend more on bottled sources (Anadu and Harding 2000).

6.3.1 Characteristics of MPs in Tap Water

MPs abundance in tap water ranged from 0.206 to over 440 MPs/L, with sizes ranging from 3 to 5,000 μm. Fragments, fibers, films, and beads of various morphologies were present in tap water, exhibiting colors such as blue, black, and transparent. Sun et al. (2024) reported that particles of size <50 μm have a higher proportion in global tap water supplies, where fragments are the dominant shape (58%), followed by fiber (38%), spheres, and films (2%), transparent is the dominant color (54%), followed by blue (13%), black (13%), red (9%), and brown (1%). In terms of polymer, polyester is the most dominant polymer (27%), followed by PP (12%), PA (11%), PVC (11%), PE (11%), and PS (6%). The concentration of MPs in tap water from developed countries was significantly lower than in their developing counterparts, at an average of 34 and 125 MPs/L, respectively. This result shows that better plastic control measures, advanced water treatment technologies, and effective municipal waste disposal and water filtration systems are being employed in developed countries (Mukotaka et al. 2021; Awad et al. 2019; Nguyen et al. 2023; Kosuth et al. 2018). The drinking WTPs have varying efficiencies at removing MPs, and particles can end up in tap water from treatment plants with lower efficiencies and subsequently be ingested by consumers (Acarer 2023; Feld et al. 2021). Drinking Water Treatment Plant (DWTP)

Table 6.2 Analytical method, abundance, shape, size range, color, and polymer in tap water

Abundance (MPs/L)	Analytical Method	Shape	Polymers	Particle Size Range (μm)	Colors	Country	Reference
5.45	RB staining, Dissection microscope	Fiber* and Fragment	NG	Range: 100–5,000	Blue, Red, Brown, Transparent, Purple, Black, and Multicolor	Global	Kosuth et al (2018)
2.181 ± 0.165 n/L	RB staining, stereomicroscope	Fiber* and Film	NG	Range: 50–4,830 150–499*	Black*, Blue, Transparent, and Yellow	China	Lam et al. (2020)
56 ± 14 (6.5–53 μm), 21 ± 7 (53–300 μm), 13 ± 5 (300–500 μm), 6.0 ± 3.0 (≥500 μm)	NR staining, Optical microscope, and ATR-FTIR	Fiber*, Fragment, and Film	PE*, PVC, PET, PP, Polytetrafluoroethylene (PTFE), PAM, and PA	Range: >6.5–5,000 6.5–53*	Blue, Red, White, and Transparent	Thailand	Kankanige and Babel (2020a)
440	NR staining, Fluorescence microscope, and μ-Raman	Fragment*, Fiber, and Sphere	PE*, PP*, copolymer of PE and PP, Polyphenylene Sulphide (PPS), PS, PET, Poly Alpha-Methyl Styrene (PMS), PTFE, PC, Poly Methyl Methacrylate (PMMA), Polybutylene terephthalate (PBT), Polybutene (PB), Ny, and PVC	Range: 3–4, 453 <50*	NG	China	Tong et al. (2020)
39 ± 44	FTIR microscope	Fragment*, Fiber, and Sphere	PS*, Styrene-Ethylene-Butylene-Styrene (SEBS), PP, PE, other	Range: 19.2–4,200 19–50*	NG	Global	Mukotaka et al. (2021)

0.88	Stereomicroscope, μ-FTIR	Fiber*, Fragment, and Film	PP, PS, PET, and ABS	Range: >10	Transparent, Black, Blue, and Pink	Denmark	Feld et al. (2021)
219 MPs/500 mL of water (North wing), 97 MPs/500 mL of water (South wing)	NR staining and Fluorescence microscope	NG	NG	Range: 6–50	NG	Brazil	Pratesi et al. (2021)
0.68 (Eastern DTWPs), 0.62 (Western DTWPs)	μ-Raman spectroscopy	Fragment*, Fiber	PE*, PET, PP, PS, and PVC	Range: 50–1,000 50–99* (Eastern DTWPs) 100–199* (Western DTWPs)	Colorless*, Color, and Black	Thailand	Chanpiwat and Damrongsiri (2021)
1.4	FTIR, SEM, Optical microscope, and μ-FTIR	Fiber*	PET* and PE	Range: >25 <40*	White, Blue	China	Zhou et al. (2023)
0.201 ± 0.113 (CT1) 0.206 ± 0.106 (CT2)	Field Emission Scanning Electron Microscopy (FE-SEM), FTIR, and μ-Raman spectroscopy	Fragment* and Fiber	PS, PE, PS, PMMA, and PVC	Range <50–>100 >90% of the particles were <50 *	white*, Transparent, Blue, Red, and Black	Iran	Taghipour et al. (2023)
14 ± 5.6	RB staining, Stereomicroscopy, SEM, μ-Raman spectroscopy	Fiber*, Fragment, Beads, and Film	HDPE*, PU, PET, Poly (hexamethylene Terephthalamide) (PA6T), and Poly(acrylamide-co-acrylic acid) (Poly(AM-co-AA))	Range: 20–5,000 100–500*	Black*, Green, Blue, Red, Brown, Transparent, Grey, and Yellow	South Africa	Ramaremisa et al. (2024)

Note: *Refers to the dominant type in the category. NG refers to Not Given.

effectively removes MPs, with removal efficiency ranging from 40–93% (Dahal and Babel 2023). The tap is allowed to run for one minute before sample collection. Various studies have employed different containers and volumes for this purpose: Ramaremisa et al. (2024) used 3L pre-washed glass jars stored in cooler boxes for transport to the laboratory, Tong et al. (2020) filled 1L HDPE bottles to the point of overflowing, and Kosuth et al. (2018) collected samples in 500 milliliter (mL) HDPE bottles, also filled to overflowing after letting the water run for a minute. Table 6.2 presents the overall scenario of bottled water, detailing its abundance, shape, size range, color, and polymer type, and the analytical method applied.

6.3.2 Source of MPs in Tap Water

In their study, Kosuth et al. (2018) reported that 81% of the tap water samples tested were contaminated with MPs. The potential source for this contamination was attributed to MPs-polluted water bodies, atmospheric deposition, and mechanical abrasion of plastic equipment during water treatment and distribution (Lam et al. 2020). Colorants such as C.I Pigment Red 1, C.I. Solvent Yellow 4, Potassium Indigo Tetrasulfonate, and C. I Pigment Black 7 was detected in a study by Ramaremisa et al. (2024). These compounds are known to be carcinogenic and mutagenic and pose risks of potential toxicity to humans. The presence of poly(AM-co-AA) in samples may be directly linked to the water treatment facility. Similarly, the detection of HDPE, PVC fragments >50 μm, and polyester fibers >100 μm in the sample suggests MPs contamination from the pipes during transportation (Ramaremisa et al. 2024; Tong et al. 2020; Mukotaka et al. 2021; Taghipour et al. 2023). The presence of PAM, an anionic polyelectrolyte, in the samples also suggests contamination from treatment plants, as it is often added to clarification tanks after alum dosing to promote flocculation. Moreover, polymers like PE, PP, PEST, and PA may end up in the water via atmospheric fallout, treatment processes, packaging materials, source contamination, and contamination along distribution networks (Zhou et al. 2023; Kankanige and Babel 2020a). Figure 6.4 shows the various sources of MPs contamination in tap water.

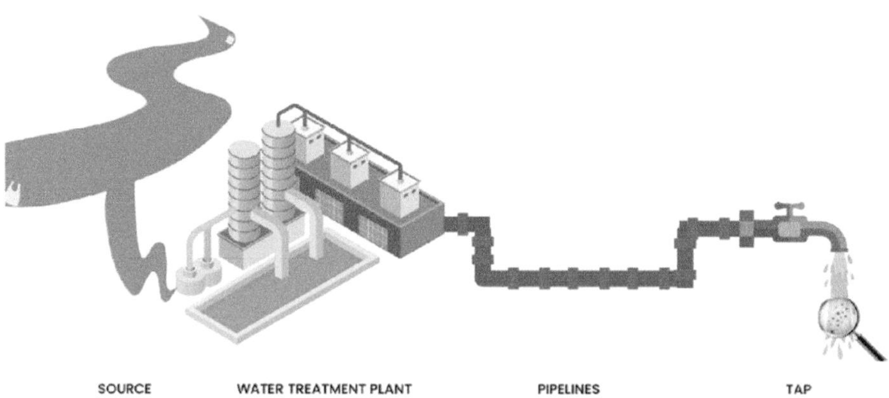

Figure 6.4 Sources of MPs contamination in tap water.

Hence, tap water, a common drinking source derived from surface or underground reservoirs, plays a crucial role in public health but faces contamination challenges, such as MPs from polluted water bodies, atmospheric deposition, and equipment abrasion during treatment. Studies reveal that a significant portion of bottled water globally originates as reprocessed tap water, highlighting concerns over water quality monitoring and regulatory standards. Contaminants such as colorants and polymers like HDPE, PVC, and polyester fibers have been detected in tap water samples, posing potential health risks due to their carcinogenic and mutagenic properties. The presence of additives like poly(AM-co-AA) suggests contamination from treatment processes, while MP fragments indicate transportation pipeline sources. Effective management of these sources is crucial to mitigate MP contamination and ensure safe drinking water for communities worldwide.

6.4 CONCLUSION

MP contamination in bottled water varies globally, with the highest levels recorded in China and the lowest in India, indicating its ubiquitous presence worldwide. The most dominant size range of MPs is less than 50 μm, with all dominant size ranges within 1,000 μm. Fragments and fibers are the prevalent morphologies, with fragments being more common. These MPs particles are typically transparent, blue, black, or green. PET is the most dominant polymer, followed by PP, PE, PA, PS, and PVC. PET contamination primarily results from the bottle due to mechanical stress, squeezing, and UV exposure.

In contrast, PE contamination often comes from bottle caps, mainly due to repeated opening and closing. Single-use plastic bottles generally have lower MP contamination compared to reusable ones. MPs have also been detected in glass-bottled water.

MPs contamination has been detected in tap water sources, with the highest levels of contamination found in China, while the lowest levels have been recorded in Iran, ranging from 3 to 5,000 μm. However, the presence of MPs in tap water is a global concern, as they are found everywhere from sources to WTP to tap water. Fibers, fragments, films, and spheres have been detected, with fibers being the most common, followed by fragments. MPs are typically transparent, black, and blue. The most dominant size range is less than 60 μm, with all dominant size ranges less than or equal to 500 μm. PE, PP, PET, PVC, and PS are the most dominant polymers. Additionally, PAM, copolymers of PE and PP, and PVC have been detected, suggesting that MPs generation increases from water treatment facilities and pipelines.

Thus, MPs' contamination should be investigated more rigorously, and guidelines should be established to mitigate its impact.

REFERENCES

Acarer S. 2023. Microplastics in wastewater treatment plants: sources, properties, removal efficiency, removal mechanisms, and interactions with pollutants. Water Sci. Technol. 87(3): 685–710.

Altunışık A. 2023. Microplastic pollution and human risk assessment in Turkish bottled natural and mineral waters. Environ. Sci. Pollut. Res. 30(14): 39815–39825.

Amato-Lourenço L.F., R. Carvalho-Oliveira, G.R. Júnior, L. dos Santos Galvão, R. A. Ando and Thais Mauad. 2021. Presence of airborne microplastics in human lung tissue. J. Hazard. Mater. 416: 126124.

Amobonye, A.E., P. Bhagwat, S. Singh and S. Pillai. 2023. Biodegradability of polyvinyl chloride. *In*: Biodegradability of Conventional Plastics. Elsevier. 201–220.

Anadu, E.C. and A.K. Harding. 2000. Risk perception and bottled water use. J. Am. Water Works Assoc. 92(11): 82–92.

Andrady, A.L. 2011. Microplastics in the marine environment. Mar. Pollut. Bull. 62(8): 1596–1605.

Arthur, C., J. Baker and H. Bamford 2009. Proceedings of the international research workshop on the occurrence, effects and fate of microplastic marine debris; 2008 Sept 9–11; University of Washington, Tacoma, WA. Silver Spring (MD): NOAA Marine Debris Division.

Awad, H., M.G. Alalm and H.K. El-Etriby. 2019. Environmental and cost life cycle assessment of different alternatives for improvement of wastewater treatment plants in developing countries. Sci. Total Environ. 660: 57–68.

Bessa, F., P. Barría, J.M. Neto, J.P. Frias, V. Otero, P. Sobral, et al. 2018. Occurrence of microplastics in commercial fish from a natural estuarine environment. Mar. Pollut. Bull. 128: 575–584.

Browne, M.A., S.J. Niven, T.S. Galloway, S.J. Rowland and R.C. Thompson. 2013. Microplastics move pollutants and additives to worms, reducing functions linked to health and biodiversity. Curr Biol. 23(23): 2388–2392.

Campanale, C., C. Massarelli, I. Savino, V. Locaputo and V.F. Uricchio. 2020. A detailed review study on potential effects of microplastics and additives of concern on human health. Int. J. Environ. Res. Public Health. 17(4): 1212.

Carpenter, E.J. and K.L. Jr. Smith 1972. Plastics on the Sargasso Sea surface. Science. 175(4027): 1240–1241.

Chanpiwat, P. and S. Damrongsiri. 2021. Abundance and characteristics of microplastics in freshwater and treated tap water in Bangkok, Thailand. Environ. Monit. Assess. 193(5): 258.

Conti, G.O., M. Ferrante, M. Banni, C. Favara, I. Nicolosi, A. Cristaldi, et al. 2020. Micro- and nano-plastics in edible fruit and vegetables. The first diet risk assessment for the general population. Environ. Res. 187: 109677.

Cox K.D., G.A. Covernton, H.L. Davies, J.F. Dower, F. Juanes and S.E. Dudas. 2019. Human consumption of microplastics. Environ. Sci. Technol. 53(12): 7068–7074.

Dahal, Y. and S. Babel. 2023. Occurrence and source of microplastics contamination in drinking water and performance of water treatment plants in removing microplastics. pp. 114–143. *In*: Krantzberg, G., S. Jetoo, V.I. Grover and S. Babel. (eds). Plastic Pollution: Nature Based Solutions and Effective Governance. CRC Press. https://doi.org/10.1201/9781003137740.

de Souza Machado, A.A., W. Kloas, C. Zarfl, S. Hempel and M.C. Rillig. 2018. Microplastics as an emerging threat to terrestrial ecosystems. Glob. Chang. Biol. 24(4): 1405–1416.

Diaz-Basantes, M.F., J.A. Conesa and A. Fullana. 2020. Microplastics in honey, beer, milk and refreshments in Ecuador as emerging contaminants. Sustainability. 12(14): 5514.

Doria, M.F. 2006. Bottled water versus tap water: understanding consumers' preferences. J. Water Health. 4(2): 271–276.

European Union. 2020. European Drinking Water Directive 2020/2184 of the European Parliament and of the Council of 16 December 2020 on the quality of water intended for human consumption (recast). Off. J. Eur. Union. 435: 1–62.

European Parliament. 2018. Microplastics: sources, effects and solutions. Available from: https://www.europarl.europa.eu/topics/en/article/20181116STO19217/microplastics-sources-effects-and-solutions.

Feld, L., V.H.D. Silva, F. Murphy, N.B. Hartmann and J. Strand. 2021. A study of microplastic particles in Danish tap water. Water. 13(15): 2097.

Galafassi, S., L. Nizzetto and P. Volta. 2019. Plastic sources: a survey across scientific and grey literature for their inventory and relative contribution to microplastics pollution in natural environments, with an emphasis on surface water. Sci. Total Environ. 693: 133499.

Gambino, I., F. Bagordo, T. Grassi, A. Panico and A. De Donno. 2022. Occurrence of microplastics in tap and bottled water: current knowledge. Int. J. Environ. Res. Public Health. 19(9): 5283.

Garside, M. 2024. Global plastic production 1950–2022. [accessed 2024 Jun 26]. Available from: https://www.statista.com/statistics/282732/global-production-of-plastics-since-1950/.

Gausepohl, H. and Nießner, N. 2001. Polystyrene and styrene copolymers. *In*: Encyclopedia of Materials: Science and Technology. 7735–7741.

Geyer, R., J.R. Jambeck and K.L. Law. 2017. Production, use and fate of all plastics ever made. Sci Adv. 3(7).

Gibbens, S. 2018. Microplastics found to permeate the ocean's deepest points. Microplastic permeates Mariana Trench and other deep sea points. 6.

Gigault, J., A. Ter Halle, M. Baudrimont, P.Y. Pascal, F. Gauffre, T.L. Phi, et al. 2018. Current opinion: What is a nanoplastic? Environ. Pollut. 235: 1030–1034.

Hossain, M.B., J. Yu, P. Banik, M.A. Noman, A.A.U. Nur, M.R. Haque, et al. 2023. First evidence of microplastics and their characterization in bottled drinking water from a developing country. Front. Environ. Sci. 11.

Huan, L.I., Z.H. Long, M.A. Mindong, W.U. Haiwen, A.N. Lihui, Y.A.N.G. Zhanhong, et al. 2023. Occurrence of microplastics in commercially sold bottled water. Sci. Total Environ. 867: 161553.

Iñiguez, M.E., J.A. Conesa and A. Fullana. 2017. Microplastics in Spanish table salt. Sci. Rep. 7(1): 8620.

Ivleva, N.P., A.C. Wiesheu and R. Niessner. 2017. Microplastic in aquatic ecosystems. Angew. Chem. Int. Ed. 56(7): 1720–1739.

Jambeck, J.R., R. Geyer, C. Wilcox, T.R. Siegler, M. Perryman and A. Andrady, et al. 2015. Plastic waste inputs from land into the ocean. Science. 347(6223): 768–771.

Jadhav, E.B., M.S. Sankhla, R.A. Bhat and D.S. Bhagat. 2021. Microplastics from food packaging: an overview of human consumption, health threats and alternative solutions. Environ. Nanotechnol. Monit. Manage. 16: 100608.

Kankanige, D. and S. Babel 2020a. Identification of micro-plastics (MPs) in conventional tap water sourced from Thailand. J. Eng. Technol. Sci. 52(1).

Kankanige, D. and S. Babel. 2020b. Smaller-sized micro-plastics (MPs) contamination in single-use PET-bottled water in Thailand. Sci. Total Environ. 717: 137232.

Karami, A., A. Golieskardi, C. Keong Choo, V. Larat, T.S. Galloway and B. Salamatinia. 2017. The presence of microplastics in commercial salts from different countries. Sci. Rep. 7(1): 46173.

Kedzierski, M., B. Lechat, O. Sire, G. Le Maguer, V. Le Tilly and S. Bruzaud. 2020. Microplastic contamination of packaged meat: occurrence and associated risks. Food Packag. Shelf Life. 24: 100489.

Koelmans, A.A., N.H.M. Nor, E. Hermsen, M. Kooi, S.M. Mintenig and J. De France. 2019. Microplastics in freshwaters and drinking water: critical review and assessment of data quality. Water Res. 155: 410–422.

Kosuth, M., S.A. Mason and E.V. Wattenberg. 2018. Anthropogenic contamination of tap water, beer and sea salt. PLoS One. 13(4).

Lam, T.W.L., H.T. Ho, A.T. Ma and L. Fok. 2020. Microplastic contamination of surface water-sourced tap water in Hong Kong—A preliminary study. Appl. Sci. 10(10): 3463.

Law, K.L. 2017. Plastics in the marine environment. Annu. Rev. Mar. Sci. 9: 205–229.

Lebreton, L. and A. Andrady. 2019. Future scenarios of global plastic waste generation and disposal. Palgrave Commun. 5(1): 1–1.

Leslie, H.A., M.J. Van Velzen, S.H. Brandsma, A.D. Vethaak, J.J. Garcia-Vallejo and M.H. Lamoree. 2022. Discovery and quantification of plastic particle pollution in human blood. Environ. Int. 163: 107199.

Liebezeit, G. and E. Liebezeit. 2013. Non-pollen particulates in honey and sugar. Food Addit. Contam. Part A. 30(12): 2136–2140.

Liebezeit, G. and E. Liebezeit. 2014. Synthetic particles as contaminants in German beers. Food Addit. Contam. Part A. 31(9): 1574–1578.

Liu, M., N. Graham, W. Wang, R. Zhao, Y. Lu, M. Elimelech, et al. 2022. Spatial assessment of tap-water safety in China. Nat. Sustain. 5(8): 689–698.

Luo, Q., Z.H. Liu, H. Yin, Z. Dang, P.X. Wu, N.W. Zhu, et al. 2018. Migration and potential risk of trace phthalates in bottled water: a global situation. Water Res. 147: 362–372.

Luo, H., Y. Xiang, D. He, Y. Li, Y. Zhao, S. Wang, et al. 2019. Leaching behavior of fluorescent additives from microplastics and the toxicity of leachate to Chlorella vulgaris. Sci. Total Environ. 678: 1–9.

Lusher, A.L., I.L.N. Bråte, K. Munno, R.R. Hurley and N.A. Welden. 2020a. Is it or isn't it: the importance of visual classification in microplastic characterization. Appl. Spectrosc. 74(9): 1139–1153.

Lusher, A.L., N.A. Welden, P. Sobral and M. Cole. 2020b. Sampling, isolating and identifying microplastics ingested by fish and invertebrates. pp. 119–148. *In*: Leo M.L. Nollet and Khwaja Salahuddin Siddiqi (eds). Analysis of Nanoplastics and Microplastics in Food. CRC Press.

Magalhães, S., L. Alves, B. Medronho, A. Romano and M.D.G. Rasteiro. 2020. Microplastics in ecosystems: From current trends to bio-based removal strategies. Molecules. 25(17): 3954.

Makhdoumi, P., A.A. Amin, H. Karimi, M. Pirsaheb, H. Kim and H. Hossini. 2021. Occurrence of microplastic particles in the most popular Iranian bottled mineral water brands and an assessment of human exposure. J. Water Process. Eng. 39: 101708.

Mason, S.A., V.G. Welch and J. Neratko. 2018. Synthetic polymer contamination in bottled water. Front. Chem. 6: 389699.

Miller, M.E., M. Hamann and F.J. Kroon. 2020. Bioaccumulation and biomagnification of microplastics in marine organisms: A review and meta-analysis of current data. PLoS One. 15(10).

Montano, L., E. Giorgini, V. Notarstefano, T. Notari, M. Ricciardi, M. Piscopo, et al. 2023. Raman microspectroscopy evidence of microplastics in human semen. Sci. Total Environ. 901: 165922.

Mukotaka, A., T. Kataoka and Y. Nihei. 2021. Rapid analytical method for characterization and quantification of microplastics in tap water using a Fourier-transform infrared microscope. Sci. Total Environ. 790: 148231.

Nacaratte, F., P. Cuevas, M. Becerra-Herrera and C.A. Manzano. 2023. Early screening of suspected microplastics in bottled water in the Santiago Metropolitan Region of Chile. Environ. Pollut. 334: 122118.

Napper, I.E., B.F. Davies, H. Clifford, S. Elvin, H.J. Koldewey, P.A. Mayewski, et al. 2020. Reaching new heights in plastic pollution—preliminary findings of microplastics on Mount Everest. One Earth. 3(5): 621–630.

Nguyen, P.D., Q.V. Tran, T.T. Le, Q.H. Nguyen, T.C. Kieu-Le and E. Strady. 2023. Evaluation of microplastic removal efficiency of wastewater-treatment plants in a developing country, Vietnam. Environ. Technol. Innov. 29: 102994.

Oßmann, B.E., G. Sarau, H. Holtmannspötter, M. Pischetsrieder, S.H. Christiansen and W. Dicke. 2018. Small-sized microplastics and pigmented particles in bottled mineral water. Water Res. 141: 307–316.

International Bottled Water Association (IBWA). n.d. Packaging—Bottled water. Available from: https://bottledwater.org/packaging/.

Patil, S., A. Bafana, K. Krishnamurthi and S. Sivanesan. 2024. Estimated exposure to microplastics through national and local brands of bottled water in Central India. Environ. Monit. Assess. 196(3): 1–20.

Pauna, V.H., E. Buonocore, M. Renzi, G.F. Russo and P.P. Franzese. 2019. The issue of microplastics in marine ecosystems: A bibliometric network analysis. Mar. Pollut. Bull. 149: 110612.

Periyasamy, A.P. and A. Tehrani-Bagha. 2022. A review on microplastic emission from textile materials and its reduction techniques. Polym. Degrad. Stab. 199: 109901.

Pratesi, C.B., M.A. Santos Almeida, G.S. Cutrim Paz, M.H. Ramos Teotonio, L. Gandolfi, R. Pratesi, et al. 2021. Presence and quantification of microplastic in urban tap water: a pre-screening in Brasilia, Brazil. Sustainability. 13(11): 6404.

Praveena, S.M., S.N.M. Shaifuddin and S. Akizuki. 2018. Exploration of microplastics from personal care and cosmetic products and its estimated emissions to marine environment: an evidence from Malaysia. Mar. Pollut. Bull. 136: 135–140.

Praveena, S.M., N.I.S. Ariffin and A.L. Nafisyah. 2022. Microplastics in Malaysian bottled water brands: Occurrence and potential human exposure. Environ. Pollut. 315: 120494.

Ragusa, A., A. Svelato, C. Santacroce, P. Catalano, V. Notarstefano, O. Carnevali, et al. 2021. Plasticenta: first evidence of microplastics in human placenta. Environ. Int. 146: 106274.

Ragusa, A., V. Notarstefano, A. Svelato, A. Belloni, G. Gioacchini, C. Blondeel, et al. 2022. Raman microspectroscopy detection and characterisation of microplastics in human breastmilk. Polymers. 14(13): 2700.

Ramaremisa, G., H. Tutu and D. Saad. 2024. Detection and characterisation of microplastics in tap water from Gauteng, South Africa. Chemosphere. 356: 4141903.

Rech, S., V. Macaya-Caquilpán, J.F. Pantoja, M.M. Rivadeneira, C.K. Campodónico and M. Thiel. 2015. Sampling of riverine litter with citizen scientists—Findings and recommendations. Environ. Monit. Assess. 187: 1–18.

Rillig, M.C. and Lehmann, A. 2020. Microplastic in terrestrial ecosystems. Science. 368(6498): 1430–1431.

Rochman, C.M. 2015. The complex mixture, fate and toxicity of chemicals associated with plastic debris in the marine environment. *In*: Marine Anthropogenic Litter. 117.

Rodríguez-Seijo, A. and R. Pereira. 2017. Morphological and physical characterization of microplastics. pp. 49–66. *In*: Teresa A.P. Rocha-Santos and Armando C. Duarte (eds). Comprehensive Analytical Chemistry, Volume 75. Elsevier.

Ronca, S. 2017. Polyethylene. pp. 247–278. *In*: M. Gilbert (ed.). Brydson's Plastics Materials. Butterworth-Heinemann. https://doi.org/10.1016/B978-0-323-35824-8.00010-4.

Roy, P., A.K. Mohanty and M. Misra. 2022. Microplastics in ecosystems: their implications and mitigation pathways. Environ. Sci. Adv. 1(1): 9–29.

Salazar-Beltrán, D., L. Hinojosa-Reyes, C. Palomino-Cabello, G. Turnes-Palomino, A. Hernández-Ramírez and J.L. Guzmán-Mar. 2018. Determination of phthalate acid esters plasticizers in polyethylene terephthalate bottles and its correlation with some physicochemical properties. Polym. Test. 68: 87–94.

Samandra, S., O.J. Mescall, K. Plaisted, B. Symons, S. Xie, A.V. Ellis, et al. 2022. Assessing exposure of the Australian population to microplastics through bottled water consumption. Sci. Total Environ. 837: 155329.

Santana, J., C. Giraudi, E. Marengo, E. Robotti, S. Pires, I. Nunes, et al. 2014. Preliminary toxicological assessment of phthalate esters from drinking water consumed in Portugal. Environ. Sci. Pollut. Res. Int. 21: 1380–1390.

Schwabl, P., S. Köppel, P. Königshofer, T. Bucsics, M. Trauner, T. Reiberger, et al. 2019. Detection of various microplastics in human stool: A prospective case series. Ann. Intern. Med. 171(7): 453–457

Schymanski, D., C. Goldbeck, H.U. Humpf and P. Fürst. 2018. Analysis of microplastics in water by micro-Raman spectroscopy: release of plastic particles from different packaging into mineral water. Water Res. 129: 154–162.

Sekar, V., S. Shaji and B. Sundaram. 2024. Microplastic prevalence and human exposure in the bottled drinking water in the West Godavari region of Andhra Pradesh, India. J. Contam. Hydrol. 264: 104346.

Senathirajah, K., S. Attwood, G. Bhagwat, M. Carbery, S. Wilson and T. Palanisami. 2021. Estimation of the mass of microplastics ingested–A pivotal first step towards human health risk assessment. J. Hazard. Mater. 404: 124004.

Silva, A.L.P. 2021. Future-proofing plastic waste management for a circular bioeconomy. Curr. Opin. Environ. Sci. Health. 22: 100263.

Sin, L.T. and B.S. Tueen. 2022. Plastics and sustainability: practical approaches. Elsevier.

Smith, M., D.C. Love, C.M. Rochman and R.A. Neff. 2018. Microplastics in seafood and the implications for human health. Curr. Environ. Health Rep. 5: 375–386.

Sruthy, S. and E.V. Ramasamy. 2017. Microplastic pollution in Vembanad Lake, Kerala, India: the first report of microplastics in lake and estuarine sediments in India. Environ. Pollut. 222: 315–322.

Sun, J., X. Dai, Q. Wang, M.C. Van Loosdrecht and B.J. Ni. 2019. Microplastics in wastewater treatment plants: detection, occurrence and removal. Water Res. 152: 21–37.

Sun, T., Y. Teng, C. Ji, F. Li, X. Shan and H. Wu. 2024. Global prevalence of microplastics in tap water systems: Abundance, characteristics, drivers and knowledge gaps. Sci. Total Environ. 929: 172662.

Taghipour, H., M. Ghayebzadeh, F. Ganji, S. Mousavi and N. Azizi. 2023. Tracking microplastics contamination in drinking water in Zahedan, Iran: From source to consumption taps. Sci. Total Environ. 872: 162121.

Tong, H., Q. Jiang, X. Hu and X. Zhong. 2020. Occurrence and identification of microplastics in tap water from China. Chemosphere. 252: 126493.

Tse, Y.T., S.M.N. Chan and E.T. Sze. 2022. Quantitative assessment of full-size microplastics in bottled and tap water samples in Hong Kong. Int. J. Environ. Res. Public Health. 19(20): 13432.

United Nations Development Programme. 2023. Microplastics and human health: how much do they harm us? United Nations Development Program Kosovo. Available from: https://www.undp.org/kosovo/blog/microplastics-human-health-how-much-do-they-harm-us.

Van Cauwenberghe, L. and C.R. Janssen. 2014. Microplastics in bivalves cultured for human consumption. Environ. Pollut. 193: 65–70.

Van Emmerik, T., R. Tramoy, C. Van Calcar, S. Alligant, R. Treilles, B. Tassin, et al. 2019. Seine plastic debris transport tenfolded during increased river discharge. Front. Mar. Sci. 6: 642.

Van Raamsdonk, L.W., M. van der Zande, A.A. Koelmans, R.L. Hoogenboom, R.J. Peters, M.J. Groot, et al. 2020. Current insights into monitoring, bioaccumulation, and potential health effects of microplastics present in the food chain. Foods. 9(1): 72.

Walker, T.R. and D.C. McKay. 2021. Comment on "five misperceptions surrounding the environmental impacts of single-use plastic." Environ. Sci. Technol. 55(2): 1339–1340.

Wei, C., Z. Yan, J. Yang and K. Huang. 2024. Occurrence, detection, and classification of microplastics in excess sludge. pp. 71–84. *In*: Sunil Kumar, Kui Huang, Sartaj Ahmad Bhat (eds). Occurrence and Behavior of Emerging Contaminants in Organic Wastes and their Control Strategies. Elsevier.

Winkler, A., N. Santo, M.A. Ortenzi, E. Bolzoni, R. Bacchetta and P. Tremolada. 2019. Does mechanical stress cause microplastic release from plastic water bottles? Water Res. 166: 115082.

World Health Organization. 2019. Microplastics in Drinking-water.

World Packaging Organisation. 2008. Market statistics and future trends in global packaging. Brazil: Formato Design-Brazil. Available from: http://www.worldpackaging.org/.

Wright, S.L. and F.J. Kelly. 2017. Plastic and human health: A micro issue? Environ. Sci. Technol. 51(12): 6634–6647.

Wu, B., X. Wu, S. Liu, Z. Wang and L. Chen. 2019. Size-dependent effects of polystyrene microplastics on cytotoxicity and efflux pump inhibition in human Caco-2 cells. Chemosphere. 221: 333–341.

Yang, Y., E. Xie, Z. Du, Z. Peng, Z. Han, L. Li, et al. 2023. Detection of various microplastics in patients undergoing cardiac surgery. Environ. Sci. Technol. 57(30): 10911–10918.

Zhou, G., Q. Wu, X.F. Wei, C. Chen, J. Ma, J.C. Crittenden, et al. 2023. Tracing microplastics in rural drinking water in Chongqing, China: Their presence and pathways from source to tap. J. Hazard. Mater. 459: 132206.

Zhou, X.J., J. Wang, H.Y. Li, H.M. Zhang and D.L. Zhang. 2021. Microplastic pollution of bottled water in China. J. Water Process. Eng. 40: 101884.

Zuccarello, P., M. Ferrante, A. Cristaldi, C. Copat, A. Grasso, D. Sangregorio, et al. 2019. Exposure to microplastics (<10 μm) associated to plastic bottles mineral water consumption: the first quantitative study. Water Res. 157: 365–371.

Chapter 7

Microplastics Intercepted in Mangrove Forests: Sources, Generation, and Mitigation Strategies

Rungpilin Jittalerk[1] and Sandhya Babel*[2]

School of Bio-Chemical Engineering and Technology,
Sirindhorn International Institute of Technology, Thammasat University,
Rangsit Center, 99 Moo 18, Khlong Luang, Pathum Thani, Thailand-12120

7.1 INTRODUCTION

The mangrove ecosystem is recognized as the implicit destination of enormous plastic debris and the potential sink of microplastics from multiple and complex sources (Luo et al. 2021; Meera et al. 2021). Several studies reported the abundance of microplastics in mangrove biota and environmental matrices (sediment and surface water) (Deng et al. 2021). In the overall majority, microplastics in mangrove forests are contributed from terrestrial, marine, and atmospheric sources (Wang et al. 2023). Both primary and secondary microplastics, including plastic debris, can reach mangrove forests through environmental events like wind, tide, and current (Martin et al. 2019; Qian et al. 2021) and dumping activities (Luo et al. 2021). A special composition of mangrove forests, pneumatophores in particular, can act as trappers that facilitate the retention of plastic waste and microplastics on the forest floor (Duan et al. 2021). In consequence, these plastics are broken

*For Correspondence: sandhya@siit.tu.ac.th

DOI: 10.1201/9781003531463-7

down into smaller pieces through abiotic and biotic degradations (Qian et al. 2021). Recently, the accumulation of microplastics in mangrove forests is more prevalent compared to non-mangrove areas (e.g., beach, sea, and seagrass) (Wang et al. 2023). This indicates the critical situation of microplastic pollution in mangrove forests, which take essential responsibilities for providing habitats, carbon storage, nutrient cycling, natural resources, and coastal defense (Deng et al. 2021). Thus, it is vital to develop and implement feasible strategies to mitigate microplastic pollution in the mangrove ecosystem (Luo et al. 2021). This chapter provides an overview of information on potential sources and generation of microplastic pollution in mangrove forests and determines its feasible mitigation strategies.

7.2 SOURCES OF MICROPLASTICS IN MANGROVE FORESTS

Plastic invaders into mangrove forests come from manifold sources that are related to their morphologies and polymer types (Meera et al. 2021; Qian et al. 2021). They are considered to originate from marine, terrestrial, and atmospheric sources (Wang et al. 2023). The mangrove ecosystem is well known as one of the potential receptacles of plastic waste and microplastics from flank ecosystems. Unequivocally, major sources are located in marine and terrestrial ecosystems. Moreover, some studies reported microplastic contamination in atmospheric deposition (Wu et al. 2019). Airborne microplastics were even found in remote areas with equivalent abundance to megacities (Allen et al. 2019). Atmospheric fallout can be another important source of microplastics. Therefore, three main sources comprising marine, terrestrial, and atmospheric sources are discussed in this section.

7.2.1 Marine Sources

Plastic waste produced in the fishery is an important source of microplastics in mangrove forests where fishing activity and aquaculture frequently occur (Qian et al. 2021). Additionally, there are some cases of plastic containers and pellets released into the ocean from shipping spills (Saliba et al. 2022). The following are potential marine-based sources of microplastic pollution.

7.2.1.1 Aquaculture

Studies have shown that the intensive use of plastic equipment in neighboring aquaculture industries encourages the high abundance of microplastics that can come into mangrove forests through ocean currents or wave directions (Deng et al. 2021). A study by Huang et al. (2021) reported the predominance of foam microplastics in the mangrove forest near the largest aquaculture area in Guangxi, China, which possesses the largest amount of foam buoy usage. Zhou et al. (2020) also found a similar result of foam microplastics in other mangrove wetlands in Guangxi, which was probably caused by the widespread usage of floating frames and foamed fishing gear in local mariculture. Li et al. (2018) indicated that Styrofoam used in mollusk rafts in Qinzhou Bay released highly contaminated foam fragments in the mangrove areas. Valsan et al. (2024) found that the proximity to aquacultures that employ polystyrene (poly(1-phenylethene-1,2-diyl)) floating frames probably caused the majority of polystyrene microplastics in the mangrove forests.

7.2.1.2 Fishing Activity

Mangrove forests can receive microplastics released from plastic waste (e.g., gears, ropes, nets, and vessels) in adjacent fishing areas (Qian et al. 2021). Fishing waste from nearby small piers and commercial ports in Chabahar Bay broke down into microplastics that can be driven to protected mangrove forests without human encroachment through wind and currents (Aliabad et al. 2019). Apart from fishing waste, other types of waste from fishermen's communities can be released into adjacent mangrove forests. Food wrapping and plastic containers from fishermen may cause the dominant presence of film microplastics in the mangrove forest located at Caño Grande in Colombia (Garcés-Ordóñez et al. 2019).

7.2.1.3 Shipping

Saliba et al. (2022) reported 30 shipping spills of containers associated with consumer-packaged goods, including plastics, that occurred in the North Sea and surrounding areas between 1917 and 2021. The containers are spread in the marine environment and may be transported at great distances through ocean currents. As a capital of plastics, Cartagena in Northern Colombia has hosted many industries for the transportation of plastic resin pellets (Acosta-Coley et al. 2019). A study by Acosta-Coley et al. (2019) reported the presence of pellet microplastics on Castillogrande and Cabrero-Marbella beaches in Cartagena. Beyond industrial wastewater, leakage during transportation may cause the distribution of plastic resin pellets in the neighboring coastal areas. Thus, it is possible that shipping spills occurring in tropical and subtropical climate regions can cause microplastic pollution in mangrove forests.

7.2.2 Terrestrial Sources

According to a study by Cordova and Nurhati (2019), the highest intensity of land-derived debris released through nine river outlets into Jakarta Bay was recorded at the highest peak of rainfall amount in the rainy season. Terrestrial runoffs can release enormous land-based waste into the marine and wetland ecosystems. Treated wastewater with remaining microplastics is discharged into the ocean or rivers (Rose et al. 2023). Then, floating plastic waste and microplastics can reach mangrove forests through the movement of seawater or natural disasters (Adyel and Macreadie 2021; Deng et al. 2021). Moreover, dumping activities directly deposit a significant amount of waste in mangrove forests (Phelan et al. 2020). The following are potential terrestrial-based sources.

7.2.2.1 Agriculture

Agricultural runoff can transfer microplastics from fields into rivers and groundwater. A study by Yoon et al. (2024) indicated that microplastics originating from agricultural lands were higher than those from other types of land uses (roadside, residential areas, parks, and forests). Microplastics can be released from mulching and fertilizers (Yoon et al. 2024). Plastic mulching for crop production leaves microplastic residues on the soil, particularly low-density polyethylene (poly(methylene)) films (Rose et al. 2023). According to a study by Zhou et al. (2019), the abundance of polyethylene microfilms in mulching farmlands was higher than in non-mulching farmlands in Hangzhou Bay. Moreover, microplastics in agricultural drainage possibly reach the nearby coastal wetland ecosystem of this bay. Jiao et al. (2022) found that upstream farmland was a main source of microplastics derived from mulching films. They indicated that microplastics from mulching were more than 44.1% of total microplastics detected in farmland soils and ranged from 9.0 to 13.7% in estuarine sediments. Besides, sewage sludge utilized as fertilizer disperses remaining microplastics from wastewater treatment plants into the soil (Rose et al. 2023). In consequence, microplastics derived from agricultural runoff can reach mangrove forests. Additionally, machines used in farming activities may produce microplastics (Yoon et al. 2024). Nevertheless, it might exist in extremely low concentrations.

7.2.2.2 Wastewater Treatment Plants

The microplastic trapping efficiency of wastewater treatment plants is 70 to 99.9% depending on different countries (Hassan et al. 2023). The remaining microplastics in treated wastewater and sewage sludge cannot be ignored because they can become an important source of enormous microplastics (Zhang and Chen 2020; Hassan et al. 2023). Domestic wastewater and industrial wastewater were separately discussed.

Domestic Wastewater: Personal care and cosmetic products (e.g., cleansing foam and scrubbers) contain deliberately manufactured microplastics, as known microbeads (Guerranti et al. 2019). Some studies have found microbeads in the mangrove forests; however, in low concentrations (Pan et al. 2020; Huang et al. 2021; Jittalerk and Babel 2024). Microfibers also contaminate domestic wastewater. Yadav et al. (2023) reported that fiber microplastics abraded from clothes during laundry remain in household wastewater. According to research conducted by De Falco et al. (2018), the washing process of a 5-kilogram load of polyester fabrics can release 6,000,000 to 17,700,000 microfibers, depending on detergent types. Moreover, a study by Huang et al. (2021) indicated that the vicinity of coastal cities influenced the abundance of fibers emanating from domestic wastewater treatment plants. The predominance of fibers was found in some studies on microplastic contamination in mangrove forests in Iran (Aliabad et al. 2019), China (Deng et al. 2020), and Brazil (Trindade et al. 2023), which was probably caused by domestic wastewater.

Industrial Wastewater: Numerous industrial productions are associated with plastic materials. Microplastics can be released through wastewater from textiles, furniture, medical devices, sports, packaging, electronic components, vehicles, garments, and plastic recycling companies (Mahon et al. 2017). Textile production is well known as one of the major sources of fiber microplastics that remain in sewage sludge and treated wastewater from wastewater treatment plants (Yadav et al. 2023). According to a study by Deng et al. (2020), the large-scale production of shoes, textiles, and garments in Jinjiang, China, tends to be a major source of microplastics in mangrove wetland sediments at Jinjiang estuary. Valsan et al. (2024) reported that the Kota mangrove forest of India, with its proximity to industries involving marine food and agrochemicals, contained the highest abundance of microplastics in post-monsoon and pre-monsoon. Moreover, primary microplastics like industrial pellets are released from the wastewater system into the adjacent environment. White pellets with a low degree of surface degradation features were mostly common on beaches in Cartagena, a well-known industrialized city in the Caribbean (Acosta-Coley et al. 2019). Pellets derived from industrial processes may be translocated to neighboring mangrove forests (Li et al. 2020). Moreover, the high demand for plastic production possibly raises the amount of microplastics discharged from industrial wastewater.

7.2.2.3 Recreational Activities

More than 3,500 mangrove attractions are situated in 93 countries and territories (Spalding and Parrett 2019). Recreational activities produce different kinds of plastic waste (e.g., bottles, caps, and food packaging) (Seeruttun et al. 2021). Plastic waste was left in the Pasir Ris and Sungei Buloh mangrove forests located in Singapore after being heavily visited by tourists (Mohamed Nor and Obbard 2014). A study by Kannankai et al. (2022) detected that low-density polyethylene microplastics were abundant in mangrove soil in Mangalavanam bird sanctuary in India, which may be due to tourist activities. Valsan et al. (2024) reported that tourism during post-monsoon could be an additional source of microplastics in the Kota mangrove forest. Furthermore, kayaking, a new recreational activity within this mangrove forest, may raise the input of microplastics through the littering behavior of tourists. Similarly, nearby boating activities were likely to be a source of microplastics detected in the Berlayar Creek mangrove forest in Singapore (Mohamed Nor and Obbard 2014). Besides, coastal recreation leaves various types of plastic debris onshore (e.g., beverage bottles and food packaging) (Wu et al. 2020). It is possible that plastic debris can be brought from coastal areas to adjacent mangrove forests. Kesavan et al. (2021) found shoreline recreational activities as the majority source of debris, especially plastic waste, on the forest floor of mangrove forests in Mumbai, India. Once plastic waste degrades, remaining plastic particles onshore that have low-setting velocity are transported seawards by swash waves (van Sebille et al. 2020). Subsequently, those plastic particles are driven to mangrove forests through wind, currents, and tides (Deng et al. 2021).

7.2.2.4 Households

Neighboring communities can serve as one of the persistent sources of microplastics in mangrove forests. According to a study by Valsan et al. (2024), the dominant abundance of polyvinyl chloride (poly(1-chloroethylene)) microplastics in the mangrove forest may be caused by pipe waste from households proximate to the forests. Poor waste management leads to the dumping of municipal waste in the environment. Garcés-Ordóñez et al. (2019) reported that solid waste was dumped into the mangrove forests in the northern and western sectors of the Ciénaga Grande de Santa Marta in Colombia. Phelan et al. (2020) found that none of the observed villages in Selayar and Wakatobi islands in Indonesia have regular waste collection. The villagers managed household waste in traditional ways. Dumping rubbish into mangrove forests was one of their disposal activities. To make matters worse, they converted wetlands to drylands by dumping rubbish. In addition, Riascos et al. (2019) indicated that the mangrove forests in Agua Dulce and Islalba, localities in Buenaventura city of Colombia, are among the most highly polluted coastal places in the world. The surrounding river is mainly responsible for transporting a vast amount of litter from urban areas in Buenaventura city to accumulate in the mangrove areas.

7.2.3 Atmospheric Sources

Atmospheric microplastics are mainly released from industrial emissions, roads, and urban dust, with additional sources like landfills and sewage sludge (Rose et al. 2023). A study by Allen et al. (2019) indicated that atmospheric transport can bring microplastics to remote areas as far as 95 kilometers (km). Gravitational force and rain can play a role in depositing microplastics floating in the air onto mangrove forests (Qian et al. 2021; Rose et al. 2023). Thus, these factors may facilitate airborne microplastics from entering the mangrove forest. Li et al. (2022) and Huang et al. (2024) validated that microplastics were detected on the surface of mangrove leaves, which can be expected as a bioindicator of atmospheric microplastic pollution. However, the accumulation and fate of airborne microplastics in the mangrove ecosystem are largely unknown. The following are potential atmospheric sources.

7.2.3.1 Industrial Emissions

Airborne microplastics may be produced during industrial processes involving plastic production. Some studies found the prevailing abundance of fibers in the mangrove forests. Jittalerk and Babel (2024) reported the predominance of fibers detected in mangrove animals and their surrounding environment in the rainy season in the Bang Pu mangrove forest, Thailand. This mangrove forest is close to many textile industries located in Bang Pu Industrial Estate (Bang Pu Industrial Estate 2022). Deng et al. (2020) observed that fiber microplastics were the most abundant in sediment in the mangrove forests at the Jinjiang estuary near the textile and apparel industries. Apart from industrial wastewater, emissions from industries into the atmosphere were possibly washed out and settled in the mangrove areas.

7.2.3.2 Vehicles

Global emissions of road microplastics can reach remote regions (e.g., Arctic and Greenland) (Evangeliou et al. 2020). The erosion of synthetic rubber tires is one of the main sources of atmospheric microplastics (Rose et al. 2023). A study by Kannankai et al. (2022) found that the proximity to the public road possibly influenced black-fragmented microplastics in mangrove soil. They suggested that detected microplastics may be generated from the deterioration of tires. In addition to street runoffs, microplastics on roads may be moved through airborne transportation into contiguous mangrove forests during dense traffic.

7.2.3.3 Urban Dust

As mentioned earlier, atmospheric microplastics can travel at great distances. Microplastics floating as atmospheric components in bustling city zones can be transported and then deposited into adjacent mangrove forests. Li et al. (2022) reported that the highest microplastic content on non-submerged mangrove leaves was found in the mangrove forests in the vicinity of the crowded urban area of Fangcheng City in China, with the predominance of fibers. Hence, the proximity to dense population areas probably raises the concentration of atmospheric microplastics entering mangrove forests.

7.3 FATE OF PLASTIC WASTE MICROPLASTICS IN THE MANGROVE FORESTS

7.3.1 Transportation

Seawater carries marine debris to the coastal ecosystem. Coastal geomorphology and specific hydrodynamics control the transportation to shore (van Sebille et al. 2020). Winds, waves, currents, and tides significantly contribute to the presence of plastic waste and microplastics in the mangrove forests (Deng et al. 2021). Martin et al. (2019) indicated that the distance to maritime traffic routes influenced the litter distribution in Red Sea mangrove forests, with the concomitant role of surface currents. They found that the strongest impact of high-intensity maritime traffic on oceanic litter can reach mangrove forests at approximately 70 km. Microplastics can even reach the secluded and uninhabited mangrove forests. Seeruttun et al. (2023) reported that microplastics were detected in sediments from the non-utilized mangrove forest and nearby commercialized mangrove forest of Mauritius, a small oceanic island. They found the highest concentration of microplastics in seaward zones of the non-utilized mangrove forest. In addition, plastic waste dominated among other anthropogenic marine debris on this island (Seeruttun et al. 2021). Besides, surface runoffs (e.g., street runoff and agricultural runoff) and wastewater discharge can carry land-based microplastics that can be released into rivers via drainage systems (Deng et al. 2021; Qian et al. 2021) or directly enter the mangrove forests (Kannankai et al. 2022). A study by Cappa et al. (2023) affirmed that the proximity to the river mouth and human settlements influenced a higher accumulation of plastic litter. Watercourses inside mangrove forests possibly facilitate the invasion of plastic debris and microplastics. Moreover, atmospheric microplastics can be deposited into mangrove forests by gravitational force, rain (Qian et al. 2021; Rose et al. 2023), and even intense tropical storms like cyclones (Adyel and Macreadie 2021) and typhoons (Li et al. 2022).

A study by Duan et al. (2021) revealed that dense pneumatophores and prop roots increased the abundance of microplastics. In contrast, some mangrove species with knee roots (e.g., *Bruguiera gymnorrhiza* (L.) Lam.) have a lesser trapping ability, so plastic waste and microplastics may be released during tidal changes (Seeruttun et al. 2023). Moreover, plastic buoyancy and settling velocity values affect the residence time in the swash zone in the coastal area (van Sebille et al. 2020). Polyethylene, or polypropylene (poly(1-methylethylene)), tends to float in surface water, while polyethylene terephthalate (poly(ethyl benzene-1,4-dicarboxylate)) is likely to sink (Deng et al. 2021). Other factors, such as steepness and sediment types, also impact the accessibility of plastic waste and microplastics in the mangrove forests (van Sebille et al. 2020).

Apart from natural drivers, plastic can be directly thrown into the mangrove forests. Valsan et al. (2024) reported that some mangrove areas are used as dumping sites for municipal waste from nearby residents. The Pichavaram mangrove ecosystem in the coastal regions of Tamil Nadu in India is utilized for plastic waste dumping (Devi et al. 2019). Moreover, the mangrove forests and wetlands in villages on Selayar and Wakatobi islands in Indonesia have become rubbish dumping places (Phelan et al. 2020).

7.3.2 Degradation

Physicochemical characteristics of plastics (e.g., crystallinity, backbone composition, molecular weight, and polymer types) differentiate the results of plastic polymers after undergoing degradation (Ali et al. 2021). Polypropylene, with a lower stability compared to polyethylene, is prone to oxygen attack (Zhang et al. 2021). Polyvinyl chloride is the most sensitive polymer to ultraviolet (UV) radiation due to rapid dehydrochlorination (Gewert et al. 2015). Moreover, in the same scenario, polyethylene (0.88–0.96 g/cm^3) and polypropylene (0.90–0.92 g/cm^3) floating on surface water may expose sunlight more than polyethylene terephthalate (1.38 g/cm^3) sinking into bottom sediments (Deng et al. 2021). Plastic waste is ubiquitous in mangrove forests with a variety of shapes, sizes, colors, and polymer types. Once disseminated in the environment, plastic waste is degraded and fragmented into secondary microplastics. In addition, the tropical climate in the mangrove ecosystem might encourage the degradation of plastic particles (Deng et al. 2021). The aging of plastic polymers is accelerated by a warmer climate, which also causes a greater generation of microplastics in the environment

(Wei et al. 2024). Furthermore, the high richness of microorganisms in the mangrove forests can alleviate microplastic pollution through biodegradation; however, most studies showed slow degradation rates (Meera et al. 2021). Plastic degradation is categorized into two types: abiotic and biotic degradation (Zhang et al. 2021).

7.3.2.1 Abiotic Degradation

Abiotic factors are solar radiation, temperature, chemicals, and mechanical forces (e.g., wind and waves) (Qian et al. 2021; Haque and Fan 2023). Photodegradation by solar irradiation is considered the initial process of plastic degradation (Zhang et al. 2021). UV radiation can form environmentally free radicals that lead to oxygen addition, hydrogen abstraction, and chain scission (Du et al. 2021). Eventually, degradation occurs. It is possible that oxidation products remaining on the outer layer of the polymer are eroded, and then the non-oxidized layer is oxidized onwards (ter Halle et al. 2017). In consequence, plastics become brittle and shatter into small pieces (Gewert et al. 2015). Zhou et al. (2020) examined the characteristics of microplastics in different mangrove sediments along the south-eastern coastal zones of China. They found that microplastics detected in the sandy beach sediment without mangroves illustrated more harsh fragments and creases compared to those in the mangrove sediment. The reason was that microplastics may receive greater UV radiation exposure on the sandy beach. However, additives are employed for reducing degradation rate (e.g., UV stabilizers) (Ali et al. 2021). Hence, degradability also depends on additives that are added to the plastic materials.

Heat at elevated temperatures can break polymeric chains, leading to free radicals that cause degradation (Haque and Fan 2023). Slow thermal degradation of plastic waste may simultaneously occur with photodegradation when highly exposed to sunlight (e.g., on beaches) (Zhang et al. 2021). An increase in temperature associated with climate change also induced thermal-oxidative degradation (Haque and Fan 2023). Nevertheless, polymeric materials like polyethylene terephthalate enhance thermal stability by the presence of heteroatoms in their main chain (Gewert et al. 2015).

Pollutants in the atmosphere are related to the chemical degradation of plastic waste and microplastics. Ozone, nitrogen dioxide, sulfur dioxide, and volatile organic compounds can act as the catalysts in photochemical reactions for radical formation, leading to plastic degradation (Zhang et al. 2021). Acid rain can encourage degradability by changing the pH of water and soil (Haque and Fan 2023).

Mechanical degradation of plastic waste and microplastics can occur through the action of external forces (Zhang et al. 2021). In the environment, plastic particles collide with rocks and are abraded with sand (Haque and Fan 2023). Moreover, mechanical stresses probably increase the abundance of microplastics torn from clothes during washing processes (De Falco et al. 2018).

7.3.2.2 Biotic Degradation

Biotic degradation of plastic waste and microplastics occurs by organisms (e.g., bacteria, fungi, algae, and insects) (Zhang et al. 2021). Microbes can utilize plastic materials as carbon and energy sources for growth (Rose et al. 2023). The mangrove ecosystem with a high biodiversity of microbial community possibly enhances microbial biodegradability (Meera et al. 2021). However, large plastic particles cannot be directly absorbed by microorganisms due to their high molecular weight (Singhal et al. 2024). They initially need to be depolymerized into smaller monomers by enzyme binding and subsequent catalysis of their hydrolytic cleavage (Yuan et al. 2020). Plastic materials are classified into hydrolysable (e.g., polyethylene terephthalate) and non-hydrolysable polymers (e.g., polyethylene and polypropylene) (Zhang et al. 2021). Functional bonds in hydrolysable plastics facilitate the attack of hydrolytic enzymes, while structural similarity in non-hydrolysable plastics makes them more resistant to enzymatic degradation (Inderthal et al. 2021). Different polymer types can be degraded by different enzymes produced by a variety of microorganisms. For example, polyethylene terephthalate-digesting enzyme produced by *Ideonella sakaiensis* Yoshida et al. is capable of degrading polyethylene terephthalate (Singhal et al. 2024). Fungi secrete strong oxidative enzymes (e.g., lignin peroxidase and manganese peroxidase) that are possibly candidates for the enzymatic degradation of non-hydrolysable polymers (Rose et al. 2023). Extracellular depolymerases are employed for breaking down macromolecular polymers into small molecules that can pass through the cell membrane, and intracellular depolymerases are used for cell metabolism (Alimi et al. 2022). Consequently, degraded products are assimilated and then mineralized into carbon dioxide and water under aerobic conditions and into methane, carbon dioxide, organic acids, water, and ammonium under anaerobic conditions (Zhang et al. 2021). Besides, bacterial and fungal consortia showed the preferable efficiency on biodegradation in different types of polymers compared to individual strains (Ali et al. 2021). A study by Ameen et al. (2015) indicated that the fungi efficiently worked together on the biodegradation of low-density polyethylene. As shown in the results, the consortium showed the highest percentage of biomass gain, carbon dioxide elution, and enzyme activity after the test (Ameen et al. 2015).

Several studies reported the potential of microorganisms from the mangrove ecosystem for plastic biodegradation. Ameen et al. (2015) demonstrated that fungal strains isolated from tidal water, sediment, and floating debris surrounding pneumatophores in the mangrove stands on the Red Sea coast of Saudi Arabia were capable of degrading low-density polyethylene films. Six fungal isolates (*Alternaria alternata* (Fr.) Keissler, *Aspergillus caespitosus* Raper and Thom, *Aspergillus terreus* Thom, *Eupenicillium hirayamae* Scott and Stolk, *Paecilomyces variotii* Bainier, and *Phialophora alba* van Beyma) and a consortium gained biomass with firm attachment on the films after co-cultivation under aeration on a rotary shaker for four weeks. According to a study conducted by Auta et al. (2018), the biodegradation ability of bacterial isolates from mangrove sediments in Peninsular Malaysia was investigated. They found that *Rhodococcus* sp. strain 36 and *Bacillus* sp. strain 27 showed the capacity to degrade microplastics, accounting for the weight loss of UV-treated polypropylene microplastics at 6.4% and 4.0%, respectively, after 40 days of incubation. Devi et al. (2019) demonstrated the potential of bacterial isolates from plastic waste dumping places in coastal environment sites of Tamil Nadu, India, in degrading high-density polyethylene films. According to the results, bacterial isolates from the Pichavaram mangrove ecosystem and Silver Beach in Cuddalore showed the highest biodegradation ability among other isolates, as shown by the highest percentage of weight loss of degrading high-density polyethylene films. Xie et al. (2021) conducted a semi-controlled *in situ* exposure experiment for different polymers of microplastics in the mangrove forest at the Zhanjiang Mangrove National Nature Reserve. The result showed that erosion pits, a sign of biodegradation by rhizosphere bacteria, were detected on the surfaces of polyethylene, polyamide 6 (poly(hexano-6-lactam)), and polyvinyl chloride microplastics after 3 months *in situ* incubation. They also found that different chemical structures can attract different microbes. Nevertheless, a few potential microbes were discovered. Limited information necessitates further study to explore the capacity of biodegradation for microplastics in the mangrove forests (Meera et al. 2021).

7.4 MITIGATION STRATEGIES

Given the manifold importance of mangrove forests, it is necessary to determine mitigation strategies for microplastic pollution. Additionally, it might be more effective if the strategies link to the management of plastic waste, which is the source of secondary microplastics. Mitigation strategies are divided into short-term and long-term strategies.

7.4.1 Short-Term Strategies

The current scenario of plastic and microplastic pollution in global mangrove forests is exacerbated by continuous encroachment. In order to mitigate severity, short-term strategies are necessary. Effective short-term strategies are focused on manual cleanups and waste trapping to reduce the accumulation and accessibility of plastic waste and microplastics. However, it should be noted that both short-term strategies are temporary solutions.

7.4.1.1 Manual Cleanups

Cleanup solutions are conducted in the furthest places from pollution sources (Wagner 2022). Despite knowing the fact, waste collection is still essential for removing plastic waste, which is an important source of future microplastics. The collection should be designed for two purposes based on the presence of plastic waste. Firstly, the long-standing accumulation of plastic waste inside the mangrove forests should be reduced. Plastic debris that is found entangling or buried in mangrove areas can be degraded into numerous secondary microplastics in the future (Mohamed Nor and Obbard 2014). To eliminate the next generation of microplastics, accumulated plastic waste has to be removed.

Secondly, waste collection along the shoreline or in adjacent rivers should be carried out to decrease additional plastic waste and microplastics (Suyadi and

Manullang 2020). Cordova et al. (2021) reported that the low concentration of microplastics in Muara Angke Wildlife Reserve was possibly caused by river and coastal cleanup programs in Jakarta. The programs were conducted in 2015 by city cleaners under the governor's regulation (Cordova and Nurhati 2019). Hence, it can be expected to reduce the influx of plastic waste and microplastics into the mangrove forests.

However, waste collection, although practical in reducing waste, is unable to deal with large-scale tasks and repetition. According to a study by Phelan et al. (2020), manual clean-up activities in localities, done mainly by hand-picking, cannot deal with enormous waste, and villagers tend to be discouraged by the vast quantity of seasonal rubbish. Continuous inputs of terrestrial waste into the ocean would encourage recurrent invasion of waste in mangrove forests. Hence, reducing or suspending the effluence of waste at the source is necessary. Furthermore, it is vital to note that the mangrove ecosystem, which provides habitats for thousands of species within the forest, should not receive much disruption. Trampling in the mangrove forests by many workers or volunteers for waste collection is probably harmful to mangrove biota (Kesavan et al. 2021). Thus, the impacts of intensive cleaning activity should be examined before implementation.

7.4.1.2 Waste Trapping

Barrier installations along coastlines are likely to trap floating debris from the ocean from entering the mangrove forests. Natural-based fences are typically used for coastal protection. Wooden or bamboo fences have been used for reducing wave energy and raising sedimentation along coastlines in Thailand (Pranchai et al. 2019), Vietnam (Van et al. 2021), and China (Huang et al. 2023). Unfortunately, the destruction of natural-based fences poses harm to the mangrove tree. Bamboo debris affects the health of mangrove forests and increases crown dieback of the mangrove trees (Pranchai et al. 2019). Although bamboo fences effectively serve the purpose of shoreline prevention (Huang et al. 2023), their limited lifespan leads to detrimental effects and mortality of the mangrove forests (Pranchai et al. 2019). To overcome this obstacle, other materials may be employed to replace the limitations of natural materials. AlAli et al. (2023) reported on research conducted on the coastline of Florida, the United States, that biomimicry of mangrove roots with concrete showed similar features to natural mangrove roots. Moreover, it was found that colonization of foundation species (e.g., oysters and tunicates) occurred within 18 months. Therefore, it is envisaged that concrete bamboo-like fences may be a good candidate. Further study should examine the ability of concrete fences to prevent the invasion of marine debris. Besides, trapped debris should be regularly collected to avoid degradation by natural abrasion and UV radiation.

Trapping is not only essential for coastlines but also for rivers. Rivers are the main drivers of enormous land-based waste into the ocean (Luo et al. 2021). Different companies widely innovate litter trappers. For example, Mr. Trash Wheel and the Great Bubble Barrier are used to trap debris before it enters the ocean (Wagner 2022). If terrestrial waste is captured before being released to the environment, the quantity of waste accumulated in the ocean and brought to the mangrove forests will be reduced.

7.4.2 Long-Term Strategies

An integral mitigation strategy is necessary to reduce plastic and microplastic pollution in the mangrove ecosystem. Any long-term solution should be employed throughout the plastic life cycle. Phelan et al. (2020) elucidated that the unavailability of essential infrastructure and poor literacy for the plastic cycle of local villages on islands in Indonesia is likely to continuously raise mismanaged plastic waste and microplastics in the ecosystem. Producers and manufacturers supported single-use products or sachets, which were also encouraged by villagers' satisfaction. The lack of formal waste management systems in the villages has led to the extensive use of traditional disposals of rubbish that ruin the environment (e.g., open-burning, burying in the garden, and dumping into the ocean and mangrove forests). Moreover, villagers had low literacy on the impacts of plastic waste and found it convenient to keep using traditional disposals, which preclude awareness and association with solving the problem. This study highlights the importance of cooperation from producers at the outset, as well as the awareness and knowledge of consumers, and the infrastructure support of related agencies at the conclusion. Thus, lenses on long-term strategies focus on plastic production, literacy and awareness, and plastic waste management.

7.4.2.1 Plastic Production

Capitalism is one of the deep-rooted causes of plastic and microplastic pollution (Wagner 2022). Polymers utilized for single-use plastics (e.g., polyethylene, polypropylene, and polyvinyl chloride) account for the largest manufacturing among other polymers in many countries (Chen et al. 2021). Single-use plastic waste is the biggest threat to mitigate plastic and microplastic pollution (Rose et al. 2023). Plastic packaging stands out for the highest portion (40%) of global single-use plastic production (Chen et al. 2021). Studies reported that plastic packaging was mostly detected in mangrove forests (Martin et al. 2019; Deng et al. 2021; Kesavan et al. 2021). Reducing the consumption of single-use plastic products definitely eases the pollution and accumulation of plastic waste and microplastics in the environment (Rose et al. 2023). Nowadays, the prohibition of single-use plastics has been implemented in several countries (Chen et al. 2021). In Mauritius, the legislation on banning plastic bags and controlling single-use plastic products was launched in 2020 (Seeruttun et al. 2023). India prohibited the use of disposable plastics in Delhi and the National Capital Region in 2017 (Rose et al. 2023). Although banning is stated, Kesavan et al. (2021) indicated that dumping waste from households or religious festivals can still lead to the accumulation of plastic waste in the mangrove regions. Thus, apart from banning, plastic manufacturing should go along with the pursuit of environmentally friendly plastic materials.

Plastic materials are versatile in many applications. It is undeniable that people are habitually familiar with the use of plastic products. Vitrimers emerged as next-generation materials with the property combination of thermoplastics and thermosets and are known for their re-processibility and recyclability (Zheng et al. 2021). Currently, some limitations require further development to be more practical. However, it can be expected that vitrimers may positively cause a change in plastic recycling. The incentive for plastic producers to adopt a circular economy may increase. Consequently, the inputs of dumping plastic waste may indirectly be reduced. Additionally, bio-based vitrimers, made from renewable materials or polyimines, are simultaneously developed to meet the requirements of the market and approach sustainability (Lucherelli et al. 2022). Ando et al. (2023) found that epoxy resin vitrimer incorporated with polyester-grafted showed 25 wt% biodegradation after exposure to seawater for 30 days. Hence, environmentally friendly production and products can be envisioned.

As the world shifts toward more sustainable practices, bioplastics are a promising alternative to conventional plastics. Typically, bioplastics are more prone to degradation by microorganisms compared to conventional ones. Poly (lactic acid) successfully enters global bioplastic commercialization, constituting 33% of all bioplastics manufactured in 2021, due to its good processability and mechanical properties (Jem and Tan 2020; Ali et al. 2023). Its monomer is derived from renewable sources (e.g., starch or sugar), enabling it to be compostable under industrial conditions, but even so, it is not completely degradable under natural conditions (Ali et al. 2023). Moreover, there are sustainable polymers from other renewable sources such as sugar cane, wheat, terpenes, soybean, oil palm, sunflower, and agricultural waste (Zhu et al. 2016). Besides, biopolymers produced by microorganisms also draw scientific attraction. According to a study by Alarfaj et al. (2015), *Bacillus thuringiensis* KSADL127 extracted from the mangrove niche in Saudi Arabia had a high capacity for poly(3-hydroxybutyrate) production. Although bioplastics are promising alternatives, bio-based productions are not cost-effective and require property improvement (Alarfaj et al. 2015; Zhu et al. 2016; Jem and Tan 2020). In addition, intentional and unintentional waste dumping into the environment would doubtless occur now and forever. Highlighting that environmentally friendly products are less harmful to the environment, but they still need proper management. Otherwise, there is no change if public awareness is not simultaneously improved.

Apart from development, legislation on reflecting hazardous impacts on nature in the cost of raw materials (e.g., levies) may decrease producers' attraction to conventional plastics, which are cheaper and more harmful (Wagner 2022). For example, the interventional action using extended producer responsibility effectively regulates appropriate implementation along the plastic life cycle. Both manufacturers and consumers should internalize responsibility for environmental impact. Eventually, a solid foundation in plastic production can efficiently alleviate plastic and microplastic pollution in the environment.

7.4.2.2 Awareness and Literacy

Overconsumption of disposable plastic products strongly influences the distribution of plastic waste and microplastics in mangrove forests. As reported in several studies, microplastics derived from commonly used polymers like polyethylene and polypropylene were mainly detected in environmental matrices in the mangrove forests (Deng et al. 2021). The convenience and cheapness of disposable plastic products can encourage traditional consumerism that can ruin the environment (Phelan et al. 2020). Ethical consumerism is introduced and supported to reduce the overconsumption of plastic products (Wagner 2022). Several regulations are applied to decrease and prohibit the demand for disposable plastic products in many countries (Rose et al. 2023). However, the enforcement of this burden on citizens would make this solution tokenistic. Plastic avoidance is sometimes criticized as it can affect low-income people (Wagner 2022). Occasionally, eco-labeled products, intending to raise awareness, can be more costly than conventional plastic products because of cost-ineffective and small-scale production (Zhu et al. 2016; Rose et al. 2023). Thus, these problems reflect customer needs and become a significant challenge to environmentally friendly production.

Although the future market facilitates consumers' purchase of eco-friendly products, there would be a need to accompany it with adequate literacy and awareness. As mentioned earlier, consumers should internalize the responsibility for their utilization of plastic products. A study by Phelan et al. (2020) showed that low literacy on plastics and a lack of awareness may lead to the distribution of plastic waste in the mangrove forests. Knowledge of plastic products and plastic waste management should be taught in schools and published in proper media for adults. Consequently, high literacy on these issues can be ideally expected to raise the awareness of people. Apart from literacy and awareness, a lack of attraction may cause a lack of involvement that becomes an obstacle to the solution. Changing consumer behavior may need the engagement of policies, campaigns, and point-of-sale interventions (Wagner 2022). Illegal waste dumping in Maharashtra's mangrove forests in India decreased after implementing regulation (Kesavan et al. 2021). Therefore, the proper behavior of consumers is a key strength of mitigation strategies for plastic and microplastic pollution in the mangrove ecosystem.

7.4.2.3 Plastic Waste Management

Policies on plastic waste management are implemented in many countries worldwide. However, it is insufficient to regulate plastic and microplastic pollution in many ecosystems, including the mangrove ecosystem. Some countries have regulations for protecting mangrove forests from plastic waste. The government of Maharashtra states that garbage cannot be dumped into the mangrove forests (Kesavan et al. 2021). Reducing waste dumping activities in the Sundarbans, Bangladesh side, is implemented by the Bangladesh Forest Department (Adyel and Macreadie 2021). Nevertheless, plastic waste and microplastics from other sources are still brought to mangrove forests through waves, tides, and surface currents (Kesavan et al. 2021). Plastic waste management of some potential sources of pollution in mangrove forests, as discussed in Sources of microplastics in mangrove forests, is not controlled under regulations or laws. Plastic waste management should be implemented not only in mangrove forests but also in potential sources of release. Notably, a lack of waste management infrastructure can cause improper alternatives that are easy and convenient for humans but detrimental to the environment (Phelan et al. 2020). Comprehensive facilities for the efficient management of plastic waste should be installed along the route of waste. In addition, the development of efficient wastewater treatment for both municipal and industrial scales is indispensable.

A roadmap for a circular economy may positively influence the reduction of mismanaged plastic waste in the environment. Recycling, considered a part of the circular economy concept, is one of the most common ways to deal with plastic waste (Chen et al. 2021). A study by Kannankai et al. (2022) revealed that the low abundance of high-density polyethylene microplastics in the Mangalavanam bird sanctuary in India may be caused by its popularity among recyclers. They mostly collected high-density polyethylene compared to other plastics because of the ease of recycling. This suggests that increasing recycling activities on plastic waste is likely to reduce the abundance of secondary microplastics in the environment. Besides, biotechnology for enzymatic depolymerization emerged as an effective alternative for the chemical recycling of plastic waste, which can turn plastic polymers into their monomers (Vollmer et al. 2020). Thus, it can be envisioned that cutting-edge technology for recycling all types of plastic polymers can ease plastic and microplastic pollution.

Nowadays, enormous plastic waste and microplastics accumulate in many mangrove forests (Deng et al. 2021). Cleanup activities can gather only plastic debris. Numerous microplastics are left in mangrove biota and environmental matrices. Microbial degradation is a reliable, environment-friendly strategy for dealing with microplastic pollution in mangrove forests (Meera et al. 2021). Several studies explored the biodegradability of mangrove-associated microbes on some types of plastic polymers (Deng et al. 2021). As discussed in Biotic degradation, many mangrove-associated microbes are able to degrade plastic particles (Ameen et al. 2015; Auta et al. 2018; Devi et al. 2019; Xie et al. 2021). In addition, Auta et al. (2022) conducted *in situ* bioremediation of mangrove soil that was intentionally contaminated with polyethylene terephthalate and polystyrene microplastics. An indigenous microbial consortium with enhanced plastic biodegradability, which was developed in the laboratory, was introduced back into the mangrove environment. According to the results, the weight loss of polyethylene terephthalate microplastics in microbes amended plots was 18% after 90 experimental days, which was higher than in non-amended plots. However, the opposite result was detected for polystyrene microplastics, which may be caused by environmental factors and the selective behavior of microbes. This study elucidated that enhancing the capacity of mangrove microorganisms probably accelerates the bioremediation of accumulated microplastics. Thus, it can be expected that engineered microbial augmentation, along with proper regulatory frameworks, is one of the potential solutions to microplastic pollution in the mangrove forest.

Besides, mangrove plants may be assistants in bioremediation. The mangrove root system facilitates the deposit of marine litter (Luo et al. 2021). This may encourage the biodegradation of mangrove rhizosphere microorganisms. Inhabiting microbiota in the mangrove rhizosphere can degrade microplastics in different polymer types (Meera et al. 2021). A study by Xie et al. (2021) reported that rhizosphere bacteria can degrade polyethylene, polyamide 6, and polyvinyl chloride. Apart from bioremediation, mangrove plants may independently decrease the accumulation of microplastics. However, the investigation of phytoremediation of mangrove plants for microplastic pollution is superficial.

7.5 CONCLUSIONS

Continuous inputs from terrestrial, marine, and atmospheric sources exacerbate microplastic contamination in mangrove forests. Among potential sources, terrestrial discharge is likely to be the most influential source. The transportation of plastic debris and microplastics to reach mangrove forests is performed by natural forces and human activities. These plastics are shattered through abiotic and biotic processes. The distribution of microplastics in a wide range of sizes is found in mangrove forests. Short-term strategies are necessary to reduce the prolonged accumulation and additional invasion of plastic debris. Waste trapping is a promising approach to protect mangrove forests from roaming debris driven by seawater and facilitate cleanup activities. In parallel, long-term strategies should be implemented to manage the deep-rooted problem of microplastic pollution in mangrove forests. The cooperation of producers is important as a starting point for plastics. Not only does the incentive in a circular economy offer a better choice of raw materials, but the strict legislation will probably shift plastic production toward sustainability. Furthermore, consumers have a crucial role in transferring plastic waste into a proper pathway. Raising awareness, literacy, and the responsibility of consumers is indispensable. Lastly, plastic waste should be efficiently managed in both mangrove forests and their potential sources and sustainably recycled to be fed back into plastic production.

7.6 ACKNOWLEDGEMENTS

The authors acknowledge the Sirindhorn International Institute of Technology, Thailand, for providing a scholarship to the first author.

REFERENCES

Acosta-Coley, I., D. Mendez-Cuadro, E. Rodriguez-Cavallo, J. de la Rosa and J. Olivero-Verbel. 2019. Trace elements in microplastics in Cartagena: A hotspot for plastic pollution at the Caribbean. Mar. Pollut. Bull. 139: 402–411.

Adyel, T.M. and P.I. Macreadie. 2021. World's largest mangrove forest becoming plastic cesspit. Front. Mar. Sci. 8: 766876.

AlAli, M., Y. Mattar, M.A. Alzaim and S. Beheiry. 2023. Applications of biomimicry in architecture, construction and civil engineering. Biomimetics. 8(2): 202.

Alarfaj, A.A., M. Arshad, E.N. Sholkamy and M.A. Munusamy. 2015. Extraction and characterization of polyhydroxybutyrates(PHB) from *Bacillus thuringiensis* KSADL127 isolated from mangrove environments of Saudi Arabia. Braz. Arch. Biol. Technol. 58(5): 781–788.

Ali, S.S., T. Elsamahy, R. Al-Tohamy, D. Zhu, Y.A.G. Mahmoud, E. Koutra, et al. 2021. Plastic wastes biodegradation: Mechanisms, challenges and future prospects. Sci. Total Environ. 780: 146590.

Ali, W., H. Ali, S. Gillani, P. Zinck and S. Souissi. 2023. Polylactic acid synthesis, biodegradability, conversion to microplastics and toxicity: A review. Environ. Chem. Lett. 21: 1761–1786.

Aliabad, M.K., M. Nassiri and K. Kor. 2019. Microplastics in the surface seawaters of Chabahar Bay, Gulf of Oman (Makran coasts). Mar. Pollut. Bull. 143: 125–133.

Alimi, O.S., D. Claveau-Mallet, R.S. Kurusu, M. Lapointe, S. Bayen and N. Tufenkji. 2022. Weathering pathways and protocols for environmentally relevant microplastics and nanoplastics: What are we missing? J. Hazard. Mater. 423: 126955.

Allen, S., D. Allen, V.R. Phoenix, G. Le Roux, P.D. Jiménez, A. Simonneau, et al. 2019. Atmospheric transport and deposition of microplastics in a remote mountain catchment. Nat. Geosci. 12(5): 339–344.

Ameen, F., M. Moslem, S. Hadi and A.E. Al-Sabri. 2015. Biodegradation of low density polyethylene (LDPE) by mangrove fungi from the Red Sea coast. Prog. Rubber Plast. Recycl. Technol. 31(2): 125–143.

Ando, S., M. Hirano, L. Watakabe, H. Yokoyama and K. Ito. 2023. Environmentally friendly sustainable thermoset vitrimer-containing polyrotaxane. ACS Mater. Lett. 5(12): 3156–3160.

Auta, H.S., C.U. Emenike, B. Jayanthi and S.H. Fauziah. 2018. Growth kinetics and biodeterioration of polypropylene microplastics by *Bacillus* sp. and *Rhodococcus* sp. isolated from mangrove sediment. Mar. Pollut. Bull. 127: 15–21.

Auta, H.S., O.P. Abioye, S.A. Aransiola, J.D. Bala, V.I. Chukwuemeka, A. Hassan et al. 2022. Enhanced microbial degradation of PET and PS microplastics under natural conditions in mangrove environment. J. Environ. Manage. 304: 114273.

Bang Pu Industrial Estate. 2022. Factory information report. https://www.ieat-bangpoo.go.th/

Cappa, P., M.E.M. Walton, M.K.O. Paler, E.B. Taboada, J.G. Hiddink and M.W. Skov. 2023. Impact of mangrove forest structure and landscape on macroplastics capture. Mar. Pollut. Bull. 194: 115434.

Chen, Y., A.K. Awasthi, F. Wei, Q. Tan and J. Li. 2021. Single-use plastics: Production, usage, disposal and adverse impacts. Sci. Total Environ. 752: 141772.

Cordova, M.R. and I.S. Nurhati. 2019. Major sources and monthly variations in the release of land-derived marine debris from the Greater Jakarta area, Indonesia. Sci. Rep. 9(1): 18730.

Cordova, M.R., Y.I. Ulumuddin, T. Purbonegoro and A. Shiomoto. 2021. Characterization of microplastics in mangrove sediment of Muara Angke Wildlife Reserve, Indonesia. Mar. Pollut. Bull. 163: 112012.

De Falco, F., M.P. Gullo, G. Gentile, E. Di Pace, M. Cocca, L. Gelabert, et al. 2018. Evaluation of microplastic release caused by textile washing processes of synthetic fabrics. Environ. Pollut. 236: 916–925.

Deng, J., P. Guo, X. Zhang, H. Su, Y. Zhang, Y. Wu, et al. 2020. Microplastics and accumulated heavy metals in restored mangrove wetland surface sediments at Jinjiang estuary (Fujian, China). Mar. Pollut. Bull. 159: 111482.

Deng, H., J. He, D. Feng, Y. Zhao, W. Sun, H. Yu, et al. 2021. Microplastics pollution in mangrove ecosystems: A critical review of current knowledge and future directions. Sci. Total Environ. 753: 142041.

Devi, R.S., R. Ramya, K. Kannan, A.R. Antony and V.R. Kannan. 2019. Investigation of biodegradation potentials of high density polyethylene degrading marine bacteria isolated from the coastal regions of Tamil Nadu, India. Mar. Pollut. Bull. 138: 549–560.

Du, H., Y. Xie and J. Wang. 2021. Microplastic degradation methods and corresponding degradation mechanism: Research status and future perspectives. J. Hazard. Mater. 418: 126377.

Duan, J., J. Han, S.G. Cheung, R.K.Y. Chong, C.M. Lo, F.W.F. Lee, et al. 2021. How mangrove plants affect microplastic distribution in sediments of coastal wetlands: Case study in Shenzhen Bay, South China. Sci. Total Environ. 767: 144695.

Evangeliou, N., H. Grythe, Z. Klimont, C. Heyes, S. Eckhardt, S. Lopez-Aparicio, et al. 2020. Atmospheric transport is a major pathway of microplastics to remote regions. Nat. Commun. 11(1): 3381.

Garcés-Ordóñez, O., V.A. Castillo-Olaya, A.F. Granados-Briceño, L.M.B. García and L.F.E. Díaz. 2019. Marine litter and microplastic pollution on mangrove soils of the Ciénaga Grande de Santa Marta, Colombian Caribbean. Mar. Pollut. Bull. 145: 455–462.

Gewert, B., M.M. Plassmann and M. MacLeod. 2015. Pathways for degradation of plastic polymers floating in the marine environment. Environ. Sci.: Process. Impacts. 17(9): 1513–1521.

Guerranti, C., T. Martellini, G. Perra, C. Scopetani and A. Cincinelli. 2019. Microplastics in cosmetics: Environmental issues and needs for global bans. Environ. Toxicol. Pharmacol. 68: 75–79.

Haque, F. and C. Fan. 2023. Fate of microplastics under the influence of climate change. iScience. 26(9): 107649.

Hassan, F., K.D. Prasetya, J.N. Hanun, H.M. Bui, S. Rajendran, N. Kataria, et al. 2023. Microplastic contamination in sewage sludge: Abundance, characteristics and impacts on the environment and human health. Environ. Technol. Innov. 31: 103176.

Huang, Y., X. Xiao, K. Effiong, C. Xu, Z. Su, J. Hu, et al. 2021. New insights into the microplastic enrichment in the blue carbon ecosystem: Evidence from seagrass meadows and mangrove forests in coastal South China Sea. Environ. Sci. Technol. 55(8): 4804–4812.

Huang, Z., Z. Dai, R. Wang, X. Zhou, W. Pang, J. Luo, et al. 2023. Dramatical hydrosedimentary changes induced by bamboo fences over mangrove tidal flat of the largest delta in Beibu Gulf, Southwestern China. Acta Oceanol. Sin. 42: 103–115.

Huang, Y., Z. Zhu, T. Li, M. Li, Z. Cai, X. Wang, et al. 2024. Mangrove plants are promising bioindicator of coastal atmospheric microplastics pollution. J. Hazard. Mater. 465: 133473.

Inderthal, H. S.L. Tai and S.T.L. Harrison. 2021. Non-hydrolyzable plastics—An interdisciplinary look at plastic bio-oxidation. Trends Biotechnol. 39(1): 12–23.

Jem, K.J. and B. Tan. 2020. The development and challenges of poly (lactic acid) and poly (glycolic acid). Adv. Ind. Eng. Polym. Res. 3: 60–70.

Jiao, M., Y. Wang, T. Li, R. Li and B. Liu. 2022. Riverine microplastics derived from mulch film in Hainan Island: Occurrence, source and fate. Environ. Pollut. 312: 120093.

Jittalerk, R. and S. Babel. 2024. Microplastic contamination in Thai Vinegar Crabs (*Episesarma mederi*), Giant Mudskippers (*Periophthalmodon schlosseri*) and their surrounding environment from the Bang Pu mangrove forests, Samut Prakan province, Thailand. Mar. Pollut. Bull. 198: 115849.

Kannankai, M.P., R.K. Alex, V.V. Muralidharan, N.P. Nazeerkhan, A. Radhakrishnan and S.P. Devipriya. 2022. Urban mangrove ecosystems are under severe threat from microplastic pollution: A case study from Mangalavanam, Kerala, India. Environ. Sci. Pollut. Res. 29: 80568–80580.

Kesavan, S., K.A.M. Xavier, G. Deshmukhe, A.K. Jaiswar, S. Bhusan and S.P. Shukla. 2021. Anthropogenic pressure on mangrove ecosystems: Quantification and source identification of surficial and trapped debris. Sci. Total Environ. 794: 148677.

Li, J., H. Zhang, K. Zhang, R. Yang, R. Li and Y. Li. 2018. Characterization, source and retention of microplastic in sandy beaches and mangrove wetlands of the Qinzhou Bay, China. Mar. Pollut. Bull. 136: 401–406.

Li, R., L. Yu, M. Chai, H. Wu and X. Zhu. 2020. The distribution, characteristics and ecological risks of microplastics in the mangroves of Southern China. Sci. Total Environ. 708: 135025.

Li, R., C. Wei, M. Jiao, Y. Wang and H. Sun. 2022. Mangrove leaves: an undeniably important sink of MPs from tidal water and air. J. Hazard. Mater. 426: 128138.

Lucherelli, M.A., A. Duval and L. Avérous. 2022. Biobased vitrimers: towards sustainable and adaptable performing polymer materials. Prog. Polym. Sci. 127: 101515.

Luo, Y.Y., C. Not and S. Cannicci. 2021. Mangroves as unique but understudied traps for anthropogenic marine debris: A review of present information and the way forward. Environ. Pollut. 271: 116291.

Mahon, A., R. Officer, R. Nash and I. O'Connor. 2017. Scope, fate, risks and impacts of microplastic pollution in Irish freshwater systems. Environmental Protection Agency (EPA) report no. 210, EPA, Wexford, Ireland.

Martin, C., H. Almahasheer and C.M. Duarte. 2019. Mangrove forests as traps for marine litter. Environ. Pollut. 247: 499–508.

Meera, S.P., M. Bhattacharyya, A. Nizam and A. Kumar. 2021. A review on microplastic pollution in the mangrove wetlands and microbial strategies for its remediation. Environ. Sci. Pollut. Res. 29: 4865–4879.

Mohamed Nor, N.H. and J.P. Obbard. 2014. Microplastics in Singapore's coastal mangrove ecosystems. Mar. Pollut. Bull. 79: 278–283.

Pan, Z., Y. Sun, Q. Liu, C. Lin, X. Sun, Q. He et al. 2020. Riverine microplastic pollution matters: A case study in the Zhangjiang River of Southeastern China. Mar. Pollut. Bull. 159: 111516.

Phelan, A.A., H. Ross, N.A. Setianto, K. Fielding and L. Pradipta. 2020. Ocean plastic crisis—Mental models of plastic pollution from remote Indonesian coastal communities. PLoS One. 15(7): e0236149.

Pranchai, A., M. Jenke and U. Berger. 2019. Well-intentioned, but poorly implemented: Debris from coastal bamboo fences triggered mangrove decline in Thailand. Mar. Pollut. Bull. 146: 900–907.

Qian, J., S. Tang, P. Wang, B. Lu, K. Li, W. Jin, et al. 2021. From source to sink: Review and prospects of microplastics in wetland ecosystems. Sci. Total Environ. 758: 143633.

Riascos, J.M., N. Valencia, E.J. Peña and J.R. Cantera. 2019. Inhabiting the technosphere: The encroachment of anthropogenic marine litter in neotropical mangrove forests and its use as habitat by macrobenthic biota. Mar. Pollut. Bull. 142: 559–568.

Rose, P.K., M. Jain, N. Kataria, P.K. Sahoo, V.K. Garg and A. Yadav. 2023. Microplastics in multimedia environment: A systematic review on its fate, transport, quantification, health risk and remedial measures. Groundw. Sustain. Dev. 20: 100889.

Saliba, M., S. Frantzi and P. van Beukering. 2022. Shipping spills and plastic pollution: A review of maritime governance in the North Sea. Mar. Pollut. Bull. 181: 113939.

Seeruttun, L.D., P. Raghbor and C. Appadoo. 2021. First assessment of anthropogenic marine debris in mangrove forests of Mauritius, a small oceanic island. Mar. Pollut. Bull. 164: 112019.

Seeruttun, L.D., P. Raghbor and C. Appadoo. 2023. Mangrove and microplastic pollution: A case study from a small island (Mauritius). Reg. Stud. Mar. Sci. 62: 102906.

Singhal, B., S. Kumar, S. Mehtab, U. Aggarwal, V. Kumar, M. Umesh, et al. 2024. Biotechnological advancements in microplastics degradation in drinking water: Current insights and future perspectives. Case Stud. Chem. Environ. Eng. 9: 100640.

Spalding, M. and C.L. Parrett. 2019. Global patterns in mangrove recreation and tourism. Mar. Policy. 110: 103540.

Suyadi and C.Y. Manullang. 2020. Distribution of plastic debris pollution and it is implications on mangrove vegetation. Mar. Pollut. Bull. 160: 111642.

ter Halle, A., L. Ladirat, M. Martignac, A.F. Mingotaud, O. Boyron and E. Perez. 2017. To what extent are microplastics from the open ocean weathered? Environ. Pollut. 227: 167–174.

Trindade, L. dos S. T.V. Gloaguen, T. de S.F. Benevides, A.C.S. Valentim, M.R. Bomfim and J.A.G. Santos. 2023. Microplastics in surface waters of tropical estuaries around a densely populated Brazilian bay. Environ. Pollut. 323: 121224.

Valsan, G., A.K. Warrier, S. Anusree, A. Tamrakar, R. Khaleel and N. Rangel-Buitrago. 2024. Seasonal variation of microplastics in tropical mangrove waters of South-western India. Reg. Stud. Mar. Sci. 69: 103323.

Van, C.M., A. Ngo, T. Mai and H.T. Dao. 2021. Bamboo fences as a nature-based measure for coastal wetland protection in Vietnam. Front. Mar. Sci. 8: 756597.

van Sebille, E., S. Aliani, K.L. Law, N. Maximenko, J.M. Alsina, A. Bagaev, et al. 2020. The physical oceanography of the transport of floating marine debris. Environ. Res. Lett. 15(2): 023003.

Vollmer, I., M.J.F. Jenks, M.C.P. Roelands, R.J. White, T. Harmelen, P. Wild, et al. 2020. Beyond mechanical recycling: giving new life to plastic waste. Angew. Chem. Int. Ed. 59(36): 15402–15423.

Wagner, M. 2022. Solutions to plastic pollution: a conceptual framework to tackle a wicked problem. 333–352. *In:* M.S. Bank. (ed.). Microplastic in the Environment: Pattern and Process. Environ. Contam. Remed. Manag. Springer, Cham, Switzerland.

Wang, Y., M. Jiao, T. Li, R. Li and B. Liu. 2023. Role of mangrove forest in interception of microplastics (MPs): challenges, progress, and prospects. J. Hazard. Mater. 445: 130636.

Wei, X.F., W. Yang and M.S. Hedenqvist. 2024. Plastic pollution amplified by a warming climate. Nat. Commun. 15(1): 2052.

Wu, P., J. Huang, Y. Zheng, Y. Yang, Y. Zhang, F. He, et al. 2019. Environmental occurrences, fate and impacts of microplastics. Ecotoxicol. Environ. Saf. 184: 109612.

Wu, X., C. Zhong, T. Wang, X. Zou, Z. Zang, Q. Li, et al. 2020. Occurrence and distribution of microplastics on recreational beaches of Haichow Bay, China. Environ. Sci. Pollut. Res. 28(5): 6132–6145.

Xie, H., J. Chen, L. Feng, L. He, C. Zhou, P. Hong, et al. 2021. Chemotaxis-selective colonization of mangrove rhizosphere microbes on nine different microplastics. Sci. Total Environ. 752: 142223.

Yadav, S., N. Kataria, P. Khyalia, P.K. Rose, S. Mukherjee, H. Sabherwal, et al. 2023. Recent analytical techniques and potential eco-toxicological impacts of textile fibrous microplastics (FMPs) and associated contaminates: A review. Chemosphere. 326: 138495.

Yoon, J.H., B.H. Kim and K.H. Kim. 2024. Distribution of microplastics in soil by types of land use in metropolitan area of Seoul. Appl. Biol. Chem. 67(1): 15.

Yuan, J., J. Ma, Y. Sun, T. Zhou, Y. Zhao and F. Yu. 2020. Microbial degradation and other environmental aspects of microplastics/plastics. Sci. Total Environ. 715: 136968.

Zhang, Z. and Y. Chen. 2020. Effects of microplastics on wastewater and sewage sludge treatment and their removal: A review. Chem. Eng. J. 382: 122955.

Zhang, K., A.H. Hamidian, A. Tubić, Y. Zhang, J.K.H. Fang, C. Wu, et al. 2021. Understanding plastic degradation and microplastic formation in the environment: A review. Environ. Pollut. 274: 116554.

Zheng, J., Z.M. Png, S.H. Ng, G.X. Tham, E. Ye, S.S. Goh, et al. 2021. Vitrimers: Current research trends and their emerging applications. Mater. Today. 51: 586–625.

Zhou, B., J. Wang, H. Zhang, H. Shi, Y. Fei, S. Huang, et al. 2019. Microplastics in agricultural soils on the coastal plain of Hangzhou Bay, East China: Multiple sources other than plastic mulching film. J. Hazard. Mater. 388: 121814.

Zhou, Q., C. Tu, C. Fu, Y. Li, H. Zhang, K. Xiong, et al. 2020. Characteristics and distribution of microplastics in the coastal mangrove sediments of China. Sci. Total Environ. 703: 134807.

Zhu, Y., C. Romain and C.K. Williams. 2016. Sustainable polymers from renewable resources. Nature. 540: 354–362.

Chapter 8

Comprehensive Approaches to Microplastic Sampling and Analysis in Drinking Water

Anh Tuan Ta[1] and Sandhya Babel*[2]
[1]Department of Sanitary Engineering, Faculty of Public Health, Mahidol University, Bangkok, Thailand-10400
[2]School of Biochemical Engineering and Technology, Sirindhorn International Institute of Technology, Thammasat University, P.O. Box 22, Pathum Thani, Thailand-12121

8.1 INTRODUCTION

Over the last century, plastics have significantly impacted various industries, offering numerous advantages across multiple sectors. Their widespread use is evident in almost every aspect of modern life, from food packaging to urban infrastructure. Nevertheless, the mass production of plastics combined with inadequate waste management practices has led to significant environmental challenges. In 2021, global plastic production exceeded 390 million tonnes (PlasticsEurope 2022). This has led to plastic waste being a major source of river and marine pollution, creating global environmental concern. Plastic debris enters aquatic environments through pathways such as mismanaged disposal practices, uncontrolled littering, industrial discharges, and insufficient waste management infrastructure. Once released, larger plastic items gradually fragment

*For Correspondence: sandhya@siit.tu.ac.th

DOI: 10.1201/9781003531463-8

into microplastics, which persist in the environment for extended periods, often spanning several centuries.

Microplastics, plastic particles measuring less than 5 mm in size, have become a growing environmental concern owing to their widespread distribution across natural and engineered ecosystems. Recent studies have detected microplastics in drinking water sources around the world, raising concerns about their potential impact on human health as they expose humans to direct plastic intake (Kankanige and Babel 2020b; Taghipour et al. 2023; Weber et al. 2021; Zhou et al. 2020). The ingestion of microplastics can lead to physical harm, chemical toxicity, and disruption of biological functions, posing significant risks (Blackburn and Green 2022; Ta et al. 2022). The physical presence of microplastics in drinking water can cause mechanical damage to tissues, while their ability to adsorb and carry harmful chemicals such as heavy metals and pesticides can result in chemical toxicity, including endocrine disruption and carcinogenicity (Ta and Babel 2023a, b). Ingested microplastics can also interfere with cellular functions, leading to oxidative stress, inflammation, and impaired immune responses. Vulnerable populations, such as infants, children, pregnant women, and individuals with compromised immune systems, may be more susceptible to these adverse effects (Sun and Wang 2023; Ta and Babel 2023b). Therefore, understanding the extent of microplastic pollution in drinking water and developing effective mitigation strategies is crucial. By assessing contamination levels and pathways through which microplastics enter drinking water, we can better protect public health and develop targeted interventions to reduce microplastic pollution. Further research and regulatory frameworks are essential to limit microplastic pollution and ensure safe drinking water.

This chapter provides detailed guidance on sampling and analyzing microplastics in drinking water. Various sampling methods offer unique benefits and considerations, including grab and passive sampling. Advanced analytical techniques such as microscopy, spectroscopy, and chromatography are highlighted for their effectiveness in detecting and characterizing microplastics across diverse environments. The chapter also emphasizes the importance of quality assurance and quality control measures to ensure the reliability and reproducibility of results.

8.2 CURRENT ABUNDANCE OF MICROPLASTICS IN DRINKING WATER

Research on microplastics in drinking water has primarily focused on regions such as Europe, Asia, and the Americas. These studies have consistently found microplastics, although the size and abundance varied significantly (Table 8.1). Bottled water samples fall into three categories: locally processed, domestically produced, and provided regionally. Most studies report microplastic levels in bottled water below 400 items/L (Kankanige and Babel 2020b; Makhdoumi et al. 2021; Nizamali et al. 2023; Weisser et al. 2021; Zhou et al. 2020; Zuccarello et al. 2019). Low levels of microplastics (2–23 items/L) were found in 23 brands of bottled water available in China, with a detection limit of 25 μm (Zhou et al. 2020). Another low value was found in bottled water from Iran (Makhdoumi et al. 2021).

Table 8.1 Selected studies on microplastic occurrence and characteristics in drinking water

Study Location	Sampling Method	Extraction Method	Identification	QA/QC	Abundance (Items/L)	Characteristics (Predominance)	Reference
Bottled water							
Glass bottles (Germany)	Volume reduction	NOAA technique and NaOH	FTIR, Micro-FTIR	Basic contamination control; blank samples; internal standards	317 ± 257	11–500 μm, Fragments, PE, PS	Weisser et al. (2021)
PET bottles (Netherlands)	Grab sample	Filtration	ATR-FTIR and LDIR Chemical Imaging	Basic contamination control; blank samples; internal standards	7 ± 3 – 90 ± 109	PET and PU	Nizamali et al. (2023)
PET bottles (Italy)	Grab sample	Filtration	SEM-EDX	Basic contamination control	3.16E + 07 – 1.1E + 08	1.28–4.20	Zuccarello et al. (2019)
Glass and PET bottles (Thailand)	Grab sample	Filtration and Nile Red staining	Optical and fluorescence microscope, FTIR and Raman spectroscopy	Basic contamination control; blank samples	52 ± 4 – 140 ± 19	6.5–20 μm, Fiber, PET	Kankanige and Babel (2020b)
PET bottles (Iran)	Grab sample	Filtration and Rose Bengal staining	Stereomicroscope and microscope FTIR	Basic contamination control	8.5 ± 10.2	Fragments PE	Makhdoumi et al. (2021)
PET bottles (China)	Grab sample	Filtration	Microscope FTIR	Basic contamination control	2 – 23	PET Fiber 100–300 μm	Zhou et al. (2020)

(Contd.)

Study Location	Sampling Method	Extraction Method	Identification	QA/QC	Abundance (Items/L)	Characteristics (Predominance)	Reference
Tap water							
Changsha (China)	Grab sample	NOAA technique	Microscope FTIR and SEM	Basic contamination control; blank samples	267 ± 11 – 405 ± 30	PE, PP Fragment 1–10 μm	Shen et al. (2021)
(Southern Brazil)	Grab sample	KOH 15%, NaCl and Nile Red staining	Fluorescence microscope	–	105.8	Fiber	Ferraz et al. (2020)
Rüsselsheim (Germany)	Volume reduction	HCl	Raman micro-spectroscopy	Basic contamination control; blank samples; internal standards	0	-	Weber et al. (2021)
(Denmark)	Volume reduction	CH_3COOH and C_2H_6O	Stereomicroscopy microscope FTIR	Internal standards	0.31 ± 0.14	PP Fiber	Feld et al. (2021)
Zahedan (Iran)	Volume reduction	NOAA technique	Micro-Raman, FTIR and SEM	Basic contamination control; blank samples; internal standards	0.201 ± 0.113 – 0.206 ± 0.106	PS, PE Fragment <50 μm	Taghipour et al. (2023)
(Thailand)	Grab sample	Filtration and Nile Red staining	Optical and fluorescence microscope and FTIR	Basic contamination control; blank samples	6.0 ± 3.0 – 56.0 ± 14.0	PE, PVC Fiber 6.5–53 μm	Kankanige and Babel (2020a)
(Saudi Arabia)	Grab sample	Filtration	FTIR	Basic contamination control	0.9 ± 1.3	PE 25–500 μm	Almaiman et al. (2021)

PE: Polyethylene, PET: Polyethylene Terephthalate, PP: Polypropylene, PS: Polystyrene, PU: Polyurethane, PVC: Polyvinyl chloride.

High levels of microplastics were found in mineral water supplied in Germany (Weisser et al. 2021). However, a particularly high value (5.42E+07 ± 1.95E+07 items/L) was observed in mineral water in Italy, where microplastics smaller than 10 μm were targeted (Zuccarello et al. 2019).

For tap water, most reviewed studies reported microplastic levels below 10 items per liter (Almaiman et al. 2021; Feld et al. 2021; Taghipour et al. 2023). However, high concentrations (100 or more items per liter) were found in tap water samples from Changsha, China, and Southern Brazil (Ferraz et al. 2020; Shen et al. 2021). In contrast, no microplastics were detected in tap water samples of 0.5–1.3 cubic meters from Rüsselsheim, Germany (Weber et al. 2021). Nonetheless, copper phthalocyanine (a pigment) was present at an average of 113 ± 83 items per cubic meter.

8.3 CURRENT SAMPLING AND ANALYTICAL TECHNIQUES OF MICROPLASTICS IN DRINKING WATER

Microplastics have been reported in drinking water supplies from diverse geographic regions worldwide. Nevertheless, substantial variation exists among studies in terms of sampling strategies and analytical methodologies. This section synthesizes the commonly adopted sampling, processing, and analytical approaches used in investigations of microplastics in drinking water. An overview of sampling methods, sample preparation procedures, and identification and characterization techniques are provided in Table 8.1.

8.3.1 Sampling

Sampling approaches for microplastics in drinking water are generally classified into grabbing sampling and volume-reduction methods. Grab sampling entails collecting the entire water sample without any reduction in volume during collection. In contrast, volume-reduction techniques involve concentrating microplastics by filtering or reducing the sample volume either in situ or during collection. Because microplastics typically occur at low concentrations in drinking water, representative sampling often requires large water volumes, in some cases reaching several hundred liters. Consequently, volume-reduction methods are more frequently applied in drinking water studies, as they allow efficient concentration of microplastics while facilitating sample handling and transport.

8.3.2 Quality Assurance and Quality Control

Quality assurance and quality control measures have been reported in many studies, particularly those conducted in developing regions; however, their application remains inconsistent. Although most investigations describe precautionary steps such as equipment cleaning and the use of protective clothing to minimize contamination, only a limited number include procedural blanks (e.g., deionized water blanks) to evaluate background contamination. Furthermore, the use of internal standards, where known quantities of microplastic polymers are introduced into control samples to assess recovery efficiency, is rarely reported. The absence of both blank controls and recovery tests may substantially undermine data reliability. In addition, the use of plastic sampling containers, nylon meshes, and synthetic laboratory garments can introduce secondary contamination, further affecting the accuracy of reported microplastic concentrations.

8.3.3 Sample Processing

Following sample collection, microplastics must be separated from coexisting organic matter and inorganic particles. One of the most widely adopted protocols is the method developed by the National Oceanic and Atmospheric Administration (NOAA) (Masura et al. 2015). Although initially designed for marine environments, this protocol has been successfully applied to freshwater and drinking water samples.

In water-based samples, organic constituents are commonly removed through oxidative digestion using hydrogen peroxide (H_2O_2), catalyzed by ferrous iron (Fe^{2+}). Following this treatment, density separation is carried out with salt solutions exhibiting densities of approximately 1.2–1.8 g/cm^3, such as zinc chloride, sodium chloride, or sodium iodide. Because the densities of most widely used plastic polymers fall within the range of 0.9–1.4 g/cm^3, lighter plastic particles remain buoyant or suspended, whereas particles with higher densities are separated by settling.

8.3.4 Identification and Characterization Techniques

8.3.4.1 Identification of Microplastics by Visual Sorting

After filtration or sieving, retained particles are enumerated and characterized using a combination of visual and instrumental techniques. Owing to the wide diversity in microplastic size, shape, and color, multiple analytical approaches are often required.

Optical or stereomicroscopy is commonly used to assess particle morphology, size, and color. However, reliable identification becomes challenging for particles smaller than 100 μm or for transparent materials. In addition, non-plastic materials such as cellulose fibers, keratin, rayon, coal ash, or paint fragments may be misidentified as microplastics, leading to false positives.

Fluorescence staining, particularly using Nile Red, has been increasingly employed to enhance the visualization of small microplastics. Nile Red selectively binds to hydrophobic polymer surfaces, allowing particles to fluoresce under appropriate illumination. While this method improves contrast and detection efficiency, it may result in false negatives and has the potential to alter surface properties of microplastics.

Scanning electron microscopy (SEM) provides high-resolution imaging of particle surfaces, enabling detailed observation of microplastic morphology and surface texture. When coupled with energy-dispersive X-ray spectroscopy (EDS), SEM can also provide elemental composition data. However, this technique alone cannot determine polymer-specific chemical composition.

8.3.4.2 Identification of Microplastics by Chemical Composition

FTIR and Raman spectroscopy are the most common analytical methods for identifying microplastic composition. These non-destructive techniques characterize microplastics by analyzing interactions with infrared radiation or laser light.

FTIR generates unique spectra as molecular fingerprints for different polymers by examining chemical bond interactions. It uses two modes: attenuated total reflectance (ATR) for larger particles (>300 μm) and focal plane array (FPA) for particles as small as 50 μm, combining FTIR with optical microscopy.

Raman spectroscopy, which is based on inelastic scattering of laser light, offers higher spatial resolution than FTIR and is capable of detecting particles in the size range of approximately 1–20 μm. This advantage makes Raman spectroscopy particularly suitable for identifying smaller microplastics, although fluorescence interference may limit its applicability in some samples.

8.4 ADVANTAGES AND LIMITATIONS OF MICROPLASTIC SAMPLING AND ANALYSIS TECHNIQUES

8.4.1 Sampling

Table 8.2 highlights the strengths and drawbacks of the microplastic sampling methods employed in the reviewed studies. For drinking water samples, on-site filtration using sieves is an effective technique for capturing microplastics from water. This approach enables the concentration of microplastics from large volumes of water into a reduced and more manageable sample volume, thereby improving analytical efficiency and lowering costs by avoiding the need for specialized or high-cost equipment. However, this method might introduce sampling bias, as smaller microplastics could escape through the sieve, leading to an underestimation of the actual microplastic concentration in the water. On the other hand, grab sampling is known for its simplicity and accessibility, providing real-time results and requiring minimal equipment and training. This makes it a straightforward and economical method for water collection. Despite its advantages, grab sampling has limitations in spatial coverage and may introduce bias, potentially failing to accurately reflect the overall microplastic distribution in the water.

8.4.2 Sample Processing

The NOAA protocol, which is widely adopted in the studies reviewed, relies on the strong oxidative capacity of hydroxyl radicals and is therefore highly effective for organic matter degradation. This approach is versatile and can be applied to a range of sample matrices and contaminant types. Its main advantages include rapid processing, operational simplicity, low cost, and the use of easily accessible reagents. Nevertheless, the efficiency of Fenton-based oxidation can be matrix-dependent, and supplementary treatment steps may be required in certain cases.

Table 8.2 Comparison of microplastic sampling and analysis techniques in river studies

Techniques	Advantages	Disadvantages
Sampling		
Grab sample	Simple and easy to implement; low operational cost; capable of retaining small particles.	Limited representativeness; restricted sample volume; results may be influenced by temporal variability.
On-site filtration technique (volume-reduced sampling)	Reduces sample transport and storage requirements; relatively cost-efficient.	Requires specialized equipment and trained personnel; potential sampling bias.
Manta trawls or plankton nets (volume-reduced sampling)	Integrates microplastic occurrence over time and space; improved representativeness; standardized approach; reduced transport and storage demands.	High equipment cost; operational complexity; skilled operators required.
Samples processing		
NOAA technique	High removal efficiency; adaptable to multiple matrices; rapid processing; simple operation; cost-effective.	Reaction efficiency can vary with sample matrix; additional treatment steps may be necessary.
Enzymatic digestion	Selective removal of organic matter; environmentally benign; preserves microplastic integrity.	Limited enzyme specificity; high cost and limited availability; longer processing time; potential interference from complex organic matrices.
Sample analysis		
FTIR	Reliable polymer identification; non-destructive technique.	Restricted to relatively large particles; requires sample preparation; limited sensitivity at low particle abundance.
Raman	Non-destructive; capable of identifying polymer composition; suitable for small microplastics.	High instrument cost and complexity; susceptibility to background fluorescence and signal interference.
SEM/EDS	High-resolution morphological imaging; elemental composition and mapping; versatile and non-destructive.	Expensive instrumentation; requires specialized expertise; time-intensive; does not directly identify polymer type.

Enzymatic digestion represents an alternative strategy that enables selective removal of organic material while preserving microplastic particles, making it an effective and environmentally benign method. Despite these benefits, its broader application is constrained by the high cost and limited availability of enzymes, extended processing durations, and possible inhibition or interference from complex organic matrices. The key strengths and drawbacks of the principal sample processing methods employed in the reviewed studies are summarized in Table 8.2.

8.4.3 Identification Techniques

As summarized in Table 8.2, visual identification using optical or stereomicroscopy remains a relatively inexpensive and accessible technique for microplastic analysis. However, it cannot provide definitive polymer identification and is highly dependent on the analyst's experience, which increases the risk of subjective interpretation and bias. In addition, the detection of very small or transparent particles is constrained by the optical resolution of the method.

Fluorescence staining techniques improve detection efficiency and enhance selectivity compared with conventional visual inspection. Despite these advantages, they are associated with several limitations, including the potential for false-positive results and more complex sample preparation procedures. Furthermore, fluorescence-based methods may not apply to all polymer types and require specialized equipment and trained personnel.

FTIR spectroscopy enables non-destructive determination of polymer composition, but its applicability is generally restricted to larger particles, and it may lack sufficient sensitivity for microplastics present at low concentrations. Raman spectroscopy, although also non-destructive and capable of analyzing smaller particles, can be affected by background fluorescence and typically requires advanced expertise and high-cost instrumentation.

Scanning electron microscopy combined with energy-dispersive X-ray spectroscopy (SEM/EDS) enables high-resolution visualization of particle surface features while simultaneously providing information on their elemental composition. Nevertheless, this technique involves expensive equipment, substantial analytical time, and specialized technical skills, particularly when processing large numbers of samples.

8.5 SUGGESTED PROCEDURE FOR SAMPLING AND ANALYSIS OF MICROPLASTICS IN DRINKING WATER

The assessment of microplastics in drinking water involves a multifaceted workflow that requires careful consideration of environmental and analytical constraints specific to potable water systems. Numerous studies have evaluated and compared sampling and analytical techniques for microplastics in drinking water and aquatic biota with the aim of improving methodological consistency and reliability. The workflow outlined below represents one of the most widely applied approaches, selected based on practical criteria such as cost efficiency and the accessibility of analytical equipment. While methodological adaptations may be required to accommodate the physicochemical characteristics of individual water sources, the proposed procedure provides a structured and reproducible framework for accurate and reliable microplastic analysis in drinking water. An overview of the complete sampling and analytical process is presented in Figure 8.1.

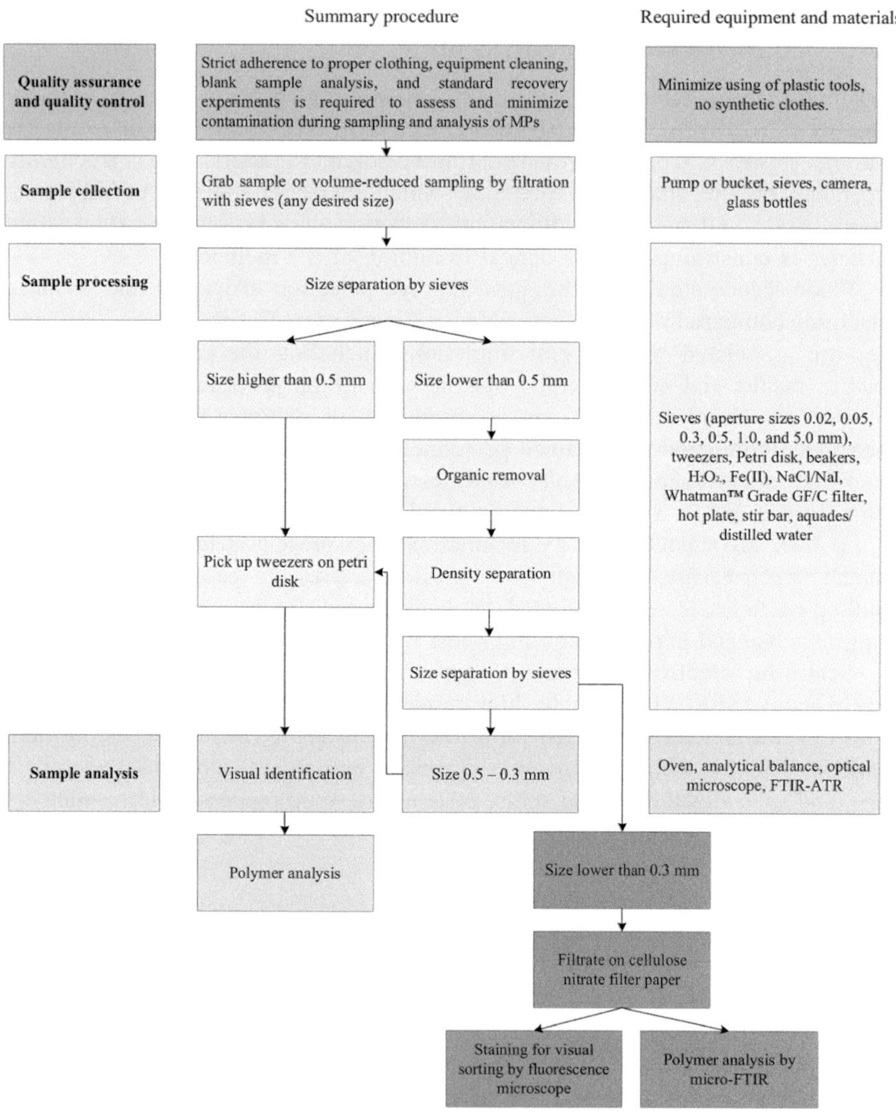

Figure 8.1 Sampling and analysis procedure of microplastics in drinking water.

8.5.1 Quality Assurance and Quality Control

8.5.1.1 Preparation and Cleaning

Quality Assurance (QA) and Quality Control (QC) are crucial in the sampling and analysis of microplastics to ensure accurate and reliable results while adhering to established standards. To avoid cross-contamination, using non-plastic equipment and instruments that have been thoroughly cleaned is vital. Furthermore, wearing synthetic clothing should be avoided during this process. The following guidelines outline the recommended procedures for preparation and cleaning:

Preparation:

- Always wear a clean white cotton lab coat and nitrile gloves.
- Choose clothing made from natural fibers, avoiding synthetic materials, even those worn underneath the lab coat.
- Minimize air movement by keeping windows and doors closed.

Cleaning:

- All equipment should be thoroughly cleaned with 70% ethanol and subsequently rinsed three times with distilled water.
- Following cleaning, glassware and sampling equipment should be covered with aluminum foil to minimize the risk of airborne contamination.
- Work surfaces should be cleaned by wiping with 70% ethanol at least three times prior to commencing any laboratory procedures.
- Petri dishes, filter papers, and forceps should be examined under a stereomicroscope before use to ensure the absence of extraneous particles.

8.5.1.2 Standard Recovery Experiment

Following sample collection, all samples should be carefully handled, clearly labeled, and stored in appropriate containers to preserve sample integrity and prevent contamination or loss. In the laboratory, evaluation of method recovery is a critical quality control step, as it reflects the efficiency of the extraction and analytical procedures in isolating microplastics from the sample matrix. Recovery performance is commonly assessed by spiking blank samples with known quantities of microplastic particles and subsequently determining the proportion recovered after processing.

To determine the recovery rate, compare the results from these spiked samples with the expected values. This comparison allows you to calculate the recovery rate, thereby validating the efficiency of the extraction and analysis procedures. Acceptable recovery rates in microplastic analysis depend on the analytical technique employed, the characteristics of the sample matrix, and applicable regulatory or quality assurance criteria. However, previous studies suggest a recovery rate above 80% is generally considered satisfactory, ensuring accurate and reliable results (Dimante-Deimantovica et al. 2022; Ta and Promchan 2024; Weber and Kerpen 2023). The following procedure is recommended for conducting the standard recovery experiment.

- **Preparation of Microplastic Standards:** Acquire microplastic particles that closely match the size, morphology, and polymer composition of those expected in the samples. These particles may be purchased from commercial suppliers or produced in the laboratory. Prepare a set of calibration standards covering a range of known microplastic concentrations from low to high levels.
- **Spike the Blank Matrix with Microplastics:** Introduce a known quantity of microplastics into the blank matrix and mix thoroughly to ensure uniform distribution. Ensure that the added microplastics fall within the concentration range of the prepared standards.
- **Extraction of the Spiked Matrix:** Process the spiked samples using the same extraction protocol applied to the actual samples. This may include chemical or enzymatic digestion, as well as physical separation techniques such as filtration, centrifugation, or density-based flotation.
- **Analysis of Recovered Microplastics:** Analyze the extracted microplastic particles using the same analytical techniques applied to the sample matrix. Depending on the study design, this may include microscopic examination, spectroscopic identification, or chromatographic analysis.
- **Calculation of Recovery Efficiency:** Recovery efficiency should be calculated as the ratio of the number of microplastic particles recovered from the spiked matrix to the number initially added, expressed as a percentage.
- **Evaluate Results:** The acceptable recovery efficiency depends on research objectives and sample type. If recovery rates fall below 80%, review the analysis procedures, identify potential errors, and make necessary adjustments before proceeding with the sample analysis.

8.5.1.3 Determination of Blank Values

Blank samples should be processed in parallel with environmental samples to evaluate potential contamination arising from sampling, handling, and laboratory procedures. Blanks must undergo identical storage, preparation, and analytical steps as the regular samples. Any microplastic particles detected in the blanks should be accounted for and deducted from the corresponding sample results. This correction helps ensure that reported concentrations accurately represent the microplastics present in the sampled environment. The procedure below describes the approach used to determine blank values.

- **Preparation of Blank Samples:** Prepare blanks that are free from microplastics and external contaminants. Suitable blank matrices may include distilled or ultrapure water, or sediment that has been pre-treated by sieving and heating at 150°C to remove residual plastic particles.
- Blank samples should be processed in parallel with environmental samples to evaluate potential contamination arising from sampling, handling, and laboratory procedures. Blanks must undergo identical

storage, preparation, and analytical steps as the regular samples. Any microplastic particles detected in the blanks should be accounted for and deducted from the corresponding sample results. This correction helps ensure that reported concentrations accurately represent the microplastics present in the sampled environment. The procedure below describes the approach used to determine blank values.

- Process the blank samples using the same procedures applied to the actual samples. This involves filtering, extracting, and analyzing these control samples.
- Analyze the blank samples for microplastics using the same techniques used for the actual samples.
- Compare the results from the blank samples with those from the actual samples to verify that any microplastics detected in the samples are not due to contamination or external sources.
- The control procedure should be repeated at regular intervals throughout the analytical workflow to ensure consistency and to verify the continued validity of the results.

8.5.2 Drinking Water Sampling

Drinking water samples may be collected using either grab sampling or volume-reduction approaches. The selection of an appropriate sampling strategy depends on several considerations, including spatial and temporal variability in microplastic occurrence, the requirement for high-resolution data, the detection limits of the analytical techniques employed, and the availability of sampling resources and equipment. Grab sampling is generally suitable for rapid screening to confirm the presence of microplastics, whereas volume-reduction methods are often required when concentrations are low or when quantitative assessment of microplastic abundance is needed. The procedure outlined below describes the steps used for sampling microplastics in tap water.

Choose a tap representing the area you want to study for tap water. Avoid taps with filters or those not used for a while.

Grab Sampling Method (Figure 8.2)

- Open the tap fully and let the water run for a few seconds to flush out any residual water in the pipes. Then, turn the tap down to a slow flow rate.
- Hold the container directly under the tap, with the opening facing upwards, and collect the water sample by filling it to the brim without any air bubbles.
- Once the container is full, carefully remove it from the tap and place the lid on tightly to avoid contamination.
- Label the container with the date, time, and location of the sampling, as well as any other relevant information.

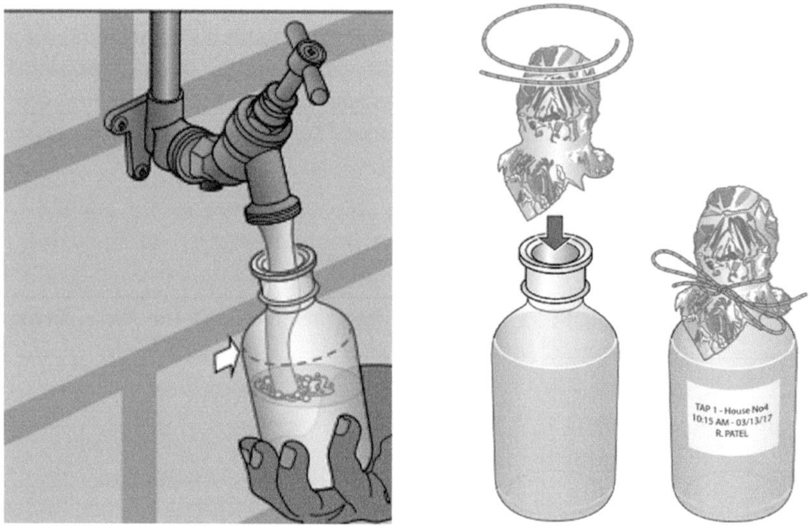

Figure 8.2 Grab sampling method for tap water.

Direct Filtration (Figure 8.3)

- Select a sieve appropriate for the size range of the microplastics you are interested in.
- Clean the sieve thoroughly to remove any debris or contaminants interfering with the microplastics analysis.
- Pour the water sample through the sieve and collect the filtrate in a container. The microplastics will be trapped on the sieve, while the filtered water will pass through.

Figure 8.3 Direct filtration of tap water.

- Rinse the sieve with deionized water to remove any remaining microplastics and transfer the rinse water to the container with the filtrate.
- Label the container with the date, time, and location of the sampling, as well as any other relevant information (Table 8.3).

Table 8.3 Examples of a datasheet for the collection of tap samples

SITE INFORMATION			
STATION DETAILS			
Sampling location			
Name of sampler			
Sampling date			
Latitude (decimal degrees)			
Longitude (decimal degrees)			
Sampling start time			
Initial flow meter			
Sampling end time			
Final flow meter reading			
Notes			

8.5.3 Sample Processing in the Laboratory

8.5.3.1 Analysis of Microplastics in Samples from WTPs

The overall procedure for analyzing microplastics in samples from WTPs is presented in Figure 8.1.

Size Separation by Sieves

- Following the implementation of contamination prevention procedures, samples are sequentially passed through stainless-steel sieves with mesh sizes of 0.05, 0.3, 0.5, 1.0, and 5.0 mm.
- Deionized water should be used to rinse all solids from the sieves.
- Material retained on the 5.0 mm sieve, including large organic debris such as leaves, insects, macroalgae, and wood fragments, should be gently rinsed with deionized water and manually removed using metal tweezers.
- This sieving process yields four particle size fractions: 0.05–0.3 mm, 0.3–0.5 mm, 0.5–1.0 mm, and 1.0–5.0 mm.
- Each resulting size fraction is collected independently and transferred into separate glass beakers using deionized water.

Particles larger than 0.5 mm:

- Particles within the 0.5–1.0 mm and 1.0–5.0 mm size fractions are suitable for visual inspection.
- Suspected plastic debris in these fractions should be manually isolated on Petri dishes using stainless-steel tweezers (Figure 8.4).

- The isolated particles are dried in an oven at 60°C for 24 hours, after which their total dry mass is measured.
- An optical microscope inspects suspected plastics.

Figure 8.4 Picking particles with a size larger than 0.5 mm.

Particles smaller than 0.5 mm:

Size fractions of 0.05–0.3 mm and 0.3–0.5 mm are challenging to assess visually due to the presence of substantial sediment and organic matter. To facilitate microplastic identification, organic material should be removed using 30% hydrogen peroxide in the presence of ferrous iron (Fe^{2+}) as a catalyst.

Organic Removal (Figure 8.5)

The procedure for organic removal is as follows:

- Add 20 mL of ferrous iron (Fe^{2+}) solution and 20 mL of 30% hydrogen peroxide to each beaker containing the sample.
- Allow the mixture to stand at room temperature for approximately 5 minutes before further treatment.
- Insert a magnetic stir bar into the beaker and cover it with a watch glass to reduce the risk of contamination and splashing.
- Heat the beakers on a hotplate to 60°C.
- When vigorous bubbling is observed, remove the beaker from the hotplate and place it in a fume hood until the reaction subsides. If excessive foaming or overflow is anticipated, add distilled water to moderate the reaction rate.
- Reheat the samples to 60°C and maintain this temperature for an additional 30 minutes.
- If visible organic residues persist, an additional 20 mL of 30% hydrogen peroxide may be introduced and the digestion procedure repeated.
- Repeat the digestion procedure as necessary until all visible natural organic material has been removed.

Figure 8.5 Organic removal of microplastic samples by hydrogen peroxide.

Density Separation

Following organic matter digestion, residual sediment and remaining organic particles can be separated using sodium iodide (NaI) solution with a density of 1.5 g/cm^3. The density separation procedure is conducted as follows:

- Prior to density separation, samples are transferred to clean beakers and dried in an oven at 60°C.
- A custom-built density separation device is prepared, consisting of a glass funnel connected to a silicone tube fitted with a control valve at the base (Figure 8.6).
- Approximately 200 mL of sodium iodide solution is added to each beaker containing the dried sample.
- The sample mixture is agitated with a magnetic stirrer for 20–30 minutes to achieve uniform particle suspension.
- The resulting suspension is carefully transferred into the glass funnels and allowed to settle undisturbed for 12–24 hours.

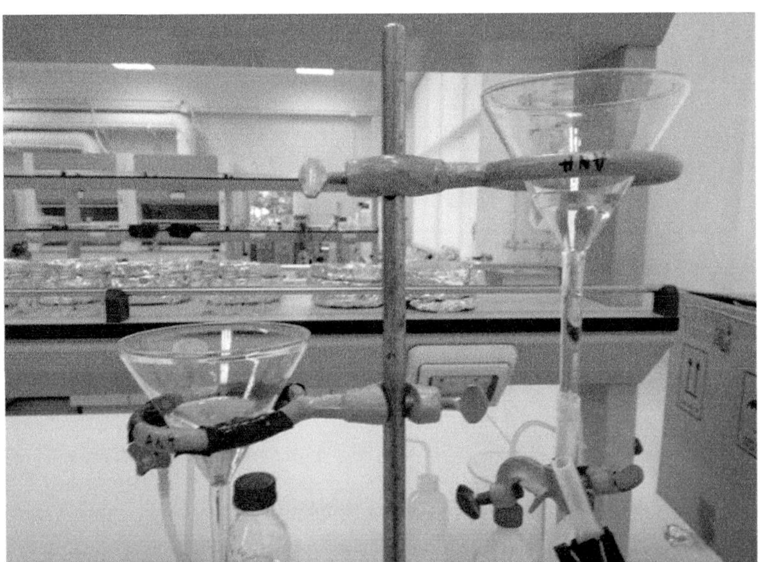

Figure 8.6 The customized apparatus for density separation.

- Low-density particles, including microplastics, remain floating at the surface, while heavier materials settle at the bottom.
- The valve at the base of the funnel is slowly opened to remove settled heavy particles without disturbing the floating fraction.
- Floating particles are gently rinsed from the funnel into clean glass beakers using deionized water.
- The recovered particles are sequentially passed through stainless-steel sieves with mesh sizes of 0.05 and 0.3 mm, using deionized water for rinsing.
- This procedure divides the particles into two size classes: 0.05–0.3 mm and 0.3–0.5 mm.
- Particles in the 0.3–0.5 mm fraction are filtered onto Whatman No. 1 filter paper and dried at 60°C.
- For the 0.05–0.3 mm size fraction, particles are placed in a beaker with 100 mL of deionized water and gently stirred for 5 minutes to obtain a homogeneous suspension.
- The suspension is then divided into two equal volumes using a graduated cylinder.
- Each aliquot is filtered through a 0.45 μm cellulose nitrate membrane filter (Whatman) under vacuum and dried at 60°C.
- One set of dried filters is subjected to Nile Red staining and analyzed by fluorescence microscopy, while the remaining set is examined using micro-FTIR spectroscopy.

Optical Microscope (Figure 8.7)

After drying, the size fraction of 0.3–0.5 mm, 0.5–1.0 mm, and 1.0–5.0 mm are inspected by an optical microscope.

- The microscope inspects parts at 10x and 20x magnification, depending on their size.

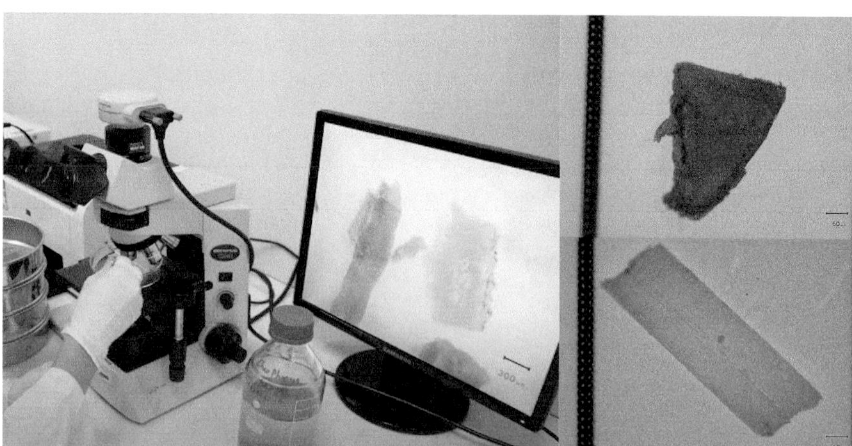

Figure 8.7 Visual sorting of microplastics by an optical microscope.

- Suspected particles are manually isolated using tweezers and placed on the microscope stage for examination.
- The abundance, color, and morphological characteristics of microplastics are then observed and systematically recorded under the microscope.

Polymer Type Identification (Figure 8.8)

Fourier-transform infrared spectroscopy

Following preliminary sorting under an optical microscope, particles within the 0.3–0.5 mm, 0.5–1.0 mm, and 1.0–5.0 mm size fractions are subjected to polymer identification using FTIR spectroscopy. The analytical procedure is conducted as outlined below:

- Configure the FTIR instrument to operate in the mid-infrared range (4,000–400 cm^{-1}) with a spectral resolution of 8 cm^{-1} and an accumulation of 8 scans per measurement.
- Clean the diamond ATR crystal thoroughly using Kimwipes moistened with ethanol to remove residual contaminants.
- Acquire a background spectrum prior to each sample measurement under identical instrumental conditions.
- Place the microplastic particle onto the ATR crystal and apply gentle pressure using tweezers or a spatula to ensure optimal contact.
- Perform background correction by subtracting the background spectrum from the sample spectrum using the instrument's compatible software.
- Identify polymer composition by comparing the resulting spectra with established reference spectral libraries.
- Polymer identification is confirmed when the spectral match factor is equal to or greater than 70.

Nile Red Staining and Fluorescence Microscope (Particles Smaller than 0.3 mm)

The organic digestion step may not be fully effective for particles in the 0.05–0.3 mm size fraction. Consequently, membrane filters containing particles smaller than 0.05 mm were subjected to Nile Red (NR) staining for fluorescence-based detection. Nile Red powder was dissolved in chloroform to obtain a working concentration of 5 mg/L (Tamminga et al. 2017). The staining and fluorescence microscopy procedure is described below.

- Once the membrane filters containing the samples are completely dried, 1–2 mL of the Nile Red solution is carefully applied to each filter.
- The filters are then covered with aluminum foil to minimize contamination and prevent particle loss.
- Stained filters are stored in the dark for 5–6 hours to allow adequate dye–polymer interaction.
- Fluorescence analysis is performed using a fluorescence microscope equipped with a DAPI excitation filter (390/18 nm) and a DAPI emission filter (435/48 nm).

- Fluorescent particles are subsequently examined, and the abundance and morphological characteristics of microplastics are recorded.

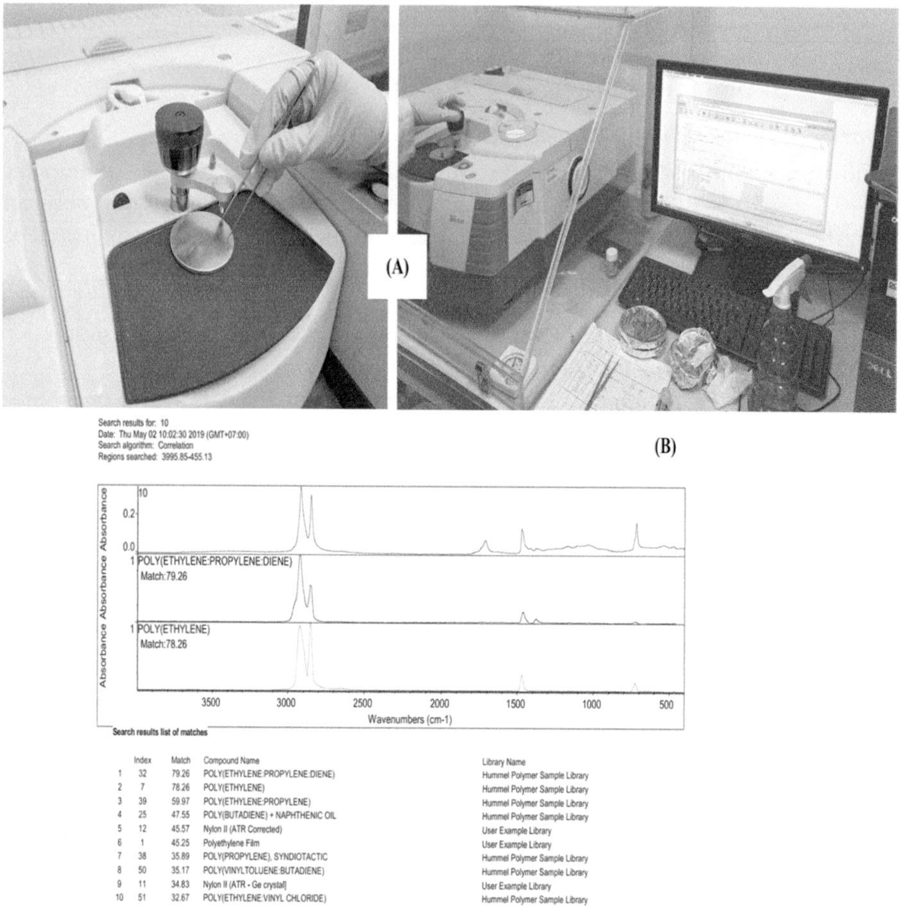

Figure 8.8 (A) FTIR analysis of microplastic particles; (B) Representative FTIR spectra of microplastic samples compared with reference spectra from polymer libraries.

Micro-Fourier-Transform Infrared Spectroscopy (Figure 8.9)

Micro-FTIR is a technique similar to ATR-FTIR that is used to analyze microplastics in small sizes. The procedure for micro-FTIR analysis is as follows:

- **Instrument Calibration:** Prior to analysis, the FTIR imaging microscope should be calibrated to ensure measurement accuracy and reproducibility.
- **Spectral Imaging:** The sample is placed on the microscope stage, and the area of interest is brought into focus. Spectral data are then acquired across multiple points on the sample surface using FTIR imaging, typically within the mid-infrared region.

- **Data Processing and Interpretation:** Acquired spectra and chemical images are processed using the manufacturer-provided Micro-FTIR software to determine polymer composition and spatial distribution of microplastics. Polymer identification is achieved through comparison with reference spectral libraries, while chemical imaging is used to assess the spatial arrangement of microplastic particles within the sample.

It should be noted that specific operational steps may vary depending on the FTIR imaging system, associated software, and characteristics of the microplastic samples. Accordingly, analyses should be performed in accordance with the manufacturer's guidelines and, where possible, under the supervision or guidance of experienced analysts.

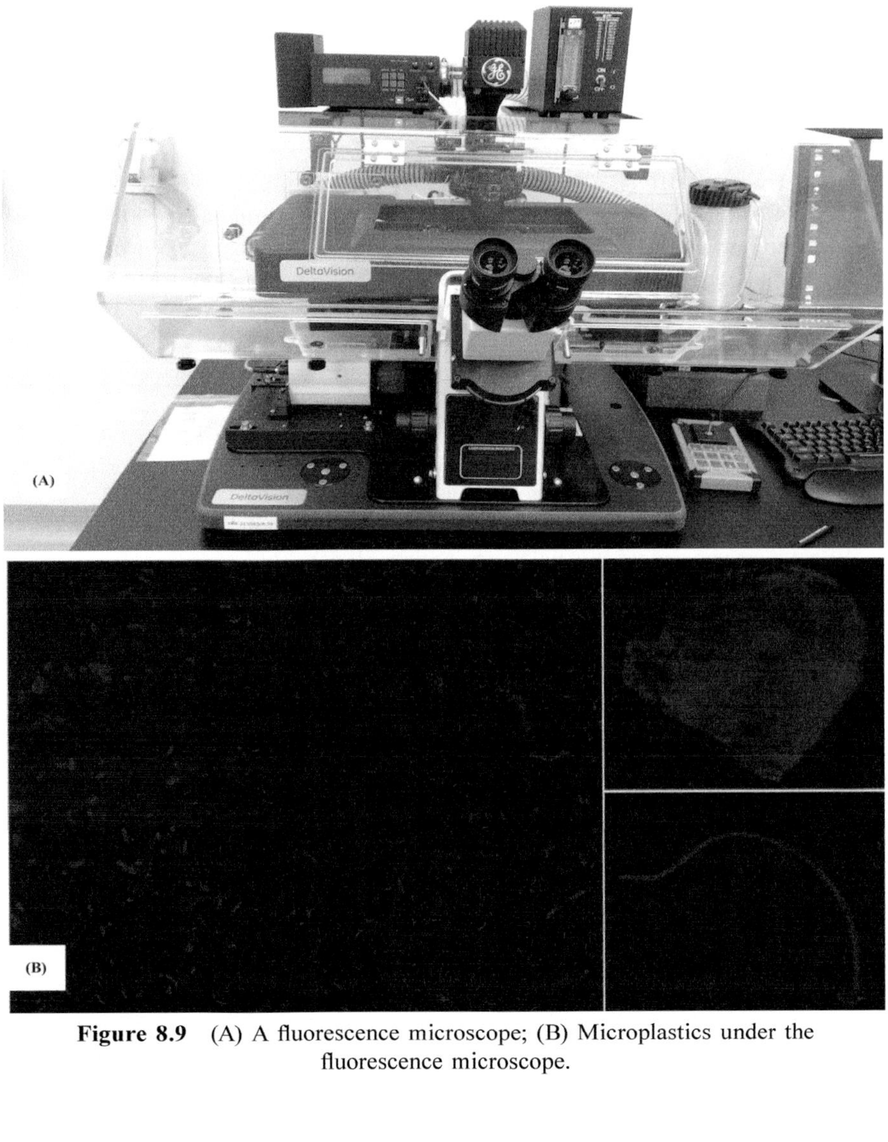

Figure 8.9 (A) A fluorescence microscope; (B) Microplastics under the fluorescence microscope.

8.6 CONCLUSION

Microplastics have emerged as a pervasive and concerning contaminant in drinking water systems worldwide, with potential implications for human health and environmental integrity. This chapter has outlined the current state of microplastic contamination in drinking water, emphasizing the variability in concentration and polymer types across regions and water sources. A detailed overview of sampling and analytical techniques has been presented, highlighting their advantages, limitations, and practical applications in both field and laboratory settings.

Standardized protocols incorporating quality assurance and control are essential to ensure the reliability and comparability of data. Methods such as grab and volume-reduced sampling, the NOAA protocol for organic removal, and advanced analytical tools like FTIR, Raman spectroscopy, and fluorescence microscopy provide a robust framework for accurately detecting and characterizing microplastics.

To advance research and public health protection, improving methodological consistency, validating techniques through recovery experiments, and developing cost-effective, sensitive, and scalable analytical approaches is critical. These efforts will support formulating regulatory standards and mitigation strategies to safeguard drinking water quality and reduce microplastic exposure risks in human populations.

REFERENCES

Almaiman, L., A. Aljomah, M. Bineid, F.M. Aljeldah, F. Aldawsari, B. Liebmann, et al. 2021. The occurrence and dietary intake related to the presence of microplastics in drinking water in Saudi Arabia. Environ. Monit. Assess. 193: 390.

Blackburn, K. and D. Green. 2022. The potential effects of microplastics on human health: What is known and what is unknown. Ambio. 51: 518–530.

Dimante-Deimantovica, I., N. Suhareva, M. Barone, I. Putna-Nimane and J. Aigars. 2022. Hide-and-seek: Threshold values and contribution towards better understanding of recovery rate in microplastic research. MethodsX. 9: 101603. https://doi.org/10.1016/j.mex.2021.101603.

Feld, L., V.H.d. Silva, F. Murphy, N.B. Hartmann and J. Strand. 2021. A study of microplastic particles in Danish tap water. Water. 13: 2097.

Ferraz, M., A.L. Bauer, V.H. Valiati and U.H. Schulz. 2020. Microplastic concentrations in raw and drinking water in the Sinos River, Southern Brazil. Water. 12: 3115.

Kankanige, D. and S. Babel. 2020a. Identification of Micro-plastics (MPs) in Conventional Tap Water Sourced from Thailand. J. Eng. Technol. Sci. 52.

Kankanige, D. and S. Babel. 2020b. Smaller-sized micro-plastics (MPs) contamination in single-use PET-bottled water in Thailand. Sci. Total Environ. 717: 137232.

Makhdoumi, P., A.A. Amin, H. Karimi, M. Pirsaheb, H. Kim and H. Hossini. 2021. Occurrence of microplastic particles in the most popular Iranian bottled mineral water brands and an assessment of human exposure. J. Water Process Eng. 39: 101708.

Masura, J., J.E. Baker, G.D. Foster, C. Arthur and C. Herring, C. 2015. Laboratory methods for the analysis of microplastics in the marine environment: Recommendations for quantifying synthetic particles in waters and sediments.

Nizamali, J., S.M. Mintenig and A.A. Koelmans. 2023. Assessing microplastic characteristics in bottled drinking water and air deposition samples using laser direct infrared imaging. J. Hazard. Mater. 441: 129942.

PlasticsEurope. 2022. Plastics – the fast Facts 2022. Plastics Europe, Brussels, Belgium.

Shen, M., Z. Zeng, X. Wen, X. Ren, G. Zeng, Y. Zhang, et al. 2021. Presence of microplastics in drinking water from freshwater sources: the investigation in Changsha, China. Environ. Sci. Pollut. Res. 28: 42313–42324.

Sun, A. and W-X. Wang. 2023. Human exposure to microplastics and its associated health risks. Environ Health. 1: 139–149.

Ta, A.T., P. Pupuang, S. Babel and L.P. Wang. 2022. Investigation of microplastic contamination in blood cockles and green mussels from selected aquaculture farms and markets in Thailand. Chemosphere. 303: 134918. https://doi.org/10.1016/j.chemosphere.2022.134918.

Ta, A.T. and S. Babel. 2023a. Microplastics and heavy metals in a tropical river: Understanding spatial and seasonal trends and developing response strategies using DPSIR framework. Sci. Total Environ. 897: 165405. https://doi.org/10.1016/j.scitotenv.2023.165405.

Ta, A.T. and S. Babel. 2023b. Occurrence and spatial distribution of microplastic contaminated with heavy metals in a tropical river: Effect of land use and population density. Mar. Pollut. Bull. 191: 114919. https://doi.org/10.1016/j.marpolbul.2023.114919.

Ta, A.T. and N. Promchan. 2024. Microplastics in wastewater from developing countries: A comprehensive review and methodology suggestions. TrAC, Trends Anal. Chem. 117537.

Taghipour, H., M. Ghayebzadeh, F. Ganji, S. Mousavi and N. Azizi. 2023. Tracking microplastics contamination in drinking water in Zahedan, Iran: From source to consumption taps. Sci. Total Environ. 872: 162121.

Tamminga, M., E. Hengstmann and E. Fischer. 2017. Nile Red staining as a subsidiary method for microplastic quantification: a comparison of three solvents and factors influencing application reliability. SDRP J. Earth Sci. Environ. Stud. 2: 165–172.

Weber, F., J. Kerpen, S. Wolff, R. Langer and V. Eschweiler. 2021. Investigation of microplastics contamination in drinking water of a German city. Sci. Total Environ. 755: 143421.

Weber, F. and J. Kerpen. 2023. Underestimating microplastics? Quantification of the recovery rate of microplastic particles including sampling, sample preparation, subsampling, and detection using μ-Ramanspectroscopy. Anal. Bioanal. Chem. 415: 2963–2973. https://doi.org/10.1007/s00216-022-04447-z.

Weisser, J., I. Beer, B. Hufnagl, T. Hofmann, H. Lohninger, N.P. Ivleva, et al. 2021. From the well to the bottle: Identifying sources of microplastics in mineral water. Water. 13: 841.

Zhou, X., J. Wang, H. Li, H. Zhang, H. Jiang and D. Zhang. 2020. Microplastic pollution of bottled water in China. J. Water. Process. Eng. 40: 101884.

Zuccarello, P., M. Ferrante, A. Cristaldi, C. Copat, A. Grasso, D. Sangregorio, et al. 2019. Exposure to microplastics (<10 μm) associated to plastic bottles mineral water consumption: The first quantitative study. Water Res. 157: 365–371.

Chapter 9

Application of Recycled Plastic Waste in the Greenest Freeway in Australia

Salman Shooshtarian*, Tayyab Maqsood and Peter S.P. Wong
School of Property, Construction and Project Management, RMIT University, Melbourne, Australia

9.1 INTRODUCTION

The Mordialloc Freeway Project (MFP) in Victoria, Australia, comprises a 9 km freeway link between Dingley Bypass and the Mornington Peninsula Freeway, including several grade-separated interchanges, six new road bridges over wetlands and Mordialloc Creek, and over 8 km of parallel shared user path along the alignment. The MFP was built between 2019 and 2021 using the delivery method design and build (D&C). This Major Road Projects Victoria (MRPV) freeway was constructed by McConnell Dowell and Decmil Joint Venture (MCDDJV). The partnership aimed to deliver a world-first freeway (Figure 9.1), one that would incorporate technological advancements in sustainability, delivering not only a long-awaited stretch of road but also contributing to the state's waste reduction goals.

The project has delivered several social, economic, and environmental benefits. In terms of the traffic outcome, according to MPRV prediction modelling, the project will save 7 and 10 minutes in peak morning and evening traffic, respectively. It will also reduce 60–70 per cent of the traffic load on Wells Rd

*For Correspondence: salman.shooshtarian@rmit.edu.au

DOI: 10.1201/9781003531463-9

Figure 9.1 Construction of Mordialloc Freeway in Melbourne, Victoria, Australia. (*Source:* Victoria Big Build 2022).

and Boundary Rd by 2031. To drive the social outcome, the project team also implemented many social procurement initiatives and achieved the following:

- 92,702 total hr for local content and employment outcomes;
- 49,242 total hr for Aboriginal employment; and
- 71,068 total hr disadvantaged Victorian commitment.

To date, the project has been recognised for its environmental sustainability, contractor performance, and design. As can be seen in Figure 9.2, the project has received awards between 2021 and 2022.

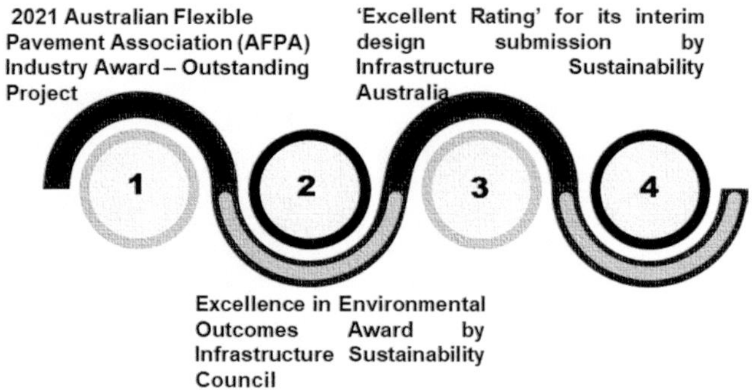

Figure 9.2 Project sustainability accreditations and awards achieved to date. (*Source:* Authors).

9.2 APPLICATION OF RECYCLED MATERIALS IN THE CASE PROJECT

The project had significant environmental impacts. Since the project's completion, the MFP has been known as 'Australia's greenest freeway.' This claim is mainly supported by a collective of implemented sustainability initiatives in response to Victoria's waste crisis, as highlighted in the report *Recycling Victoria: A New Economy* (Victoria State Government 2022). The project saved over 300,000 tonnes of materials from landfills or 3 ha of land space that may be required for stockpiling. The use of these products compared to comparable products on the market has saved over 4,000 t of greenhouse gas emissions.

The Recycled materials quantity, type, and application areas in this project include the following:

- 75% recycled plastic rotationally moulded panel (RMP) noise walls (10 km of noise walls using 600 tonnes of plastic waste equalling 32,000 m^2)
- 100% recycled subbase pavement materials and 44% recycled materials in asphalt pavements (over 270,000 t of pavement material incorporating the maximum allowable recycled content)
- 10% crushed glass and 30% reclaimed asphalt pavement in pavements
- 100% recycled polypropylene plastic (PP) concrete reinforcing mesh (30 t of plastic waste)
- 100% recycled high-density polyethene (HDPE) stormwater drainage pipe (4.6 km installed, totalling 75 t of plastic waste)

The project applied 407 VicRoads Standard Specifications (VicRoads 2022) to ensure the quality of recycled materials to be used in this project met the minimum acceptable range. This specifications document was updated in collaboration between the industry and government agencies to allow for more recycled asphalt products to be used in all pavements in the future.

9.2.1 Recycled Plastic Rotationally Moulded Panel (RMP) Noise Wall

The use of recycled plastic RMO noise walls in this project occurred for the first time in the world. These walls were built using over 570 tonnes of plastic waste (Figure 9.3), with half of it sourced from kerbside recycling, including items like milk and soft drink bottles (Victoria State Government 2021). The remaining recycled content consisted of soft plastics, such as bread bags, food wrappers, and bubble wrap, which are challenging to recycle and often end up in landfills.

Figure 9.3 Recycled plastic RMP noise walls used in MFP, Melbourne, Australia. (*Source:* Victoria Big Build 2022).

In a technical note drafted by the National Transport Research Organisation (NTRO), four major advantages of using recycled plastic noise walls over other more traditional materials (including concrete, timber, and still) were highlighted (NTRO 2022). These include RMP being lighter and safer to install and maintain. They can be manufactured in any colour or pattern, possess better acoustic performance due to the void created within the panel, and are made using recycled post-consumer plastic. Conversely, additional considerations for RMP noise walls include factors such as impact resistance, UV stability, fire resistance, light reflection, graffiti resistance, and various other criteria.

9.3 RESEARCH APPROACH

A case study analysis was conducted to explore the process of utilising plastic waste in the building and construction sector. The selected case project, MFP, met the two primary selection criteria: significant use of recycled materials and recent or imminent completion. After identifying the project, the perceptions of four stakeholder groups playing an essential role in using recycled materials were sought. These groups included the designer, client, head contractor, and supplier (Figure 9.4).

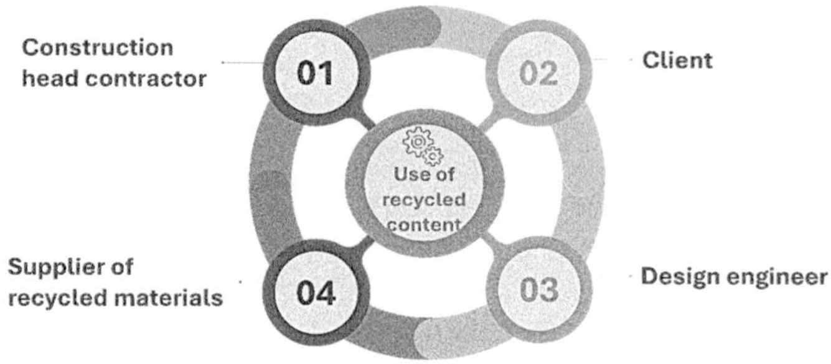

Figure 9.4 Stakeholders who were interviewed in the case study. (*Source:* Authors).

During the interviews, the participants responded to questions on their demographic details, working experience, and relevant expertise in the use of recycled materials in the construction project. Furthermore, they were asked about the main motivations, challenges, and opportunities for the increased application of these materials in the case project.

9.4 PARTICIPANTS' PROFILE

The interviewees were experienced individuals with a significant track record of involvement in building public infrastructure in Victoria (Table 9.1). The client representative (P_1) was extensively engaged in the Mordialloc Freeway Project in two capacities: project director at the pre-construction and procurement phases, and senior project manager responsible for the delivery of the southern part of the project. The participant representing the head contractor (P_2) has worked in the building and construction sector for the last 20 years and has experienced different roles, from site engineer and project engineer to project manager in various road infrastructure and mechanical projects. As the project manager, P_2 led a group of

Table 9.1 Summary of participants' profiles in MFP

Stakeholder Type	Experience/Expertise	Participant Code
Client	Senior project manager with extensive experience in project managing public infrastructure projects	P_1
Head-contractor	20 years of experience in the building and construction sector as a site engineer, project engineer, and manager	P_2
Designer	15 years of experience working as a design consultant	P_3
Supplier	Highly experienced corporate communicator with a background in government, corporate, industry, and community organisations, with a four-year employment history in the organisation	P_4

engineers working on different parts of the project. This participant's involvement in C&D waste management was limited to engaging waste operators to take the on-site waste to landfills.

The third research participant (P_3), who represented the design team, had more than 15 years of experience as a design consultant, primarily on road infrastructure projects such as the removal of PTV level crossings and the construction of the West Gate Tunnel. This participant's primary experience with recycled materials was related to the case study project. In this project, P_3 managed the main freeway alignment packages, the freeway crossings with major local roads, and the design coordination in the role of civil design lead. The final participant (P_4) worked as the organisation's communication manager and participated in the project development process at an early stage. While the participant was not directly involved in managing the project's recycled materials supply, they had a good understanding of the organisation's supply activities due to their participation in some communications with other stakeholders.

9.4.1 Organisations' Involvement in the Project

The client collaborated with recycled materials suppliers to increase their application in the project, leading to the revision of materials specifications. The client had meetings with the project's potential contractors and suppliers to review the project scope and requirements. Furthermore, the client project team established the sustainability committee to monitor and report the use of recycled materials in this project. The head contractor, representing a joint venture, was responsible for conducting construction activities according to the design. This organisation also procured recycled materials from various suppliers.

The design team was responsible for the design work mainly in the civil roads' drainage, structural design, geometric highway, traffic management, and alignment of the other roads. The supplier had a key role in suggesting reforms to the existing recycled materials specifications in this project. The main materials they supplied included reclaimed asphalt pavement, crushed glass, crushed rock, and concrete. A summary of the stakeholders' main involvement areas in the project is presented in Table 9.2.

Table 9.2 Main involvement areas of case study organisations in MFP

Organisation	Involvement/Activities
Client	Working with stakeholders to develop new materials specifications and establishing a sustainability committee to monitor the sustainability outcome
Head-contractor	Conducting the construction work according to the design
Design	Providing design work for the civil roads' drainage, structural design, geometric highway, traffic management and alignment of the other roads
Supplier	Working with the client to revise recycled materials specifications, supplying reclaimed asphalt pavement, crushed glass, crushed rock, and concrete

9.4.2 Stakeholders' Interaction Modelling in MFP

Figure 9.5 shows the interaction model between the four study stakeholders in MFP. The modelling was done based on the responses captured during the interviews. The interaction model delineates the collaborative efforts of these stakeholders to incorporate recycled materials effectively into the construction project.

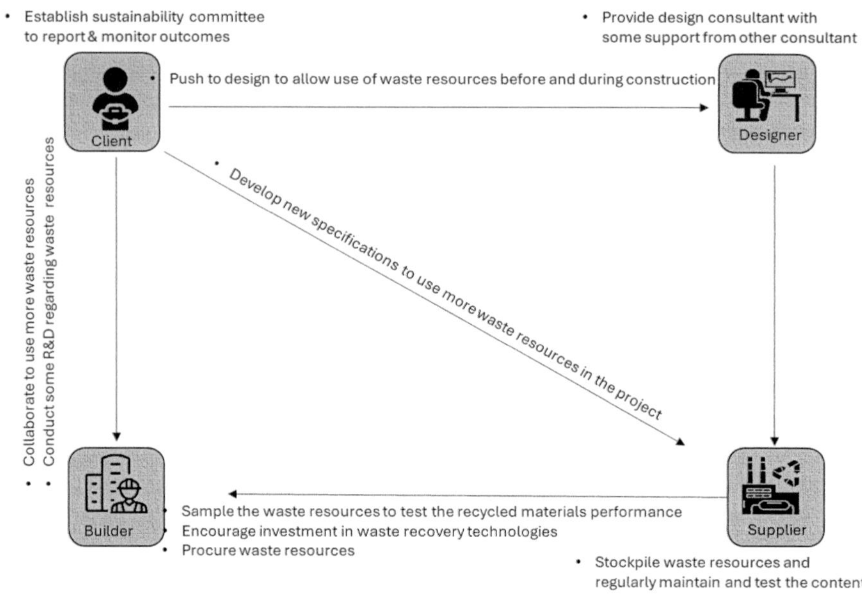

Figure 9.5 Stakeholders' Interaction Model in MFP. (*Source:* Authors).

9.5 QUALITATIVE FINDINGS

In this section of the chapter, we present the qualitative findings derived from the interviews conducted. The responses from the interviewees shed light on both individual and organizational motivations for utilising recycled materials, as detailed in Section 9.5.1. Furthermore, it outlines the primary challenges faced in procuring and applying these resources, as discussed in Section 9.5.2, along with the strategies proposed to overcome these challenges, as elaborated in Section 9.5.3.

9.5.1 Major Motivations for Using Recycled Materials

The four stakeholders interviewed in this case study provided 20 reasons why they have been engaged in this project and why they were motivated to use recycled materials. The most frequent reason was to contribute to the environmental sustainability of the construction activities, which the client, designer and supplier of recycled materials indicated. Figure 9.6 illustrates these motivations for each interviewee.

Figure 9.6 Main motivations for using recycled materials in MFP. (*Source:* Authors).

The client representative reported that support from the organisation's executive management, revising the materials specifications, financial incentives, and the project's aspiration to be green were strong motivations to use recycled materials in this project. P_1 highlighted that their executive management team was a strong supporter of using these materials as it would support recyclers, reduce the quantity of waste landfilling, and contribute to the state's economy:

> *"...to be fair, if I had been just the project team, we might not have done that, but we were supported and encouraged to do that by our executive. And that was done on the basis of supporting recycled materials, but also supporting the Victorian economy".* [P_1]

Furthermore, according to this participant, the DoT specifications were revised in collaboration with the supplier to allow more usage of recycled materials in this project, notably asphalt and plastic noise walls. The client organisation also provided financial incentives for the head contractor to achieve the targets set for using recycled materials in this project, which was based on a 70% ISC rating:

> *"Ultimately, we worked with the DoT and industry...to develop a new specification, and we used it on the Mordialloc freeway project. And that allowed for more recycled asphalt products to be used in all pavements going forward...that opened up a lot more recycled products into the Victorian market. And that was the big deal with the Mordialloc project."* [P_1]

> *"What we also did is we also provided a financial incentive in that contract to achieve a 70 rated ISC outcome."* [P_1]

> *"I think it was a project goal to build the greenest freeway that we could. So there was a motivation, from the organisation, from the individuals who planned that to achieve that...there's benefits in relation to the Victorian economy to provide all the support producers that we're in Victoria. So the recycled first element of that, you know, to reduce the, you know, the, the amount of product going to landfill."* [P_1]

P_2 identified three key factors that serve as the major enablers or motives for using recycled materials in this project: contract obligations, building organisational capacity to work with recycled materials, and preparing to respond to current and future client needs. In the contract awarded to the head contractor organisation, the use of recycled materials was required to meet the ISC sustainability rating requirements. Furthermore, according to this participant, with the emerging trend in the state's public projects to use more recycled materials, the participant's organisation willingly adopted these materials to build the organisational capacity to address the market needs and secure the kinds of projects in which the use of recycled materials is necessary:

> *"Because it's good for business, you need to differentiate yourself from the others, and then you need to respond to your clients to know what they want, what they need out of projects. This is when we sell ourselves for the next project. What we're used to as our selling point is we say, well, look at all this great stuff, we are a company that does this type of thing as part of our business models."* [P_2]

P_3, the representative of the design team, highlighted seven factors as the main motives behind using recycled materials. These include responding to the client's request, providing design solutions that satisfy all stakeholders involved in the project, developing sustainable design, the superior physical characteristics of recycled materials, being less energy intensive, being recyclable, minimising the need for virgin materials, and certification for their quality and performance. The participant explained:

> *"...we definitely sort of wanted to provide a good sort of sustainable design. That was one factor of it. It was just about really keeping the client happy, as well as the contractor, and coming up with a solution that worked for everyone."* [P_3]

This participant emphasised the environmental benefits of using recycled materials in their project. The benefits realised included reducing the need for virgin materials, using materials with lower energy to be produced, and a higher level of recyclability at the end of their service life, which further contributes to a CE in the building and construction sector. The participant expanded on this, stating:

> *"For the noise wall panels, for example, I know the benefits of using the recycling plastics are that they're lightweight, they've got the certification behind them in terms of testing for UV resistance and chemical resistance, they're less sort of energy-intensive to make and manufacture. And then you can also easily replace sections of the wall and recycle them at the end of their useful life as well. So you can do, all those things also with other materials such as concrete and steel. But you're already starting with a recycled product, to begin with, rather than creating something, and then going on to recycle it later down the line."* [P_3]

P_4 also highlighted three main reasons for being involved in this project. These reasons were business purposes, and to make a greater impact, set examples for other smaller council and private construction projects, make procurement of recycled materials business-as-usual, and contribute to lowering the sector's carbon footprint. The participant's organisation had been supplying recycled materials for construction projects for many years, particularly in Victoria and Queensland. Hence, their engagement in this project was primarily based on business grounds:

> *"...we are probably the largest recycler, by weight and volume, in the country; we recover materials to manufacture heavy materials for roads and infrastructure. So this is our bread and butter. And we would always seek to supply major projects with these materials because that's where we can make the biggest impact."* [P_4]

Furthermore, according to P_4, by engaging in this project, they were able to make the greatest impact due to the scale of supply, which further resulted in building confidence in the market to use these materials, particularly in smaller-scale construction projects. This eventually should result in using these materials as business as usual. This project allowed them to set an example for the safe application of recycled materials in the Australian context. Lastly, the use of recycled materials contributes to reduced carbon emissions resulting from construction activities. They explained:

> *"...we want to make sure that major projects especially, are getting, creating, or paving the way for councils and smaller construction, and setting an example. We know that through the use of our materials, these projects can reduce their carbon footprint by up to about 65%. So having these materials used in major projects is a really great way to communicate to the rest of the market that these are really viable materials to use".* [P_4]

> *"...but we want to make our motivation for this...to make this part of business as usual for the entire sector. Purchasing and utilising recycled materials and infrastructure should just be the norm."* [P_4]

9.5.2 Major Challenges to Using Recycled Materials

In response to the question regarding the primary challenges associated with the use of recycled materials, the respondents identified several factors that inhibit the widespread adoption of these products. As illustrated in Figure 9.7, unsupportive

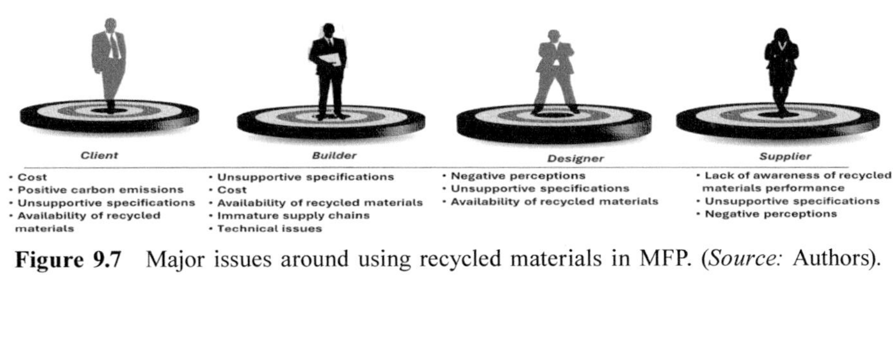

Figure 9.7 Major issues around using recycled materials in MFP. (*Source:* Authors).

specifications and the availability of recycled materials are the major challenges reported. These two factors were indicated by all participants as the two main major challenges.

Cost and positive carbon emission, as well as specifications and availability, were cited by the client as the most significant challenges in this project. P_1 indicated that the cost of acquiring recycled materials was greater than that in the case of new materials due to the small scale of production:

> *"...the cost was one [challenge] and we did take the choice to pay a premium to move from a very efficient product in the noise walls, to the recycled plastic product."* [P_1]

In addition, the participant noted that the use of recycled materials would not necessarily result in a reduction in carbon emissions, which influenced the decision to use some recycled products. Lastly, according to P_1, while the organisation was able to overcome some of the obstacles imposed by the Department of Victoria (DoT) material specifications, other restrictions following specifications prevented the use of additional recycled materials:

> *"...we weren't able to do that, ultimately, because the emissions for the trucks to pick up the water and take them to the site was greater than the benefit of actually using the recycled water [recycled materials]. So yeah, there's a balance in sometimes you get environmental benefits in one area, but you're harming the environment in another area. So, there were trade-offs that sometimes did not work in favour of recycled materials."* [P_1]

> *"...so we had the contractor who wanted to achieve it [the target of using recycled materials], because they had commercial benefits. And then we had suppliers who were reluctant. And they were reluctant because the DoT had these specifications...So whilst we were able to overcome some of those specification challenges with the asphalt, there are probably other areas that we didn't even try."* [P_1]

P_2 mentioned five factors in response to this question: traditional specifications, the premium cost of recycled materials, their availability, immature supply chains, and technical challenges. According to this participant, the problem with conventional specifications is that they do not recognise the use of new materials in the common elements of infrastructure projects, such as the use of recycled plastic instead of concrete in pipes. It was also noted that the high cost of recycled materials prevented the use of these materials in this project. The participant further connected this obstacle to the immaturity of the waste recycling industry and its inability to produce products at reasonable prices and the required scales:

> *"First and foremost, traditional specifications don't allow it...the pipe, for instance. Well, the specification said it had to be the concrete pipe. So if I were just to follow the specification that I can't do it all the recycled asphalt well, there was no testing regime, there was no accepted product. So the challenge is to get the specifications to*

be invited to innovation specifications traditionally. They prevent innovation." [P_2]

"Some of these products come at a premium cost. And probably just because the industry is a bit immature, like they don't have the volume and the scale on some of these products yet, so I think we need to get more volume and scale and to get to drive the price down...there was only one supplier up in Mildura [that could produce sleepers with recycled plastic], and they could only produce like five a day or something which meant we could only do a very short section of the work, they just didn't have the capacity in their workshop to produce the amount that we needed. And they didn't have the money to invest in that production scale because not enough jobs were using it." [P_2]

The last challenge was related to the technical aspect of the application of these materials. The participant noted that the characteristics of recycled materials are different from those of new products, which created issues for their application in this project:

"...for instance, the thickness of the panel causes any issue. So there's just that the maturity of the engineering that goes into them, so to know how to make the product work with all the rest of the systems that fit around it...And these noise walls were quite a lot thicker than traditional ones. So all of a sudden, you've got to know, from a design point of view, our designers left whatever was a two-and-a-half metre space to fit all this in." [P_2]

The third participant, P_3, indicated that the three major challenges were negative perceptions, specifications, and the availability of recycled materials. They described:

"...there's a perception that the newer materials are a bit sort of untested. So getting the road authorities to approve the use of new materials can be a challenge. So obviously, with roadside infrastructure, it needs to be sort of performance tested, you don't want to instal something that is going to sort of fail or fall apart in the next 10 years, when it should be lasting for much longer...so there is some trepidation around that until it's been demonstrated that the product or the recycled material actually performs a similar function to traditional materials." [P_3]

"...availability of recycled materials as well, it's probably easier in metropolitan Melbourne to get access to these alternative materials... but maybe some other parts of Australia, it's more difficult when you've only got a limited supply chain to deliver materials for projects." [P_3]

The last participant in this case study (P_4) referred to the lack of awareness of recycled materials' performance, unsupportive material specifications, and people with a sceptical mindset about recycled materials as the main three challenges, stating:

"I think the challenging part for us is that there's a lot of talk about recycled materials being new or innovative or cutting edge...people wanting to do trials and things like that to research [application of] recycled materials [in] projects. The fact of the matter is that we've been manufacturing these materials for decades, and they have been used for decades in projects across Victoria and Queensland. So we're actually past the point where this is new and newsworthy." [P_4]

"...the specifications are probably the key thing if the specifications allow for the use of recycled material, and that takes away a huge barrier and makes it far easier to get acceptance for it in the market. So you know, it's open to the regulators and Vic Roads and major projects and people like that to really, you know, step up and make sure that there aren't these, you know, ridiculous invisible barriers in the way of people taking up the use of "recycled material." [P_4]

"So if you have individual projects, engineers, or people working on the project that have old ideals around the use of materials that don't have experience with recycled materials, I might be a little bit resistant or reluctant to take that on." [P_4].

9.5.3 Major Strategies to Overcome Challenges

Different strategies were suggested by the interviewees to remove challenges to the use of recycled materials in this project. As depicted in Figure 9.8, 12 strategies were captured in the interviews that can be clustered under three categories: financial, regulatory support, and technical.

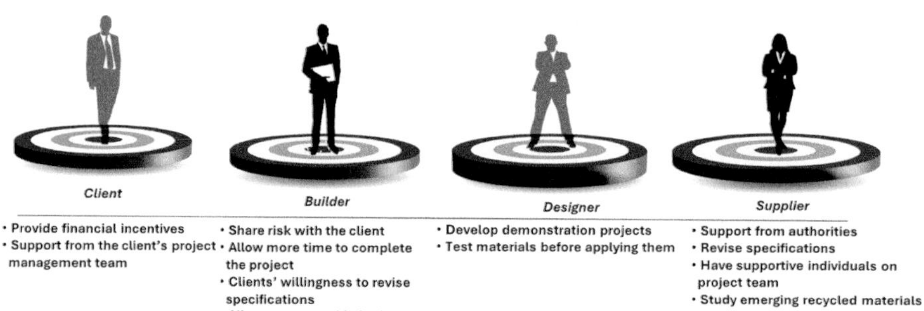

Figure 9.8. Major strategies to remove barriers to using recycled materials in MFP. (*Source:* Authors).

The first participant (P_1) noted two major strategies: financial incentives and a supportive client project management team to monitor the sustainability progress of the project. They explained:

"...we had a sustainability committee for the project, which met monthly, we reported and monitored outcomes in that timeframe as well...[we also] monitored the success in this area. So yeah, it was

not treated as a side matter, it was treated as a significant element of the project that needed to be monitored and managed to be able to be achieved." [P_1]

The second participant named the main strategies to overcome the reported challenges as sharing the risk with the client, allowing more time than usual, the client's efforts in revising/updating materials specifications, and the client's preparation to allocate a reasonable budget. They believed that taking away some risks from using these materials could result in the contractor's motivation to use more of them. Furthermore, the time allowed for exploring options and testing innovative ideas was seen as a way to solve the technical issues that contractors may face during construction. They stated:

"I think the clients need to take away the risk from the contractor on the performance of these innovations. So that would motivate us to use them a lot more. They need to give us the time to explore the options properly to look for the answer." [P_2]

According to P_2, the client should be prepared to work with the contractor to revise the specifications that facilitate the utilisation of a wider range of recycled materials with different characteristics. As the cost of recycled materials is somewhat higher than that of the usual new materials, the client should consider the premium cost in budget planning:

"...they will also deal with the specification issue around the spec explicitly not to use this type of thing. So if they can, then the can be overcome predominantly. You know, maybe they need to be prepared to pay for it as well". [P_2]

The participant from the design team P_3 indicated two strategies to overcome the challenges they faced during the design stage of this project. These included developing demonstration projects in which recycled materials are used and testing the properties of the material before specifying it for the project:

"...just having a precedent set from this project about what types of applications you can use for the recycled materials is one...they can be a bit of a risk in being the first project to do something in that way, because it can be a little bit untested...And then you may see sort of poor outcomes in another project, if they haven't been installed properly or selected appropriately. So, materials such as plastic wood will need to be considered in terms of how UV resistant they are, how they are dealt with, I guess, impacts loadings, fire as well." [P_3]

The participant from the recycled materials supplying organisation, P_4, suggested some strategies to deal with challenges that are faced when using recycled materials in these kinds of projects. These included support from authorities, revision of material specifications in favour of using recycled materials, having supportive individuals on the project team, and studying the application of emerging materials. Supportive authorities and up-to-date specifications could significantly remove many challenges to using recycled materials in the project, because, according to this participant:

> "*...the specifications are probably the key thing if the specifications allow for the use of recycled material, and that takes away a huge barrier and makes it far easier to get acceptance for it in the market... So you know, it's up to the regulators and Vic Roads and Major Projects and people like that to really step up and make sure that there aren't these ridiculous invisible barriers in the way of people taking up the use of recycled material.*" [P_4]

> "*[and] then there'll be emerging materials that come onto the market, which we'll need to find. Place to use as well. And we need to be open to those sorts of things and be agile and responsive with specifications so that we can broaden the scope of recycled materials and construction.*" [P_4]

9.6 ACKNOWLEDGMENTS

This book chapter is based on research developed by Australia's Sustainable Built Environment National Research Centre (SBEnrc). The SBEnrc develops projects informed by industry partner needs, secures national funding, project manages collaborative research, and oversees research into practice initiatives. Current Core Members of the SBEnrc include ATCO Australia, BGC Australia, Government of Western Australia, Queensland Government, Sunshine Coast Council, Curtin University, Griffith University, Western Sydney University, and RMIT University. The industry-driven research described in this publication would not have been possible without the valuable support of our core industry, government, and research partners.

REFERENCES

NTRO. 2022. Recycled plastic builds noise walls in Victoria. National Transport Research Organisation. Melbourne, Australia. Available from https://bit.ly/3MroUWU

VicRoads. 2022. Standard sections - 400 series - Asphalt and surface treatments 407 dense graded asphalt. Melbourne, Australia. Available from https://bit.ly/3XqErtW

Victoria State Government. 2021. Mordialloc Fwy recycled plastic noise walls a world first. Melbourne, Australia. Available from https://bit.ly/4g2hMxR

Victoria State Government. 2022. Recycling Victoria: A new economy. Melbourne, Australia. Available from https://bit.ly/42OT0KP

Chapter **10**

Beyond Disposal: Unleashing the Potential of Plastic and Metal Waste in Producing Metal Organic Frameworks (MOFs) for Environmental Revitalization

Preeti Waribam[1], Thilina Rajeendre Katugampalage[1], Wanida Chooaksorn[2] and Paiboon Sreearunothai*[1]

[1] School of Integrated Science and Innovation (ISI), Sirindhorn International Institute of Technology (SIIT), Thammasat University, Thailand

[2] Department of Environmental Science, Faculty of Science and Technology, Thammasat University, Thailand

10.1 INTRODUCTION

Improper disposal of single-use plastics and heavy metals leads to significant environmental pollution, contaminating both water and soil, and posing risks to ecosystems and human health. The integration and repurposing of organic and inorganic wastes for the development of highly porous metal-organic frameworks

*For Correspondence: paiboon@siit.tu.ac.th

DOI: 10.1201/9781003531463-10

(MOFs) offer a promising approach to waste management by conserving resources and recycling hazardous materials. MOFs, known for their tunable structures and diverse properties such as high surface area, thermal stability, and chemical resistance, hold great potential for environmental applications. Polyethylene terephthalate (PET), a prevalent plastic waste, can be depolymerized into benzene-dicarboxylic acid (BDC) to create waste-derived MOFs. Simultaneously, metal precursors sourced from inorganic waste—such as electroplating sludge, spent batteries, aluminum dross, industrial wastewater, and biological materials like bones and eggshells—can provide the necessary metal ions for MOF synthesis. This approach diverts plastic and metal waste from the environment, transforming it into valuable, functional materials. This chapter explores the use of various waste sources, synthesis methods, and types of waste-derived MOFs, highlighting their applications in environmental remediation.

Metal-organic frameworks (MOFs) are a class of porous materials composed of metal ions or metal ion clusters coordinated to organic linkers, forming a highly ordered, three-dimensional structure (Figure 10.1). These materials are renowned for their exceptionally high surface area, tunable pore sizes, and versatile chemical functionalities, making them highly effective in a range of applications such as gas storage, adsorption, catalysis, and drug delivery. MOFs show great promise in environmental remediation, particularly in removing heavy metals and organic pollutants from wastewater, due to their ability to selectively adsorb and trap organic molecules and metal ions in their structures (Yusuf et al. 2022). The versatility and tunability of MOFs make them promising candidates for addressing key challenges in materials science and environmental remediation, paving the way for developing next-generation functional porous materials.

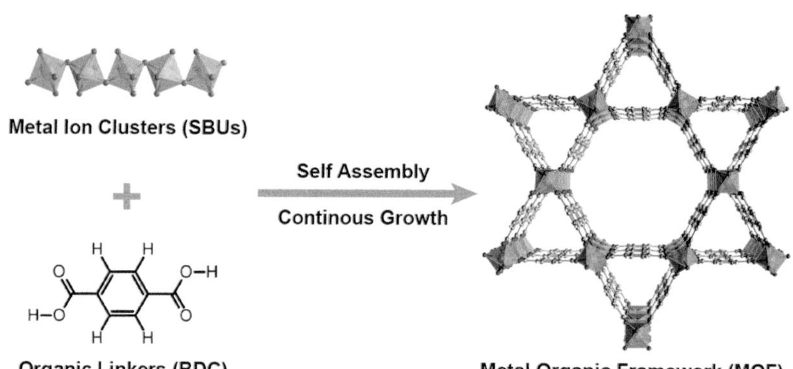

Figure 10.1 Formation of metal-organic frameworks (MOFs).

In the synthesis of MOFs, one of the most commonly utilized organic linkers is the benzene-1,4-dicarboxylic acid (BDC), which is also a monomer of polyethylene terephthalate (PET), a major consumer plastic and the most popular beverage packaging material worldwide (Sarda et al. 2022). BDC linkers can be combined with various metals such as Al, Cr, Fe, Zn, Co, Ca, Zr, and Ni to produce BDC-based MOFs. The recycling of post-consumer PET waste into specialty MOFs that can be used, for instance, in toxic pollutant capture, seems practical and corresponds also with the concept of circular economy (Karam et al. 2021).

The release of toxic heavy metals into the environment poses a significant threat, given their prolonged presence in the atmosphere and ability to accumulate in living organisms through bioaccumulation (Mitra et al. 2022). Waste sources like electroplating sludge (EPS), spent batteries, and wastewater contain high concentrations of heavy metals, demanding immediate attention before the situation is exacerbated. EPS, a byproduct of electroplating processes, contains various heavy metals such as chromium, nickel, zinc, and copper. Additionally, industrial wastewater often contains heavy metals like cadmium, lead, and mercury, which contaminate water bodies and pose risks to human health through bioaccumulation in the food chain (Wuana and Okieimen 2011). Similarly, the waste produced from batteries, particularly spent lithium-ion batteries (LIBs) and alkaline batteries, contains several heavy metals, such as lithium (Li), zinc (Zn), manganese (Mn), cobalt (Co), nickel (Ni), iron (Fe), copper (Cu), and aluminum (Al) (Du et al. 2022). Proper collection of heavy metals and their treatment before disposal could produce valuable metal resources for MOF synthesis. The presence of these heavy metals in EPS, wastewater, and batteries highlights the importance of implementing effective waste management practices and developing sustainable solutions for their treatment and disposal to safeguard the environment and public health.

Thus, harnessing organic ligands from spent PET plastic and inorganic metals from waste to produce MOFs for environmental cleanup emerges as an efficient approach for crafting sustainable materials with applications in environmental remediation. It is an approach that enhances resource efficiency while simultaneously reducing waste accumulation and pollution. This chapter extensively explores the primary methods of producing valuable metal-organic frameworks (MOFs), emphasizing the innovative use of plastic and metal waste. By repurposing these waste materials, the approaches discussed below contribute to sustainable practices and offer a cost-effective solution for MOF synthesis. The transformative potential of these methods highlights the intersection of waste management and advanced material production, showcasing the environmental and economic benefits of recycling waste into high-value MOFs.

10.2 PLASTIC WASTE TO VALUABLE PRODUCTS

10.2.1 Importance of Addressing Improper Disposal of PET Plastic and Related Environmental Pollution

Plastics derive their name from the Greek term "plastikos," meaning "able to be molded into different shapes." Plastics are manufactured from a blend of organic and inorganic elements, including carbon, silicon, hydrogen, nitrogen, oxygen, and chlorine. The foundational components for producing synthetic plastics are sourced from oil, coal, and natural gas. Over the past five decades, synthetic polymers have progressively supplanted natural materials, becoming indispensable in everyday life. Plastic pollution has emerged as a critical environmental issue, with the surge in disposable plastic goods overwhelming waste management systems and leading to plastic accumulation in the environment (Ghosh et al. 2013). When PET plastic breaks down in the environment, it can release harmful chemicals and microplastics. This pervasive microplastic can enter the food chain, water bodies, and oceans, harming wildlife habitats and human populations (Geyer et al. 2017; Rocha-Santos and Duarte 2015).

Polyethylene terephthalate (PET), the most widely used plastic daily, is a semi-crystalline, thermoplastic polyester known for its high strength, transparency, and safety. It is synthesized from pure terephthalic acid (TPA) and ethylene glycol (EG) derived from crude oil. PET is commonly used in plastic bags, packaging, clothes, and bottles, contributing to significant environmental issues, especially with the rise of single-use PET products. Improper disposal contributes to resource depletion and increases the demand for new plastic production. The production and incineration of plastics, including PET, contribute to greenhouse gas emissions. Addressing improper disposal can reduce the need for virgin plastic production, thereby mitigating the carbon footprint associated with plastic manufacturing. Chemical recycling and upcycling (Al-Sabagh et al. 2016) of PET into valuable products have been suggested as one of the most effective strategies for managing and reducing environmental impact associated with the consumption of the versatile PET plastics.

10.2.2 Recycling and Depolymerization of PET

PET waste can be recycled through both chemical and mechanical methods. Mechanical recycling deals with challenges arising from the contaminated nature of PET waste. Despite its simplicity and cost-effectiveness, mechanical recycling is limited by a decline in product quality caused by factors like heat-induced photo-oxidation and mechanical stress. Chemical recycling enables the recovery of the petrochemical constituents of PET wastes (Park and Kim 2014), which can be utilized for remanufacturing PET products or other synthetic chemicals. Therefore, chemical recycling is one of the most effective techniques for recycling PET into its monomeric units.

Chemical recycling of PET involves breaking down the polymer chain into monomers or oligomers, which can then be used to produce new PET as well as other functional materials. Chemical recycling enhances sustainability by reducing greenhouse gas emissions by up to 50–80% compared to virgin PET production, providing a carbon-neutral approach to PET manufacturing. Additionally, chemical recycling allows for the recycling of contaminated or partially degraded PET waste, which is unsuitable for mechanical recycling. Besides energy efficiency, chemical recycling can yield high-quality PET monomers with the available chemical recycling technologies, including glycolysis, hydrolysis, methanolysis, and aminolysis, as shown in Figure 10.2 (Babaei et al. 2024).

Among various chemical recycling methods, PET hydrolysis is a promising depolymerization technique that produces BDC as a major product. PET hydrolysis is categorized into alkaline, neutral, and acid hydrolysis, depending on the pH of the aqueous reaction medium. Neutral hydrolysis of PET involves depolymerization in an excess of water or steam in a high-pressure autoclave. Reaction temperature and pressure range between 200 and 400°C and 40–300 atm, and depolymerization was carried out for 1–6 hours (Colnik et al. 2021; Zope and Mishra 2008), producing BDC monomer and EG. BDC can be recovered from unreacted PET through acidification and filtration of hydrolysis products. It is considered more environmentally friendly since it eliminates unwanted inorganic salts. However, this method does not effectively remove impurities from PET waste, leading to higher impurity levels in the produced BDC. Consequently, a more rigorous purification process is also required. The operational costs of neutral hydrolysis are high, primarily due to the elevated reaction temperature and pressure requirements.

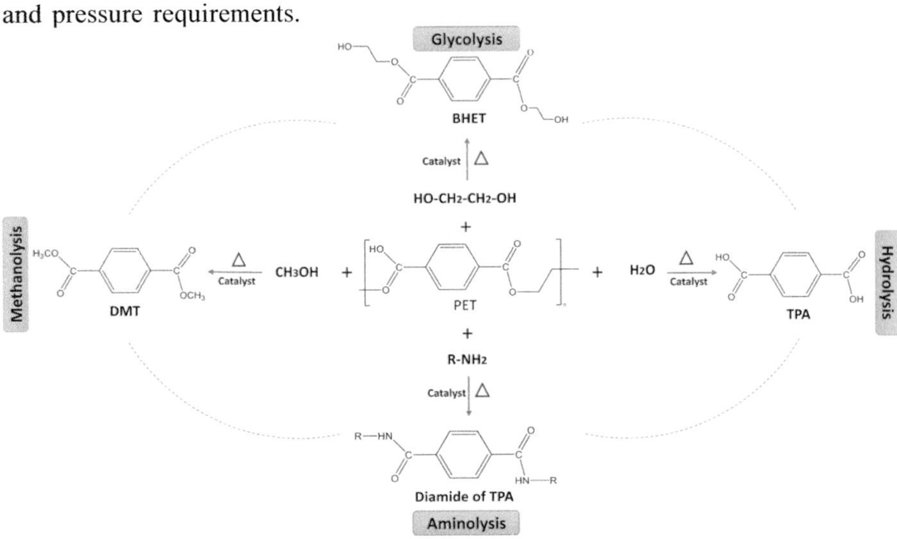

Figure 10.2 Chemical recycling of PET and types of products (Babaei et al. 2024).

Another common method for PET recycling is alkaline hydrolysis, which allows the reaction of PET waste with a base catalyst in an aqueous solution. Typically, a metal hydroxide catalyst like NaOH or KOH is used at 4–20% w/w concentrations, and depolymerization is carried out around 110°C between 1 and 5 hr (Corak et al. 2022; Wang et al. 2022a). Nevertheless, depolymerization using microwave irradiation achieved degraded products within 30 minutes at 180°C (Siddiqui et al. 2010). In the case of using NaOH, the resulting product is the sodium form of BDC (Na^{2+}–BDC), and BDC monomers can be precipitated by adding a strong acid like sulfuric acid or hydrochloric acid to adjust the solution pH to 2. Moreover, Na-BDC can be collected as a solid product via rotary evaporation, and it can be utilized as a linker to prepare MOF in a water-based system. Since Na-BDC completely dissolved in water, it will further coordinate with metal ions, leading to the MOF growth.

Lastly, acid hydrolysis has been used to depolymerize waste PET to BDC and EG in the presence of mineral acids, particularly sulfuric acid, as the acid catalyst. PET is hydrolyzed in concentrated acid solutions exceeding 90% of their weights to mitigate the need for high reaction temperatures and pressures. Following complete hydrolysis, the produced BDC is recovered through a crystallization process. Unlike alkaline hydrolysis, acid hydrolysis does not necessitate an additional acidification step for BDC recovery in cases of complete PET conversion. The concentration of the mineral acids, reaction temperature, pressure, particle size, and shape (Abedsoltan 2023) of waste PET are essential to achieve complete depolymerization in acid hydrolysis. Acid hydrolysis is a straightforward depolymerization technique, unlike neutral and alkaline hydrolysis, which require an additional step to precipitate out BDC monomer via acidification. Acid hydrolysis, however, is less popular due to the more complicated equipment in handling concentrated acid solutions.

10.3 EXTRACTION OF METALS FROM INORGANIC WASTES

10.3.1 Overview of Heavy Metals Discharge from Industries and Impact on the Environment and Public Health

Naturally occurring heavy metals in the environment are essential for survival and can become hazardous when they accumulate in organisms. The release of harmful organic and inorganic wastes exacerbates rapid industrial expansion to meet the needs of the growing population. Ultimately, these pollutants contaminate water, soil, and air, adversely affecting human health. Health risks from toxic metals depend on their concentration and duration of exposure. Even at low levels, long-term exposure can cause neurotoxicity, nephrotoxicity, carcinogenicity, hepatotoxicity, immunotoxicity, cardiovascular toxicity, skin toxicity, reproductive and developmental toxicity, and genotoxicity (Mitra et al. 2022).

Some commonly found heavy metals like chromium, cobalt, nickel, copper, and zinc are ideal candidates, especially in MOF synthesis. Chromium (Cr) is a toxic, carcinogenic heavy metal with two stable oxidation states: the less hazardous Cr (III) and the more hazardous Cr (VI). Among various industries, the ferrochrome industry has the highest chromium discharge to the environment (Coetzee et al. 2020). Cobalt (Co) and its compounds are widespread and play crucial roles in various human activities. While cobalt is essential as a component of vitamin B_{12}, excessive exposure can result in adverse health effects. Exposure to cobalt can occur through occupational, environmental, dietary, and medical exposure (Leyssens et al. 2017). Nickel (Ni) is found in the environment due to natural and anthropogenic sources. The atmospheric nickel that originates from the weathering of rocks and soils, volcanic activities, and human activities causes various adverse effects on humans upon exposure (Genchi et al. 2020). Copper (Cu) is an essential trace element. However, high doses of copper can induce toxic responses, particularly liver toxicity (Georgopoulos et al. 2006). Lastly, zinc (Zn) is a fundamental metal crucial for its role as a cofactor in numerous enzymatic reactions, essential for various physiological processes. Smelting and mining activities are major sources of environmental zinc contamination, significantly impacting ecosystems and living organisms (Zhang et al. 2012).

Therefore, recycling or treating heavy metals to prevent their accumulation in the environment is essential, which can lead to significant ecological damage. Effective management and remediation techniques are necessary to mitigate the release of toxic substances from industrial processes, thereby reducing their harmful impact on water bodies, soil, air, and human health.

10.3.2 Extraction of Metal Ions from Various Types of Inorganic Wastes

Waste reclamation is a demanding research topic for regenerating valuable materials from spent contaminants. Electroplating sludge (EPS) is a heavy metal-containing byproduct produced by the electroplating industry. This sludge often comprises metals such as copper, nickel, zinc, chromium, and iron in a complex liquid-solid mixture. It is formed as a residue after the chemical precipitation of heavy metals from acidic or alkaline solutions and rinse waters used in electroplating processes. It is classified as hazardous waste; the storage and disposal of this sludge are costly, leading to the loss of substantial amounts of valuable metals (Li et al. 2010).

The three common methods for extracting low-grade and complex heavy metals from EPS include the selective ammonia treatment process (SAP), efficient acid treatment process (EAP), and roasting-acid treatment process (RAP). EAP is reported as a superior technique for extracting heavy metals from actual EPS compared to SAP and RAP (Zhao et al. 2025).

Industrial wastewater, a major source of heavy metals, has been increasing due to industrial and human activities. Industries including plating, electroplating, battery production, pesticide manufacturing, mining, rayon production, metal rinsing, tanning, textiles, metal smelting, petrochemicals, paper manufacturing, and electrolysis have extensively contributed to this problem (El-Sherif et al. 2013; Zou et al. 2016). Various methods have been developed to eliminate heavy metal ions from wastewater, categorized into adsorption, membrane filtration, chemical treatment, electrochemical processes, and photocatalysis (Qasem et al. 2021). The precipitation of heavy metals using lime, sulfide, and caustic soda is a simple and widely used method for metal recovery. Meanwhile, stepwise extraction of Fe, Al, Ca, and Zn through acid leaching could successfully extract metals via precipitation (Qu et al. 2021).

Another primary industry utilizing heavy metals is the battery industry, which is experiencing rapid growth, driven by increasing demand for portable electronic devices, electric vehicles, and renewable energy storage solutions. Alkaline, lithium, and zinc-carbon spent batteries have often ended up in landfills or been incinerated instead of recycled (Pistoia 2005). Given the environmental impact of heavy metals, recycling these devices is of utmost importance to mitigate pollution and conserve valuable resources. Spent lithium-ion batteries usually contain 5–20% cobalt (Co), 5–10% nickel (Ni), 5–7% lithium (Li), 5–10% of other metals (e.g. copper (Cu), aluminum (Al), iron (Fe), etc.), 15% organic compounds, and about 7% plastic. Recycling valuable metals like lithium, nickel, cobalt, and manganese from spent lithium-ion batteries presents considerable economic benefits (Zheng et al. 2018).

The two primary recycling processes for recovering metals from spent batteries are pyrometallurgy and hydrometallurgy (Makuza et al. 2021; Yao et al. 2018). The pyrometallurgy process involves the selective volatilization of metals at elevated temperatures, followed by condensation. Hydrometallurgical processes utilizing aqueous chemistry for metal recovery are recognized for their environmental friendliness and cost-effectiveness, particularly for treating materials with low metal grades on a small scale (Chen et al. 2017).

Therefore, waste materials such as electroplating sludge (EPS), industrial wastewater, or spent batteries can significantly benefit the environment. Recycling these inorganic wastes to recover metals helps eliminate harmful heavy metals from the environment while conserving valuable resources.

Considering the immense potential of plastic waste recycling and heavy metal recovery, an innovative approach is to harness these recycled materials to produce porous metal-organic frameworks (MOFs), which can be used in environmental remediation. This approach facilitates waste management and promotes the sustainable production of valuable materials, offering a greener solution for addressing both heavy metal pollution and plastic waste management. The subsequent content of this chapter delves into the synthesis and utilization of metal-organic frameworks (MOFs) derived from waste materials. Additionally, it covers a range of synthetic methods and explores their practical applications in environmental remediation.

10.4 MOFs DERIVED FROM ORGANIC AND INORGANIC WASTES

10.4.1 MOFs Prepared Using BDC as the Organic Linker

The main goal in MOF synthesis is to establish conditions that create defined inorganic building blocks without decomposing the organic linker while ensuring that the crystallization kinetics allow for proper nucleation and growth of the desired phase. Generally, MOFs can be synthesized using various methods, including room temperature synthesis, conventional electric (CE) heating, microwave (MW) heating, electrochemistry (EC), mechanochemistry (MC), and ultrasonic-assisted (US) techniques (Stock and Biswas 2012). MOF-5 is one of the earliest MOFs published in 1999, with the framework consisting of zinc acetate ($Zn_4O(COO)_6$) clusters and the well-known ditopic linker benzene dicarboxylate (BDC) (Li et al. 1999). Further studies on this have expanded to form a series of isoreticular MOFs (IRMOFs), which can be functionalized with the organic groups such as –Br, $-NH_2$, $-OC_3H_7$, $-OC_5H_{11}$, $-C_2H_4$, and $-C_4H_4$ (Eddaoudi et al. 2002). Furthermore, various BDC-based MOFs have been extensively synthesized using diverse metal sources. Furthermore, there is a growing interest in bimetallic MOFs, where more than one metal is present in the secondary building units (SBUs). These bimetallic MOFs offer versatile functionalities and can be fine-tuned for specific properties, particularly in catalysis and across various other applications such as gas storage and separation, advanced oxidation, sensing, and photocatalysis (Masoomi et al. 2019; Katugampalage et al. 2024).

As stated earlier, the synthesis of MOFs requires organic linkers, such as BDC, and various sources of metal ions or clusters to form intricate 3D network structures. However, developing MOFs using commercial chemicals often requires significant investment, making them less cost-effective, especially for applications such as environmental remediation. Therefore, there is a growing interest in utilizing organic wastes as a source of organic linkers and inorganic wastes as raw materials for metal clusters in MOF synthesis, promoting a greener and more sustainable approach. For instance, upcycling PET waste, prevalent in plastic bottles, could be the primary source of BDC linkers. Meanwhile, the recovered metals from various waste sources, such as EPS, wastewater, batteries, and eggshells, offer a cost-effective and efficient approach to producing MOFs. This approach preserves available resources and repurposes waste materials to create valuable products, advancing green chemistry practices and fostering a circular economy.

10.4.2 MOFs Synthesized Using Waste PET as the Source of BDC Linker

Polyethylene terephthalate (PET) waste has been successfully utilized as the source of the organic linker in the MOFs synthesis via depolymerizing it into its monomeric unit, commonly known as terephthalic acid (TPA) or 1,4-benzene dicarboxylic acid (BDC). PET-derived MOFs have been synthesized through a one-pot or two-step synthesis protocol. During the one-pot synthesis, PET flakes were loaded along with the metal precursors in the reaction vessel, where PET depolymerization to BDC linkers and coordination with the metal SBUs took place simultaneously during the hydrothermal process. However, in the two-step process, PET was depolymerized to BDC monomer, then reacted with the hydrolyzed metal salts in the second step, forming MOFs. Both the synthesis protocol method has advantages as well as drawbacks. In the two-step process, a stepwise synthesis pathway is required, involving the isolation of linkers, purification, and subsequent reaction with metal salts to form the MOF. In the single-step process, PET hydrolysis usually occurs in the presence of acids as an additive along with metal salts, requiring high temperatures around 200–240°C. As of today, the MIL series and UiO-66 MOFs are commonly synthesized through straightforward one-pot methods, while various types of MOFs are typically produced using a two-step reaction approach.

10.4.2.1 Single-Step Conversion of PET into MOFs

The single-step conversion of PET to MOFs is an intriguing technique, as it allows for the direct transformation of PET waste into valuable MOFs within a short timeframe. In recent years, significant advancements have been made in synthesizing trivalent metal-based BDC MOFs, including MIL-101(Cr), MIL-53 (Cr, Al, Ga), and MIL-47(V) by substituting organic linkers with PET waste via hydrothermal process at temperatures between 160–210°C. The depolymerization of the PET has been achieved via acid catalysts like HF, HCl, and formic acid present in the reaction mixture, leading to BDC monomers. One-pot synthesis of PET-derived MIL series MOFs has shown comparable PXRD patterns and N_2 adsorption properties to those synthesized using pure chemicals (Lo et al. 2016). Furthermore, several studies confirmed that no PET remained in the MOF products or the supernatant after one-step synthesis of MOF from PET, evidencing the effectiveness of the conversion process. Figure 10.3 shows various MOFs synthesized via one-pot synthesis.

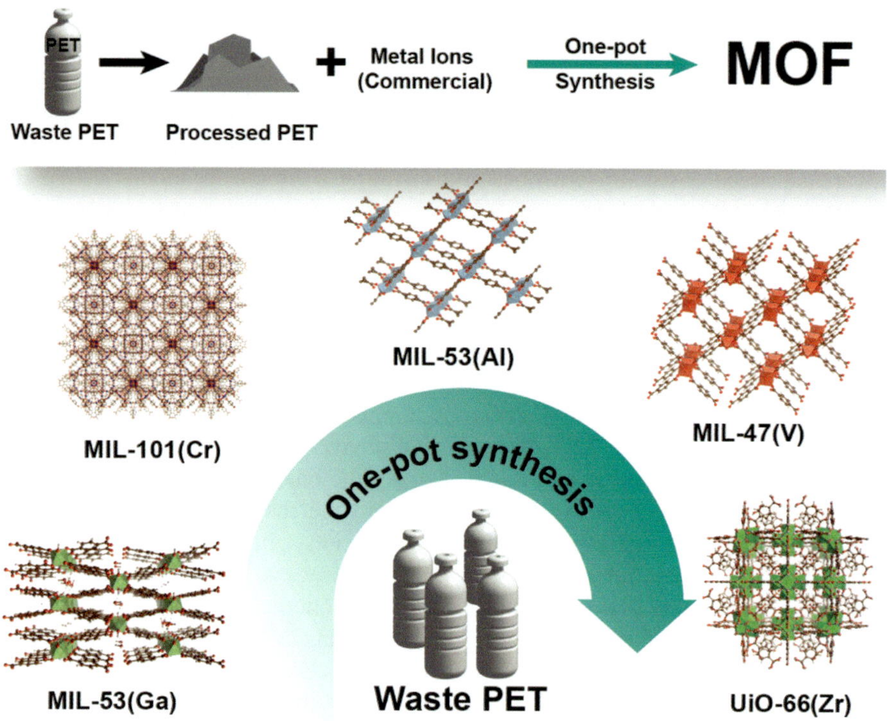

Figure 10.3 Illustration of one-pot synthesis of waste PET-derived MOFs. (The images were generated using databases as referenced in the following publications: (Lebedev et al. 2005; Lutton-Gething et al. 2023; Meilikhov et al. 2009; Oien et al. 2014; Vanpoucke et al. 2014).

Ren et al. (2016) successfully prepared one of the widely reported MOFs, MIL-101(Cr), from waste PET via a one-pot synthesis following a slightly modified protocol from above. This synthetic protocol was much greener, utilizing formic acid as a catalyst for the synthesis of MIL-101(Cr), thereby avoiding the commonly used toxic hydrofluoric acid (HF) and making it a more environmentally friendly method for MOF preparation. The reaction was carried out at 210°C for 8 h using $CrCl_3$ as the metal source, resulting in a solid green MOF (Ren et al. 2016). Furthermore, a similar reaction using a commercial BDC linker revealed that both materials exhibited comparable PXRD patterns. However, the crystal sizes of the waste-derived Cr-MOF were relatively smaller than those obtained from commercial chemicals.

MIL-53 is one of the widely studied MOF candidates, well known for its heat stability and breathing behavior of its structure. The work carried out by Deleu et al. shows the potential of a microwave-assisted one-pot synthesis process for the synthesis of MIL-53 (Al) and MIL-47 (V). The synthesis was carried out at 200°C for 12 hr in a microwave oven, using $AlCl_3$ and VCl_3 as the metal salts and PET as the source for the linker. During the hydrothermal synthesis, HCl was added to accelerate PET depolymerization, leading to the in-situ production of BDC linkers, which then coordinated with the hydrolyzed metal nodes to form the MOFs (Deleu et al. 2016).

UiO-66 is one of the highly stable Zr-based MOF candidates, and it has been successfully prepared using Zr metal salt ($ZrCl_4$) and PET bottle chips in a one-pot synthesis process. The reaction was conducted using formic acid and acetone mixed solvent at 160°C for 24 hr in a hot air oven. Interestingly, the synthesized waste-derived UiO-66(Zr) exhibits a defective hcp-UiO-66 structure, containing 15 BDC linkers instead of 18 BDC linkers. Generally, the **hcp** (hexagonal closed packed) topology of UiO-66 consists of $Zr_{12}O_8(OH)_8(OH)_6$ core coordinated by 18 BDC ligands. The PET-derived hcp-UiO-66 is the thermodynamically favorable phase, featuring hexagonal prism-shaped microcrystals with an average size of 0.49 μm. The defects were generated in the structure due to the abundance of formic acid, which acted as a modulator during the synthesis. The defective PET-derived **hcp**-UiO-66 shows high thermal stability at 350°C (in air) and hydrothermal stability at 200°C. The PXRD patterns of the MOF remain unchanged after exposure to these high temperatures, and the structure does not dissolve under harsh hydrothermal conditions (Deleu et al. 2016).

The microwave-assisted one-pot synthesis of PET-derived UiO-66, reported by Waribam et al. (2023), achieved complete depolymerization of PET and simultaneous growth of highly defective UiO-66 MOF in 30 minutes at 200°C. The reaction was carried out in an aqueous system where HCl was used as an acid catalyst to accelerate PET depolymerization. Acetic acid served as a modulator to induce defects into the UiO-66 structure. The PET-derived UiO-66 exhibited the PXRD pattern of ***reo***-topology with the general formula $3[Zr_6(\mu_3\text{-}O)_4(\mu_3\text{-}OH)_4L_4(Mod)_4]$, formed by 8 coordinated BDC linkers and a missing Zr cluster along with its connected linkers. The PET-derived UiO-66 exhibited a more pronounced ***reo***-topology than the one prepared using pristine BDC linkers, indicating that byproducts generated during PET depolymerization significantly create more structural defects, which can be desirable for higher reactivity. Interestingly, PET-derived defective UiO-66 is cuboctahedron-shaped and micron-sized, with an average diameter of approximately 5–10 μm, exhibiting a higher BET surface area of 1,206 m^2/g, comparable to the reported pristine UiO-66 (Waribam et al. 2023).

10.4.2.2 Two-Step Conversion of PET into MOFs

Various types of MOFs have been synthesized using PET as the source of BDC linkers by first depolymerizing PET into its monomeric units and then utilizing the purified PET-derived BDC in the MOF synthesis (Figure 10.4). Several studies have reported the synthesis of MIL-68(Al), Cu-BDC MOF, Cu/Zn-BDC MOF, and Ni/Co-BDC MOFs using these purified BDC from PET processing.

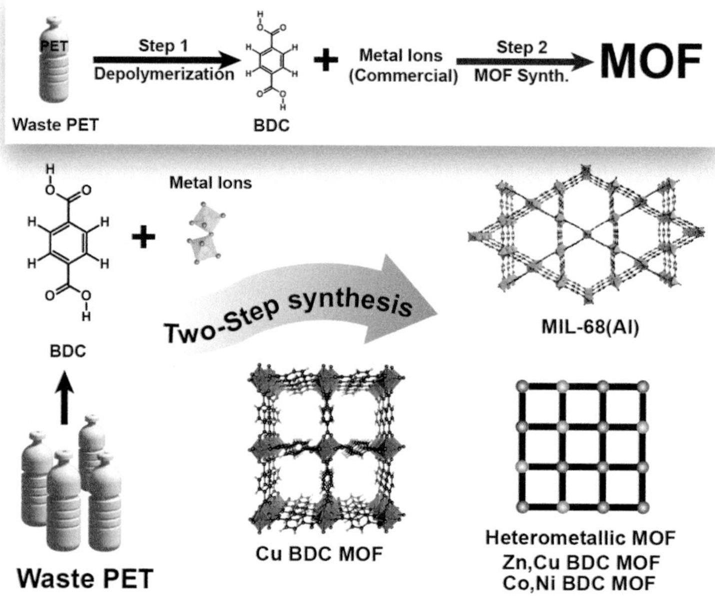

Figure 10.4 Two-step synthesis of waste PET-derived MOFs. (The images were generated using databases as referenced in the following publications: (Carson et al. 2014; Volkringer et al. 2008).

PET-derived MIL-68(Al) has been synthesized using extracted BDC from PET waste, reacting with Al^{3+} metal salts under solvothermal conditions. Although the MIL-68(Al) has been synthesized using purified BDC linkers from PET waste, the produced MOF exhibited low crystallinity with a higher degree of defects compared to the pristine MIL-68(Al), which showed fewer defects with higher crystallinity. Further characterization revealed that both MIL-68(Al) candidates had microporous structures, whereas PET-derived MIL-68(Al) exhibited a larger diameter pore of about 1.95 nm. In addition, morphological changes were also observed, where PET-derived MIL-68(Al) exhibited an irregular rod-like morphology compared to the regular rod-like structure of its pristine counterpart due to the generated defects (Heng et al. 2023).

Similarly, Cu-BDC MOF was successfully synthesized using waste-derived BDC obtained through alkaline hydrolysis of PET. The classical solvothermal synthesis could produce well-defined cubic-shaped Cu-BDC MOF with sizes ranging from 9–24 μm (Doan et al. 2020). Farshchi et al. also reported the formation of recycled PET-derived Cu-BDC and Cu-NH_2-BDC MOFs through solvent-free thermo-mechano synthesis via ball milling. The formed waste-derived Cu MOFs demonstrated a layered structure, where Cu-BDC-NH_2 exhibited rough surfaces due to the formation of thin layers compared to the Cu-BDC MOFs. The specific surface area of the Cu-BDC MOF and Cu-NH_2-BDC MOF was 726.1 and 1,025.6 m^2/g, where the increased surface area was due to the steric hindrance caused by the amino functional group (Ebrahimi et al. 2023). Both types of Cu-BDC MOFs have been used in environmental remediation, with the Cu-BDC MOF synthesized via the solvothermal process serving as an adsorbent for removing methylene blue dye, while the MOF produced via the ball milling process is used for the catalytic degradation of 4-nitrophenol.

Beyond monometallic MOFs, the realm of MOFs extends to bimetallic MOFs prepared using linkers derived from PET waste. Bimetallic MOFs feature two distinct metal ions within their secondary building units (SBUs), interconnected by BDC linkers, yielding 3D porous structures. Among various bimetallic MOFs, the Cu/Zn BDC MOF and Co/Ni BDC MOF produced using BDC linkers originating from PET waste show a circular approach to MOF synthesis. The bimetallic Cu/Zn BDC MOF was synthesized using BDC produced via PET glycolysis employing $ZnCl_2$: urea as a catalyst (Yarahmadi et al. 2023). The PET glycolysis process was conducted for 30 minutes at 170°C. Subsequently, the resulting BDC was further reacted with a mixture of $Zn(OAc)_2{\cdot}2H_2O$ and $Cu(NO_3)_2{\cdot}3H_2O$ at 110°C for 36 hr, yielding a vibrant blue precipitate of Cu/Zn MOF. The PXRD patterns and FTIR spectra of the bimetallic Cu/Zn MOF were used to characterize the bimetallic MOF product, where a cubic-shaped product was observed via FE-SEM. The PET waste-derived Cu/Zn BDC MOF has been employed as a heterogeneous catalyst in the chemical degradation of organic pollutant dyes.

Moreover, the Co/Ni BDC MOF has been synthesized using linkers derived from the hydrothermal depolymerization of PET using ethylene glycol (EG) and H_2O at 210°C for 8 hr. The obtained BDC linkers with higher purity were characterized by ^{1}H-NMR and ^{13}C-NMR analysis, where further purification was not required. The Co/Ni MOF synthesis occurred under solvothermal conditions at 150°C for 24 hr in a mixed solvent of DMF, ethanol, and water. The PXRD patterns of PET-derived Co/Ni-BDC MOF are isostructural to the previously reported Ni-based MOFs (No. 985792, CCDC), and it depicted a micrometer-sized dandelion-like morphology, composed of bulks with a length of 6–10 μm. XPS analysis further confirmed the existence of Ni, Co, C, and O elements in the bimetallic MOF, showing Ni^{2+} and Co^{2+} cation coexistence in the structure. The waste-derived bimetallic MOF was used as the anode material for Li-ion batteries, contributing to the production of active materials in a cost-effective manner for the battery industry (Wang et al. 2022b).

In summary, using plastic waste presents a viable pathway for developing both monometallic and bimetallic MOFs. The foregoing discussion has focused on MOFs synthesized using PET-derived linkers that react with commercially obtained metal ions. A greener and more sustainable approach could involve sourcing both the BDC linkers and metal ions from waste sources. However, achieving high-purity metal ions from waste sources remains challenging, potentially leading to MOFs with multiple coordinated metal ions. Henceforth, the following section discusses MOFs synthesized by sourcing BDC linkers and metal ions from waste sources.

10.4.3 Metals Extracted from Waste as a Metal Precursor for MOF Synthesis

The extraction of metals such as Zn^{2+}, Mn^{2+}, Fe^{3+}, Ca^{2+}, and Cr^{3+} for MOF synthesis represents a highly sustainable approach. Typically, these metals are generated as waste by various industries and are often released into the environment. Several studies have documented the successful production of MOFs utilizing metals from various waste sources such as wastewater, sludges, spent batteries, and more. Notable examples of MOFs derived from metal waste sources include MOF-5(Zn) from spent batteries, MIL-88(Fe), Ni-MOF, and MIL-53(Cr) from electroplating sludge, V-MOFs from carbon black, and MIL-53(Al) from aluminum cans are depicted in Figure 10.5.

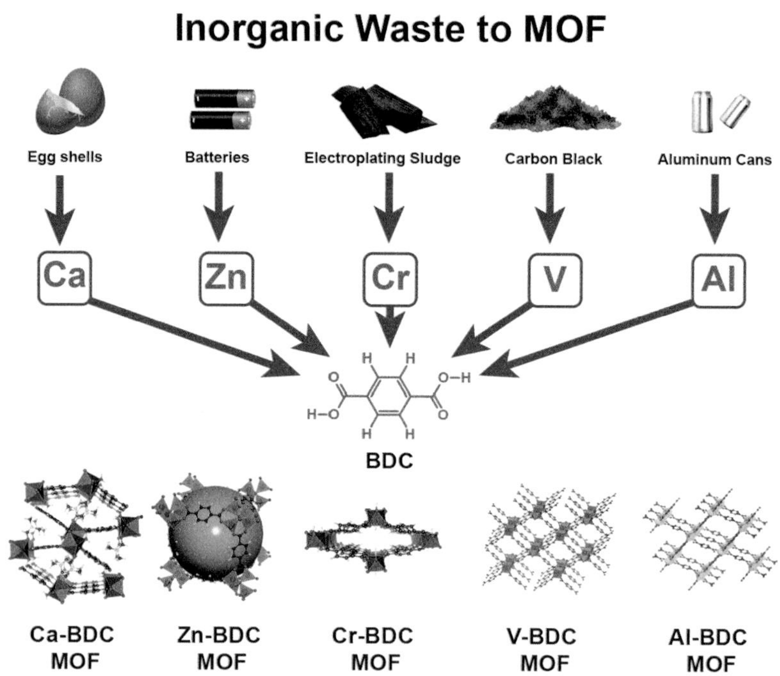

Figure 10.5 Example of MOFs synthesized using metals obtained from various inorganic wastes as the source for metal precursors. (The images were generated using databases as referenced in the following publications: (Kim et al. 2011; Liang et al. 2011; Meilikhov et al. 2009; Serre et al. 2002; Vanpoucke et al. 2014).

A well-known divalent zinc MOF known as MOF-5 has been obtained from spent Zn-MnO_2 dry alkaline batteries, an industrially attractive high-value product. The metal ions were extracted via acid leaching using HNO_3, and subsequently separating Zn^{2+} from co-existing metals like Mn^{2+}, Fe^{2+}, and Fe^{3+} via the base precipitation technique. Waste-derived MOF-5 formation involved reacting zinc leachate and BDC linker in the presence of ascorbic acid as a reducing agent to eliminate interfering ions during MOF formation. The waste MOF-5 has shown similar crystallinity with the pristine MOF-5 and diverse crystal structure with tetragonal/hexagonal shapes (Vellingiri et al. 2018).

Electroplating sludge (EPS) contains a significant quantity of various metals, and effective extraction and separation techniques could facilitate their use as SBUs in MOF synthesis. The synthesis of Fe and Ni MOFs using EPS presents intriguing insights into the subsequent separation of metal ions during MOF synthesis. The initial leaching of Fe/Ni EPS with sulfuric acid and subsequent reaction with fumaric acid could yield MOFs exclusively containing Fe ions, with no Ni-MOF observed. The selective coordination of fumaric acid with Fe^{3+} ions in the leachate solution facilitated the formation of Fe-MIL-88(A). Furthermore, reacting the remaining leachate with 2, 6-napthalenedicarboxylate tetrahydrate after separating the solid Fe-MIL-88(A) could produce Ni-MOFs. Thus, this work successfully demonstrates the production of highly pure monometallic MOFs from a mixed metal ion leachate, showcasing the potential of gradient recovery of metal ions during MOF synthesis (Han et al. 2021).

The MIL (Materials of Institute Lavoisier) series is one of the most popular groups of MOFs, known for their high stability, flexible structure, and exceptionally large specific surface area, making them ideal candidates for a wide range of applications (Zhang et al. 2022). Several studies have highlighted the potential of synthesizing MIL series MOFs using inorganic metal ion waste, resulting in the creation of MIL-53(Cr), MIL-53(Al), and MIL-96(Al). The study conducted by Kabtamu et al. demonstrates the synthesis of MIL-53 (Cr) using chromium-containing electroplating sludge. This process is noteworthy for its straightforward approach, which avoids the need for acid leaching. Pre-treated EPS has been reacted with BDC via classical hydrothermal process at 220°C for 18 hr to produce MIL-53 (Cr), with and without using hydrofluoric acid (HF) in the reaction medium. The MOFs underwent post-synthetic activation by heating in DMF at 150°C for 12 hr, followed by washing with methanol, rendering the overall process slightly less environmentally friendly. It was noticed that adding a small amount of HF during the synthesis led to the formation of highly crystalline and relatively pure MIL-53 (Cr) micro rods with large porosity and surface area (Kabtamu et al. 2020).

MIL-53(Al) and MIL-96(Al) production using household aluminum waste, such as aluminum foil and beverage cans, is a fascinating and innovative approach. One-pot hydrothermal reaction between BDC and non-dissolved aluminum foils resulted in needle-like MIL-53(Al) nanocrystals supported on the aluminum foil. The homogenous growth of one-directional needle morphology differs from the typically heterogeneous and rectangular-shaped MIL-53(Al) (Joshi et al. 2019). Generic non-supported MIL-53(Al) was also synthesized using Al foil, which was solubilized using hydrochloric acid (0.5 M) before the MOF synthesis. Furthermore, using a similar synthetic process, MIL-96(Al) and MIL-53-NH_2(Al) MOFs have also been successfully prepared using trimesic acid

(BTC) and ammonium terephthalate as organic linkers, respectively. Following the successful use of aluminum foil to prepare various Al-MOFs with different linkers, the authors further explored the potential of using waste beverage cans as an aluminum source for Al-based MOFs.

Waste beverage cans were successfully transformed into unsupported MIL-53(Al) and MIL-96(Al) MOFs using terephthalic acid (BDC) and 1,3,5-benzenetricarboxylate acid (BTC), respectively. The various impurities coexisting with the aluminum metal, such as epoxy coating and plastic labeling, did not significantly affect the formation of Al-MOFs. Coca-Cola can derive highly microporous MOFs, as measured by N_2 physisorption, displayed XRD patterns characteristic of the reported MIL-53(Al) and MIL-96(Al). However, the BET surface area for MIL-53(Al) generated from Coca-Cola cans was 800 m^2/g, which was approximately 30% lower than pristine MIL-53(Al) (S_{BET} 1,140 m^2/g). Overall, this evidence suggests that the purity of metallic aluminum does not hinder the formation of aluminum MOFs, indicating significant flexibility in material selection for Al-MOF production from solid wastes (Joshi et al. 2019).

Carboni and co-workers have developed an open-loop recycling process for Li-Ion batteries to produce three different types of MOFs. Spent Li-ion batteries, also known as NMC batteries, provided by industries were used as the metal source without any pretreatment by dissolving them in aqueous HCl and HNO_3 solutions. A solution containing Al, Mn, Ni, Co, Li, F, and P has been obtained by dissolving the components of waste batteries. The interesting protocol developed in this work shows the sequential synthesis of different MOFs from the metal ion solution by reacting it with organic ligands in the presence of various acids. MIL-96(Al) MOF was prepared by reacting the bulk metal ions with a benzene-tricarboxylic acid (BTC) linker in the presence of a strong acid (HCl or HNO_3) and DMF at 90°C for 24 hr. Then, the residual metal ion mixture from Al-MOF synthesis was further heated at 90°C for 24 hr. Square-shaped Ni-Mn-BTC MOF was obtained in the presence of HCl, while HKUST-1(Cu MOF) was obtained when HNO_3 was introduced. Furthermore, a promising result was shown during their scale-up process since 15 g of waste could be converted to 10 g of MOFs, demonstrating an effective and interesting method for upcycling spent Li-ion batteries (Cognet et al. 2020).

Vanadium-based MOF is an interesting example of MOFs synthesized using carbon black waste. The leftover carbon black from crude oil processing contains a high vanadium (V) concentration, approximately 1,400 ppm. Vanadium could be extracted from carbon black using NaOH as the chemical leaching agent with an efficiency of ~97.2%. Vanadium was then reduced using sodium dithionite from the V^{5+} state to its favorable V^{3+} state prior to the MOF synthesis (Selwyn and Tse 2008). Subsequently, the hydrothermal synthesis of V-MOFs was performed at 200°C for 15 h in the presence of BDC linkers and cetyltrimethylammonium bromide (CTAB) as a capping agent, giving rise to V-BDC-i MOF nanofibers. Similarly, replacing the BDC linker with 1,4-naphthalenedicarboxylic acid (NDC) has given rise to V-NDC-i MOF nanorods. The capping agent, CTAB, was used to control the morphological shape of the V-MOF crystals, which also aided the solubility of the organic linker. This greener synthetic approach can produce highly crystalline V-BDC and V-NDC MOFs, serving as an environmentally sustainable source of metal ions through the chemical leaching of industrial wastes (Liu et al. 2012; Zhan et al. 2018).

MOFs from eggshells highlight the creativity of scientists in producing valuable materials from common waste. Eggs, a major protein-rich food consumed globally, generate significant waste in the form of eggshells, which are rich in calcium carbonate ($CaCO_3$) and have been successfully used to produce Ca-BDC MOF. Van Niekerk and Langmi (2023) have innovatively utilized $CaCO_3$ from the waste eggshells, where 94.92% comprises $CaCO_3$. Ca-BDC MOF was prepared by reacting finely ground eggshell powder with BDC using DMF as the solvent under solvothermal conditions. The reaction temperature played a significant role in producing Ca-BDC MOFs with different phases. The reaction temperatures of 110°C and 150°C produced triclinic and orthorhombic phases of Ca-BDC MOF, respectively, where 130°C led to the formation of a mixed phase of triclinic and orthorhombic phases. The PXRD patterns and SEM analysis further evidenced the phase conversion of the synthesized Ca-MOF. This study demonstrates a sustainable and greener method that potentially eliminates energy-intensive processes required for converting limestone, dolomite, and aragonite into commercial $CaCO_3$ (Van Niekerk and Langmi 2023).

The above-mentioned MOF products were used in various applications, with waste being the source only for the metal ions in their synthesis. The subsequent section of this chapter delves into synthesizing MOFs using both waste metal ions and organic linkers derived from waste.

10.4.4 MOFs Prepared Using Inorganic and Organic Wastes

Utilizing organic and inorganic waste materials as resources for organic linkers and metal precursors to create high-value MOFs offers a more environmentally friendly and cost-effective route within the circular economy framework. This approach promotes sustainability while minimizing environmental impact. Due to the presence of contaminants and impurities in the waste-derived metal ions and organic linkers, the structure of the final MOF products slightly deviates from their pristine form. MIL-53(Al), Ni-MOF, Ca-BDC MOF, MIL-101(Cr), and Cu-BDC MOF are notable examples synthesized using organic and inorganic waste sources, as shown in Figure 10.6.

The synthesis of MIL-53(Al) discussed thus far has obtained metal ions or organic linkers from commercial sources. Combining BDC monomers from alkaline hydrolysis with the Al metal ions extracted from Coca-Cola cans under hydrothermal conditions (220°C for 72 hr) could produce MIL-53(Al) MOFs. It is important to mention that the Al cans were processed to remove paint from their surfaces and dissolved in a 6M HCl solution. The aluminum nitrate crystals were obtained using a concentrated nitric acid solution, resulting in aluminum nitrate with a purity of 93% (w/w). The waste-derived MIL-53(Al) was further activated by heating it at 330°C for 24 hr to remove trapped solvents and linkers from the

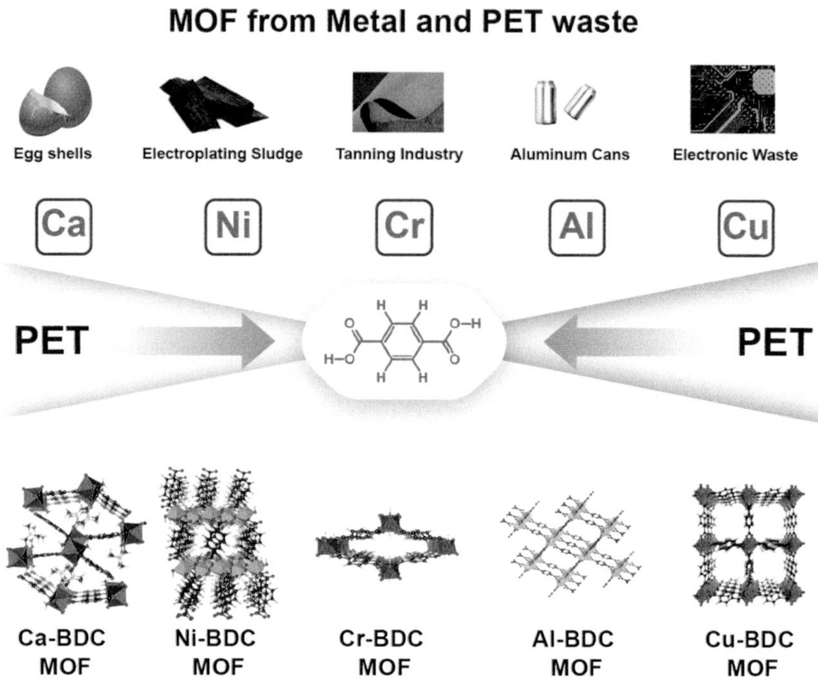

Figure 10.6 Various MOFs have been synthesized using both metal and PET wastes. (The images were generated using databases as referenced in the following publications: (Carson et al. 2014; Carton et al. 2007; Liang et al. 2011; Meilikhov et al. 2009; Serre et al. 2002).

structure. The FTIR and PXRD analysis confirmed the successful preparation of the MOF since it demonstrated a micro-crystalline structure with narrow pores similar to the pristine MIL-53 (Al) (Van Niekerk and Langmi 2023).

Similarly, the work carried out by Panda et al. reports the synthesis of MIL-53(Al) PET waste bottles (beverage bottles and water bottles), aluminum cans, and Al foils. The PXRD pattern of MIL-53(Al) shows diffraction planes (110), (011), and (211) corresponding to a well-defined crystalline structure of commercial MIL-53(Al) with an aggregated rod-like structure with a length of 1 to 2 μm. Most importantly, the cost of producing kilogram-scale MIL-53(Al) from waste PET bottles and aluminum nitrate is approximately one-third of that of commercial MIL-53(Al) from Sigma-Aldrich. Hence, this protocol provides a cost-effective and sustainable method to overcome the large-scale production of MOFs by utilizing solid wastes (Panda et al. 2020).

Combining electroplating sludges (EPS) with PET waste bottles presents another viable approach to producing waste-derived MOFs. For instance, Ni-MOF has been successfully produced using Ni-rich EPS and PET waste. Alkaline hydrolysis of PET and acid leaching of EPS could produce crystalline Ni-BDC MOFs under hydrothermal conditions (140°C for 24 hr). The PXRD patterns of the as-synthesized waste-derived Ni-MOF-W were similar to those of the MOF prepared using commercial chemicals (Ni-MOF-C), exhibiting 2D MOF layers with lengths ranging from several hundred nanometers to micrometers. It was observed that the size of Ni-MOF-W was considerably shorter compared to Ni-MOF-C, likely due to the influence of co-existing ions (such as Na^+) in the leaching solution. Overall, the upcycling of waste PET bottles and EPS to porous material was facile and cost-effective, and the conversion rate of the sources could reach ~100% (Song et al. 2021).

Crickmore et al. recently developed Ca-BDC MOF by utilizing PET waste and chicken eggshells as organic linkers and Ca^{2+} sources. Waste Ca-BDC MOF was prepared using two different approaches. In the first step, ball-milled eggshell powder and PET flakes were heated at 210°C for 15 hr in a mixture of DI water and ethylene glycol. The resulting solid was $Ca(BDC)(H_2O)_3$, as confirmed by TGA, PXRD, and SEM. The second approach, a direct mechanochemical synthesis of MOF, was carried out by ball-milling eggshells with PET flakes, which was unsuccessful. Therefore, a two-step process was employed: depolymerized BDC was mixed with eggshells and ball-milled with a small amount of water, successfully forming the Ca-MOF. This technique also offers a low-cost, sustainable method with minimal solvent use (Crickmore et al. 2021).

The tanning industry significantly contributes to chromium (Cr) waste pollution, generating substantial amounts of Cr-containing effluents, which, if not properly treated, can lead to severe environmental contamination. The study providing a sustainable solution by producing Cr MOFs using wastewater and PET waste is particularly inspiring in this context. The tannery effluent contains ~263.7 ± 28.2 mg/L of Cr ions, further concentrated by evaporation, and the BDC linkers obtained via alkaline hydrolysis of PET. The hydro/solvothermal reaction between Cr metal liquor and organic linker could produce MIL-101(Cr) and MIL-53(Cr). However, the one-pot synthesis approach of reacting tannery effluent liquor and PET flakes resulted in an amorphous product of unknown phases. Both waste-derived MIL-101 and MIL-53 (Cr) have shown characteristic PXRD patterns like those reported in the literature, which confirmed the successful transformation of wastes to porous material. The waste-derived Cr-BDC MOFs also demonstrated high thermal and hydrolytic stability, as their PXRD pattern remained unchanged after 3 days in water. N_2 BET analysis showed that waste-derived MIL-53(Cr) (890 m^2/g) and MIL-101(Cr) (2,546 m^2/g) have high BET surface areas, comparable to those prepared with commercial chemicals, MIL-53(Cr) (996 m^2/g) and MIL-101(Cr) (2,849 m^2/g). This approach could generate low-cost waste-MOFs by reducing the cost from ~\$450 to ~\$50 (estimated amount to produce 1 kg of MOF) (Delhali et al. 2023).

With the global demand for electric and electronic equipment steadily rising, the production of PCBs has soared, leading to a staggering increase in PCB waste accumulation, estimated to exceed 2 million tons. This surge in waste highlights the urgent need for sustainable solutions to manage wastewater and mitigate environmental impacts. Therefore, Zhang and colleagues have discovered an intriguing method for recovering copper (Cu) from PCB wastewater, where copper is found in the highest concentration, at 94.50 g/L. The alkali reduction

(AR) of wastewater, a byproduct of textile manufacturing, contains a significant concentration of BDC (6.03 g/L). Waste-derived Cu-BDC MOF was developed utilizing PCB wastewater as a source of Cu metal ions and AR wastewater as a source for BDC linkers. The synthesis of Cu-BDC MOF involved stirring a specified amount of PCB and AR wastewater in a water/ethanol mixture, followed by adding trimethylamine (TEA) as a base. Cu-BDC-W (waste-derived) and Cu-BDC-C (commercial sources) have shown similar 2D laminar structures, but Cu-BDC-W depicted few-layered platelets with a thinner structure. Moreover, the XRD patterns of the as-synthesized Cu-BDC-W agree with the simulated result and exhibit a type-I isotherm with a S_{BET} of 69.06 m^2/g. The economic evaluation of this study revealed that producing Cu-BDC MOF from PCB and AR wastewater leads to notably higher gross profits compared to current strategies for treating AR wastewater and recovering Cu from PCB wastewater in practice. Thus, this pathway provides a facile, simple, and cost-effective method to produce value-added MOF from wastewater in a more sustainable and greener approach (Zhang et al. 2020).

10.5 ENVIRONMENTAL REMEDIATION USING WASTES DERIVED MOFs

Employing waste-derived MOFs as adsorbents and catalysts for environmental remediation offers a promising approach to conserving resources while effectively removing hazardous wastes and contaminants from the environment. The following section of this chapter explores the application of waste-derived MOFs in environmental remediation, involving the removal of organic pollutants, CO_2 capture/conversion, and mitigating microplastics in the environment. This approach not only aids in waste management but also plays a pivotal role in fostering a cleaner and healthier living environment.

10.5.1 Removal of Organic Pollutants Using Wastes Derived MOFs

MOFs are promising adsorptive materials for various organic contaminants and greenhouse gases due to their high surface area, large porosity, and tunable structures. It has been used as an adsorber and a catalyst to treat wastewater by eliminating organic pollutants through adsorption and catalytic degradation. The MIL-53(Cr) fabricated from EPS has been used as an adsorbent for the adsorptive removal of dibenzothiophene (DBT) from liquid fuel. When synthesized with HF, MIL-53(Cr) exhibits a higher desulfurization capacity (40.11 mg/g) compared to synthesis without HF (32.80 mg/g). The higher desulfurization capacity is mainly due to the production of highly crystalline and relatively pure MIL-53(Cr) with large porosity and surface area, facilitated by the HF added during synthesis. Even after five cycles, the waste-derived MIL-53(Cr) synthesized with HF retained over 94% of its initial sulfur adsorption capacity (Kabtamu et al. 2020).

Catalytic reduction of methyl orange (MO) using waste PET-derived Cu-BDC MOF with $NaBH_4$ shows interesting properties compared to reported cobalt, gold, silver, and platinum nanoparticles. Thermal activation of Cu-MOF at 160°C and 225°C was crucial for catalysis, resulting in significant methylene orange reduction, whereas no activation led to minimal reduction. Both activated waste-derived Cu-BDC MOFs achieved a dye removal efficiency of approximately 85% in just 5 min, surpassing reported results for cobalt, gold, silver, and platinum nanoparticles in terms of speed. As reported by the authors, in the catalytic reduction of methyl orange, the activated Cu-BDC MOF increases the hydrogen generation rate by $NaBH_4$ and transfers the obtained reducing agents to dye molecules. Then, the liberated hydrogen reduces the azo group of MO, producing sodium p-aminobenzenesulphonate and N,N-dimethyl-p-phenylenediamine as products. The reduction was performed for five consecutive cycles using Cu-BDC MOF (160°C) without reactivation to assess its recyclability, with the removal efficiency remaining constant, where the reduction rate dropped with each reuse (Rahmani et al. 2018).

The adsorptive removal of dimetridazole (DMZ) using defective D-MIL-68(Al) shows an ultrafast and outstanding adsorption capacity where 97% of DMZ was removed in 10 s, exhibiting the record adsorption kinetic constant of 2.84 g/mg/min. The adsorption capacity of PET-derived D-MIL-68(Al) was 3 times higher than that of the pristine MIL-68(Al). The generation of multiple active sites in PET-derived D-MIL-68(Al) has accelerated mass transfer diffusion, enhancing the adsorption of DMZ. Additionally, the π-π stacking interactions, Al-OH sites, and strong H-bonding interactions between DMZ and the active sites in the D-MIL-68(Al) further facilitated the adsorption. The regeneration study revealed that the efficiency of DMZ removal remained nearly constant for five cycles without causing structural deformation, showcasing the outstanding stability of D-MIL-68(Al) (Heng et al. 2023).

4-nitrophenol, a persistent toxic contaminant used in the large-scale synthesis of acetaminophen, poses significant environmental and health risks when released into the environment (Xu et al. 2021). PET-derived Cu-BDC and Cu-NH_2-BDC MOFs have been employed for the adsorptive removal of 4-nitrophenol (4-NP). Cu-NH_2-BDC MOF exhibited superior adsorption of 4-NP compared to Cu-BDC MOF, with a Q_{max} of 161.3 mg/g (Langmuir isotherm), attributed to hydrogen bonding between the amino and nitro functional groups of the MOF and 4-NP. Electrostatic and π-π interactions also play a minor role in the adsorption process, further enhancing the adsorption capacity of Cu-NH_2-BDC MOF. Furthermore, Cu-NH_2-BDC showed potential simultaneous adsorption and rapid reduction of 4-nitrophenol(4-NP) to 4-aminophenol (4-AP) in the presence of $NaBH_4$ as the reducing agent. Higher adsorption capacity observed in the MOF enhances the reduction of 4-NP. Consequently, the catalytic process facilitates the efficient transfer of free electrons, facilitating the conversion of 4-NP into 4-AP. The recyclability of Cu-NH_2-BDC MOF as a catalyst for the reduction of 4-NP was assessed over twelve reduction cycles, demonstrating good stability with minimal structural changes observed through XRD and FTIR analysis (Ebrahimi et al. 2023).

Furthermore, waste-derived MOF-5, synthesized from PET, was used as an adsorbent for tetracycline (TC) removal at a maximum adsorption capacity of 2,325 mg/g (25°C). The adsorption capacity exhibited temperature sensitivity, with higher capacities of 2,653 mg/g and 3,199 mg/g achieved at 35°C and 45°C, respectively. Further studies carried out loading MOF-5 into hydrogel beads were also used as an adsorbent for TC removal (Cheng et al. 2024).

These studies demonstrate the significant potential of waste-derived MOFs as effective adsorbents for environmental remediation, showcasing their ability to transform waste materials into valuable solutions for tackling pollution and promoting sustainability.

10.5.2 Waste-Derived MOFs for CO_2 Capture and Conversion

MOFs are a promising group of adsorbents for CO_2 capture due to their unique properties, including high porosity, large surface area, and ease of functionalization, which enable them to provide numerous adsorption sites. Some well-known MOFs, including NH_2-MIL-53, Ni-MOF-74, Zn-MOF-74, UiO-66, HKUST-1, and MIL-101, have been synthesized from organic or inorganic wastes for use in CO_2 capture (Ding et al. 2019). Moreover, MOFs exhibit high selectivity towards CO_2, making them ideal for gas separation applications. Additionally, MOFs have been successfully used as catalysts to convert CO_2 into syngas or other valuable products. Thus, this demonstrates the promising adsorption and catalytic properties of waste-derived MOFs in CO_2 capture and conversion applications comparable to those of pristine MOFs.

Waste derived UiO-66, Cu-BDC-W, MIL-53(Al), MIL-53(Cr), and MIL-101(Cr) have been used in CO_2 capture, with adsorption capacities of 2.06 mmol/g (Waribam et al. 2023), 5.75 cm^3/g (Zhang et al. 2020), 2.10 mmol/g (Panda et al. 2020), 3.64 mmol/g, and 4.88 mmol/g (Lo et al. 2016), respectively. In UiO-66, the increased CO_2 adsorption capacity of the highly defective UiO-66 MOF was attributed to the presence of missing cluster defects, which consequently led to increased microporosity and an abundance of open metal sites (Waribam et al. 2023).

In the case of waste-derived Cr-based MIL-101(Cr) (Lo et al. 2016), though the MOFs were not activated to remove residual contaminants trapped in the pores, a CO_2 adsorption capacity of 4.88 mmol/g was achieved. The study demonstrates the potential of attaining high-purity MOFs from waste, avoiding typical activation processes, and suggests thermal activation of waste-derived MOFs could further enhance their performance. The Cu-BDC-W MOF derived from printed circuit boards (PCB) and alkali reduction (AR) wastewater has demonstrated excellent CO_2 capture properties (5.75 cm^3/g), which was slightly higher than the Cu-BDC-C (4.05 cm^3/g) prepared using commercial resources (Zhang et al. 2020). The slightly higher CO_2 gas uptake observed in Cu-BDC-W has been attributed to the contribution of interlamellar voids in the structure. This approach offers insights into developing MOFs from various waste sources, potentially creating a feasible platform for using waste-derived MOFs in industrial applications in the future.

Furthermore, the waste-derived MOFs that have been used as catalysts in CO_2 conversion further expand the potential of these materials. MIL-47 (V) and MIL-101 (Cr), synthesized from waste PET, have demonstrated excellent catalytic activity in converting CO_2 into valuable products. MIL-47 (V) was tested for the chemical fixation of CO_2 by converting CO_2 and epoxides into cyclic carbonates, allowing the reaction between propylene oxide (PO) and CO_2 to form propylene carbonate (PC), using tetra-n-butylammonium bromide (n-Bu_4NBr) as a co-catalyst. MIL-47(V) achieved a 95% conversion rate in 24 hr under mild reaction conditions (50°C, 150 psi CO_2). Without the co-catalyst, the yield of PC was significantly lower (15% with MIL-47 and 9% with n-Bu_4NBr), highlighting the efficiency of the combined catalytic system. Similarly, MIL-101(Cr), synthesized from waste PET in the presence of n-Bu_4NBr, was tested for the coupling reaction of cyclohexene oxide (CHO) and CO_2 to produce cyclohexene carbonate (CHC). At 80°C and 2 MPa CO_2, this system achieved over 90% conversion of CHO with a turnover number (TON) of 469 in 72 h. Notably, the MIL-101(Cr)/n-Bu_4NBr system was highly selective, yielding over 99% of the cis-isomer of CHC. Hence, this research demonstrates that MIL-47(V) and MIL-101(Cr), in combination with n-Bu_4NBr as a cocatalyst, are promising catalytic systems for CO_2 cycloaddition reactions under mild conditions (Lo et al. 2016).

Fe and Ni MOFs, derived from Fe/Ni-rich EPS, have been used as catalysts for the photoreduction of CO_2 to syngas in a mixed solution of acetonitrile and H_2O under 5 W LED irradiation (400–1,000 nm) with $[Ru(bpy)_3]Cl_2 \cdot 6H_2O$ as the photosensitizer and triethanolamine as the electron donor. The main products detected were CO and H_2. Fe MOF exhibited a CO production rate of 2.3 $\mu molh^{-1}$ and an H_2 evolution rate of 11.4 $\mu molh^{-1}$ in pure CO_2, with a CO selectivity of 16.8%. Ni MOF demonstrated a higher CO production rate of 13.6 $\mu molh^{-1}$ and a lower H_2 generation rate of 0.42 $\mu molh^{-1}$, resulting in a CO selectivity of 97%. When tested with 10% CO_2, typical for fire power plant exhaust, Ni MOF CO production rate slightly decreased to 11.9 $\mu molh^{-1}$, and CO selectivity dropped to 93.5%. In contrast, Fe MOF showed an increased H_2 production rate of 12.2 μmol/h and a reduced CO rate of 0.81 μmol/h under low-concentration CO_2. These results indicate that low-concentration CO_2 conditions favor H_2 production and decrease CO selectivity in Fe MOF, particularly highlighting the superior performance of Ni MOF for CO_2 conversion in both pure and low-concentration CO_2 environments (Han et al. 2021). Thus, this shows that EPS waste could be a useful resource for preparing waste-derived MOFs, addressing the challenge of producing tunable syngas via low-concentration CO_2 photoreduction. This approach is expected to open new opportunities for a sustainable future in the industry.

Similarly, research on the photoconversion of CO_2 to H_2 and CO using a layered Ni MOF-W synthesized from Ni EPS and PET waste has shown promising results. The Ni-MOF-W nano catalyst achieved a CO evolution rate of ~10.0 mmol/h/g and an H_2 production rate of ~0.4 mmol/h/g, with a CO selectivity of 96.7%. The apparent quantum yield at 420 nm was 1.36%. Considering the environmental benefits and economic advantages, Ni-MOF-W derived from EPS and PET wastes is a highly promising material for various practical applications (Song et al. 2021).

10.5.3 MOFs to Remove Microplastics from Water or Wastewater

Microplastics (MPs) or nano-plastics (NPs) are tiny plastic particles in the range of millimeters or nanometers in diameter that have emerged as a substantial environmental pollutant with widespread implications for ecosystems and human health. These particles originate from various sources, including cosmetics, polythene bags, plastic containers, electrical appliances, goods packaging, glass, microbeads in personal care products, and synthetic fibers from clothing. Some of the major plastic components of MPs/NPs include polyethylene terephthalate (PET), polyethylene (PE), polypropylene (PP), polyvinyl chloride (PVC), polylactic acid (PLA), polyamide (PA), polystyrene (PS), polyurethane (PU), and polycarbonates (PC). These are often mixed with functional additives, such as organic dyes (Crawford and Quinn 2017). Secondary MPs/NPs originate from the breakdown of larger plastics into nano, micro, and macro sizes due to weathering processes such as wind abrasion, wave action, photodegradation, biodegradation, hydrolysis, and ultraviolet radiation (Gewert et al. 2015). Their small size allows them to infiltrate water bodies, soil, and air, making them ubiquitous in the environment. Additionally, microplastics can absorb and transport hazardous chemicals, exacerbating their environmental impact. Consequently, micro/nano plastic pollution poses a growing threat that requires urgent and coordinated global mitigation efforts (Lamichhane et al. 2023).

Hence, several techniques have been employed to eliminate MPs/NPs from the environment, including (1) sorption and filtration methods, such as adsorption on green algae, membrane technologies, and advanced filtration technologies in wastewater treatment plants; (2) chemical treatments for microplastics; and (3) biological removal and ingestion (Padervand et al. 2020). Therefore, MOFs are considered one of the potential materials for removing MPs/NPs from water and wastewater through adsorption, catalytic degradation, and membrane filtration.

ZIF-67 has been utilized as an adsorbent to remove polystyrene (PS) MPs from water, achieving an adsorption ratio of up to 92.1% and a maximum adsorption capacity of 11.6 mg/g, reaching equilibrium within 20 minutes. The effectiveness of ZIF-67 spans a pH range of 3 to 10, with optimal adsorption observed around pH 8. Adsorption isotherms indicate that the process involves multilayer adsorption, fitting well with the Freundlich model. This study highlights the efficiency of the treatment and cost-effectiveness, demonstrating the potential of MOFs in removing PS MPs from water (Wan et al. 2022).

MIL-101 (Cr) MOF has been investigated as a promising adsorbent for removing polystyrene nano plastics (PSNPs) from Milli-Q water at various pH levels, achieving a 96% removal rate at pH 5 for initial concentrations of 5 and 70 ppm. The adsorption isotherm at room temperature best fits the Freundlich model, with a maximum adsorption capacity of 800 mg/g. The effective removal is due to electrostatic interactions between the negatively charged PSNPs and the positively charged MIL-101 (Cr) MOF. Additionally, MIL-101 demonstrated acceptable regeneration capacity, supporting its potential for sustainable water purification (Modak et al. 2023).

Recent studies have shown the effectiveness of nanofiltration in removing microplastics (MPs) from textile wastewater by incorporating MIL-100 (Fe) into polysulfone (PSF) to create a mixed matrix membrane. The optimized membrane (M0.5), containing 0.5 wt% hydrophilic MIL-100 (Fe), demonstrated the highest rejection rate for methylene blue (MB), superior permeability, enhanced pore size, and excellent anti-fouling performance. Methylene blue was not adsorbed onto the membrane mainly due to electrostatic repulsion. It performed well across various dye and MP concentrations, showed good salt rejection, and could be reused for six cycles of MB dye rejection. These improvements are due to size exclusion, electrostatic interaction, and repulsion, making the PSF/MIL-100 (Fe) membrane a promising solution for treating textile wastewater (Gnanasekaran et al. 2021).

Given the substantial contamination of the ocean with microplastics, there is a significant demand for technology to remove these particles from seawater. The composite material developed by combining melamine porous foam (MF) with Zr-based UiO-66-X, featuring a range of functional groups (H, NH_2, OH, Br, and NO_2), has shown impressive results. The composites were tested on suspensions of polyvinylidene fluoride (PVDF), poly(methyl methacrylate) (PMMA), and polystyrene (PS), simulated microplastic containing water, and simulated seawater. UiO-66-OH@MF-3 stood out, achieving up to 95.5% MPs removal efficiency and maintaining performance over several filtration cycles. The high efficiency of UiO-66-OH@MF-3 is due to the functional groups of MOF nanoparticles and structural defects, which facilitate adsorption through hydrogen bonds, electrostatic, and Van der Waals interactions. This MOF composite foam shows great potential for effectively removing MPs and NPs from water (Chen et al. 2020).

Furthermore, UiO-66(Zr) MOF was used in PET degradation, producing terephthalic acid and mono-methyl terephthalate in a solvent-free process at 260°C in 24 hr. The effectiveness of the UiO-66 catalyst, in terms of activity and selectivity for PET degradation, remained high even in the presence of polypropylene (PP) or polyethylene (PE), achieving conversion yields up to 93%. The study further revealed that UiO-66 was transformed into MIL-140A during PET degradation. These findings highlight the versatility of MOFs as an innovative class of catalysts for the degradation of PET MPs and NPs, offering solutions to environmental challenges (Wu et al. 2022).

10.6 CONCLUSION

This chapter highlights the pathway of upcycling organic and inorganic wastes to transform them into valuable functional materials, commonly known as metal-organic frameworks (MOFs). The widely used plastic, polyethylene terephthalate (PET), can be depolymerized to its BDC monomer, which can then be combined with metals extracted from various waste sources to form waste-derived MOFs. The waste-derived MOFs showed promising properties as adsorbents for removing harmful pollutants such as dyes, antibiotics, heavy metals, and microplastics, as well as for capturing toxic gases from the environment. Waste-derived MOFs have also demonstrated good catalytic properties and have been used as heterogeneous catalysts for converting CO_2 into valuable products. This chapter highlights that both plastic and inorganic wastes, notorious for causing environmental pollution, can be innovatively transformed into MOFs with significant potential for various applications. It is also presented as one of the options in the holistic approach toward the circular economy, where waste streams are converted into useful functional materials.

10.7 ACKNOWLEDGMENTS

PW gratefully acknowledges the Thammasat Postdoctoral Fellowship (TUPD 6/2567). TK gratefully acknowledges the TU-NSTDA Excellent Research Graduate Scholarship (No. T2.1-6.-01) from the National Science and Technology Development Agency (NSTDA), Thammasat University, and the EFS Scholarship from Sirindhorn International Institute of Technology. WC gratefully acknowledges research support from Thammasat University, TUFT 033/2563. PS gratefully acknowledges research support from the Program Management Unit for Human Resources and Institutional Development, Research, and Innovation grant (no. B40G660035) and general support from the Center of Excellence in Digital Earth and Emerging Technology (DEET), Thammasat University.

REFERENCES

Abedsoltan, H. 2023. A focused review on recycling and hydrolysis techniques of polyethylene terephthalate. Polym. Eng. Sci. 63: 2651–2674.

Al-Sabagh, A.M., F.Z. Yehia, G. Eshaq, A.M. Rabie and A.E. ElMetwally. 2016. Greener routes for recycling of polyethylene terephthalate. Egypt. J. Pet. 25: 53–64.

Babaei, M., M. Jalilian and K. Shahbaz. 2024. Chemical recycling of Polyethylene terephthalate: A mini-review. J. Environ. Chem. Eng. 12: 112507.

Carson, C.G., G. Brunnello, S.G. Lee, S.S. Jang, R.A. Gerhardt and R. Tannenbaum. 2014. Structure solution from powder diffraction of copper 1,4-benzenedicarboxylate. Eur. J. Inorg. Chem. 2014: 2140–2145.

Carton, A., A. Mesbah, T. Mazet, F. Porcher and M. Francois. 2007. Ab initio crystal structure of nickel(II) hydroxy-terephthalate by synchrotron powder diffraction and magnetic study. Solid State Sci. 9: 465–471.

Chen, W.S., C.T. Liao and K.Y. Lin. 2017. Recovery zinc and manganese from spent battery powder by hydrometallurgical route. Energy Procedia. 107: 167–174.

Chen, Y.J., Y. Chen, C. Miao, Y.R. Wang, G.K. Gao, R.X. Yang, et al. 2020. Metal–organic framework-based foams for efficient microplastics removal. J. Mater. Chem. 8: 14644–14652.

Cheng, S., Y. Li, Z. Yu, R. Gu, W. Wu and Y. Su. 2024. Waste PET-derived MOF-5 for high-efficiency removal of tetracycline. Sep. Purif. Technol. 339: 126490.

Coetzee, J.J., N. Bansal and E.M.N. Chirwa. 2020. Chromium in environment, its toxic effect from chromite-mining and ferrochrome industries, and its possible bioremediation. Expo. Health. 12: 51–62.

Cognet, M., J. Condomines, J. Cambedouzou, S. Madhavi, M. Carboni and D. Meyer. 2020. An original recycling method for Li-ion batteries through large scale production of metal organic frameworks. J. Hazard. Mater. 385: 121603.

Colnik, M., Z. Knez and M. Skerget. 2021. Sub- and supercritical water for chemical recycling of polyethylene terephthalate waste. Chem. Eng. Sci. 233: 116389.

Corak, I., A. Tarbuk, D. Dordevic, K. Visic and L. Botteri. 2022. Sustainable alkaline hydrolysis of polyester fabric at low temperature. Mater. 15: 1530.

Crawford, C. and B. Quinn. 2017. Physiochemical properties and degradation. 57–100. *In*: C.B.C.a.B. Quinn (ed.). Microplastic Pollutants. Elsevier Science, Netherlands.

Crickmore, T.S., H.B. Sana, H. Mitchell, M. Clark and D. Bradshaw. 2021. Toward sustainable syntheses of Ca-based MOFs. Chem. Comm. 57: 10592–10595.

Deleu, W.P.R., I. Stassen, D. Jonckheere, R. Ameloot and D.E. De Vos. 2016. Waste PET (bottles) as a resource or substrate for MOF synthesis. J. Mater. Chem. A. 4: 9519–9525.

Delhali, A., A.H. Assen, A. Mohammed, K. Adil and Y. Belmabkhout. 2023. Enabling simultaneous valorization of tannery effluent and waste plastic via sustainable preparation of Cr-BDC MOFs for water adsorption. Sci. Rep. 13: 14653.

Ding, M., R.W. Flaig, H.L. Jiang and O.M. Yaghi. 2019. Carbon capture and conversion using metal–organic frameworks and MOF-based materials. Chem. Soc. Rev. 48: 2783–2828.

Doan, V.D., T.L. Do, T.M.T. Ho, V.T. Le and H.T. Nguyen. 2020. Utilization of waste plastic pet bottles to prepare copper-1,4-benzenedicarboxylate metal-organic framework for methylene blue removal. Sep. Sci. Technol. 55: 444–455.

Du, K., E.H. Ang, X. Wu and Y. Liu. 2022. Progresses in sustainable recycling technology of spent Lithium-Ion batteries. Energy Environ. Mater. 5: 1012–1036.

Ebrahimi F.M., N. Madadian Bozorg, A. Ehsani, H. Aghdasinia, Z. Chen, S. Rostamnia, et al. 2023. Green valorization of PET waste into functionalized Cu-MOF tailored to catalytic reduction of 4-nitrophenol. J Environ Manage. 345: 118842.

Eddaoudi, M., J. Kim, N. Rosi, D. Vodak, J. Wachter, M. O'Keeffe, et al. 2002. Systematic design of pore size and functionality in isoreticular MOFs and their application in methane storage. Sci. 295: 469–472.

El-Sherif, I.Y., S. Tolani, K. Ofosu, O.A. Mohamed and A.K. Wanekaya. 2013. Polymeric nanofibers for the removal of Cr(III) from tannery waste water. J. Environ. Manage. 129: 410–413.

Genchi, G., A. Carocci, G. Lauria, M.S. Sinicropi and A. Catalano. 2020. Nickel: Human health and environmental toxicology. Int. J. Environ. Res. Public Health. 17: 679.

Georgopoulos, P.G., S.W. Wang, I.G. Georgopoulos, M.J. Yonone-Lioy and P.J. Lioy. 2006. Assessment of human exposure to copper: A case study using the NHEXAS database. J. Expo. Sci. Environ. Epidemiol. 16: 397–409.

Gewert, B., M.M. Plassmann and M. MacLeod. 2015. Pathways for degradation of plastic polymers floating in the marine environment. Environ. Sci.: Process. Impacts. 17: 1513–1521.

Geyer, R., J.R. Jambeck and K.L. Law. 2017. Production, use, and fate of all plastics ever made. Sci Adv. 3: e1700782.

Ghosh, S.K., S. Pal and S. Ray. 2013. Study of microbes having potentiality for biodegradation of plastics. Environ. Sci. & Pollut. Res. 20: 4339–4355.

Gnanasekaran, G., Arthanareeswaran G. and Y.S. Mok. 2021. A high-flux metal-organic framework membrane (PSF/MIL-100 (Fe)) for the removal of microplastics adsorbing dye contaminants from textile wastewater. Sep. Purif. Technol. 277: 119655.

Han, B., X. Ou, Z. Zhong, S. Liang, X. Yan, H. Deng, et al. 2021. Photoconversion of anthropogenic CO_2 into tunable syngas over industrial wastes derived metal-organic frameworks. Appl. Catal. B: Environ. 283: 119594.

Heng, Y., Z. Fang, J. Li, L. Luo, M. Zheng and H. Huang. 2023. Defective metal–organic framework derived from the waste plastic bottles for rapid and efficient nitroimidazole antibiotics removal. J. Colloid Interface Sci. 650: 836–845.

Joshi, J.N., C.M. Moran, H.P. Feininger, J.M. Dow and K.S. Walton. 2019. Household aluminum products as insoluble precursors for directed growth of metal–organic frameworks. Cryst. Growth Des. 19: 5097–5104.

Kabtamu, D.M., Y.N. Wu, Q. Chen, L. Zheng, K.I. Otake, L. Matovic, et al. 2020. Facile Upcycling of hazardous Cr-containing electroplating sludge into value-added metal–organic frameworks for efficient adsorptive desulfurization. ACS Sustain. Chem. Eng. 8: 12443–12452.

Karam, L., A. Miglio, S. Specchia, N. El Hassan, P. Massiani and J. Reboul. 2021. PET waste as organic linker source for the sustainable preparation of MOF-derived methane dry reforming catalysts. Mater. Adv. 2: 2750–2758.

Katugampalage, T.R., P. Waribam, P. Opaprakasit, C. Kaewsaneha, S.H. Hsu, W. Chooaksorn, et al. 2024. Bimetallic Fe:Co metal–organic framework (MOF) with unsaturated metal sites for efficient Fenton-like catalytic degradation of oxytetracycline (OTC) antibiotics. Chem. Eng. J. 479: 147592.

Kim, H., S. Das, M.G. Kim, D.N. Dybtsev, Y. Kim and K. Kim. 2011. Synthesis of phase-pure interpenetrated MOF-5 and its gas sorption properties. Inorg. Chem. 50: 3691–3696.

Lamichhane, G., A. Acharya, R. Marahatha, B. Modi, R. Paudel, A. Adhikari, et al. 2023. Microplastics in environment: global concern, challenges, and controlling measures. Int. J. Environ. Sci. 20: 4673–4694.

Lebedev, O.I., F. Millange, C. Serre, G. Van Tendeloo and G. Ferey. 2005. First direct imaging of giant pores of the metal–organic framework MIL-101. Chem. Mat. 17: 6525–6527.

Leyssens, L., B. Vinck, C. Van Der Straeten, F. Wuyts and L. Maes. 2017. Cobalt toxicity in humans—A review of the potential sources and systemic health effects. Toxicology. 387: 43–56.

Li, H., M. Eddaoudi, M. O'Keeffe and O.M. Yaghi. 1999. Design and synthesis of an exceptionally stable and highly porous metal-organic framework. Nature. 402: 276–279.

Li, C., F. Xie, Y. Ma, T. Cai, H. Li, Z. Huang, et al. 2010. Multiple heavy metals extraction and recovery from hazardous electroplating sludge waste via ultrasonically enhanced two-stage acid leaching. J. Hazard. Mater. 178: 823–833.

Liang, P.C., H.K. Liu, C.T. Yeh, C.H. Lin and V. Zima. 2011. Supramolecular assembly of calcium metal–organic frameworks with structural transformations. Crystal Growth & Design. 11: 699–708.

Liu, Q., L.N. Jin and W.Y. Sun. 2012. Facile fabrication and adsorption property of a nano/microporous coordination polymer with controllable size and morphology. Chem. Comm. 48: 8814–8816.

Lo, S.H., D.S. Raja, C.W. Chen, Y.H. Kang, J.J. Chen and C.H. Lin. 2016. Waste polyethylene terephthalate (PET) materials as sustainable precursors for the synthesis of nanoporous MOFs, MIL-47, MIL-53(Cr, Al, Ga) and MIL-101(Cr). Dalton Trans. 45: 9565–9573.

Lutton-Gething, A.R.B.J., L.T. Nangkam, J.O.W. Johansson, I. Pallikara, J.M. Skelton, G.F.S. Whitehead, et al. 2023. Breathing behaviour modification of gallium MIL-53 metal–organic frameworks induced by the bridging framework inorganic anion. Chem. Eur. J. 29: e202203773.

Makuza, B., Q. Tian, X. Guo, K. Chattopadhyay and D. Yu. 2021. Pyrometallurgical options for recycling spent lithium-ion batteries: A comprehensive review. J. Power Sources. 491: 229622.

Masoomi, M., A. Morsali, A. Dhakshinamoorthy and H. Garcia. 2019. Mixed-metal MOFs: Unique opportunities in metal-organic framework functionality and design. Angew. Chem. 58: 15188–15205.

Meilikhov, M., K. Yusenko and R.A. Fischer. 2009. The adsorbate structure of ferrocene inside [Al(OH)(bdc)]x (MIL-53): a powder X-ray diffraction study. Dalton Trans. 600–602.

Mitra, S., A.J. Chakraborty, A.M. Tareq, T.B. Emran, F. Nainu, A. Khusro, et al. 2022. Impact of heavy metals on the environment and human health: Novel therapeutic insights to counter the toxicity. J. King Saud Univ. Sci. 34: 101865.

Modak, S., M. Kasula and M.R. Esfahani. 2023. Nanoplastics removal from water using metal–organic framework: Investigation of adsorption mechanisms, kinetics, and effective environmental parameters. ACS Appl. Eng. Mater. 1: 744–755.

Oien, S., D. Wragg, H. Reinsch, S. Svelle, S. Bordiga, C. Lamberti, et al. 2014. Detailed structure analysis of atomic positions and defects in zirconium metal–organic frameworks. Cryst. Growth Des. 14: 5370–5372.

Padervand, M., E. Lichtfouse, D. Robert and C. Wang. 2020. Removal of microplastics from the environment. A review. Environ. Chem. Lett. 18: 807–828.

Panda, D., S. Patra, M.K. Awasthi and S.K. Singh. 2020. Lab cooked MOF for CO_2 capture: A sustainable solution to waste management. J. Chem. Educ. 97: 1101–1108.

Park, S.H. and S.H. Kim. 2014. Poly (ethylene terephthalate) recycling for high value added textiles. Fash. Text. 1: 1.

Pistoia, G. 2005. Spent battery collection and recycling. 225–245. *In*: G. Pistoia (ed.). Batteries for portable devices. Elsevier Science, Netherlands.

Qasem, N.A.A., R.H. Mohammed and D.U. Lawal. 2021. Removal of heavy metal ions from wastewater: A comprehensive and critical review. npj Clean Water. 4: 36.

Qu, Z., T. Su, S. Zhu, Y. Chen, Y. Yu, X. Xie, et al. 2021. Stepwise extraction of Fe, Al, Ca and Zn: A green route to recycle raw electroplating sludge. J. Environ. Manage. 300: 113700.

Rahmani, A., H. Rahmani and A. Zonouzi. 2018. Cu(BDC) as a catalyst for rapid reduction of methyl orange: Room temperature synthesis using recycled terephthalic acid. Chem. Pap. 72: 449–455.

Ren, J., X. Dyosiba, N.M. Musyoka, H.W. Langmi, B.C. North, M. Mathe, et al. 2016. Green synthesis of chromium-based metal-organic framework (Cr-MOF) from waste polyethylene terephthalate (PET) bottles for hydrogen storage applications. Int. J. Hydrogen Energy. 41: 18141–18146.

Rocha-Santos, T. and A.C. Duarte. 2015. A critical overview of the analytical approaches to the occurrence, the fate and the behavior of microplastics in the environment. TrAC, Trends Anal. Chem. 65: 47–53.

Sarda, P., J.C. Hanan, J.G. Lawrence and M. Allahkarami. 2022. Sustainability performance of polyethylene terephthalate, clarifying challenges and opportunities. J. Polym. Sci. 60: 7–31.

Selwyn, L. and S. Tse. 2008. The chemistry of sodium dithionite and its use in conservation. Stud. Conserv. 53: 61–73.

Serre, C., F. Millange, C. Thouvenot, M. Nogues, G. Marsolier, D. Louer, et al. 2002. Very large breathing effect in the first nanoporous chromium(III)-based solids: MIL-53 or $Cr^{III}(OH)\cdot\{O_2C-C_6H_4-CO_2\}\cdot\{HO_2C-C_6H_4-CO_2H\}x\cdot H_2Oy$. J. Am. Chem. Soc. 124: 13519–13526.

Siddiqui, M.N., D.S. Achilias, H.H. Redhwi, D.N. Bikiaris, K.A.G. Katsogiannis and G.P. Karayannidis. 2010. Hydrolytic depolymerization of PET in a microwave reactor. Macromol. Mater. Eng. 295: 575–584.

Song, K., X. Qiu, B. Han, S. Liang and Z. Lin. 2021. Efficient upcycling electroplating sludge and waste PET into Ni-MOF nanocrystals for the effective photoreduction of CO_2. Environ. Sci. Nano. 8: 390–398.

Stock, N. and S. Biswas. 2012. Synthesis of Metal-Organic Frameworks (MOFs): Routes to various MOF topologies, morphologies, and composites. Chem. Rev. 112: 933–969.

Van Niekerk, G.R. and H.W. Langmi. 2023. Modulator-free green synthesis of the calcium-terephthalate metal-organic framework derived from waste eggshells. Polyhedron. 245: 116618.

Vanpoucke, D.E.P., J.W. Jaeken, S. De Baerdemacker, K. Lejaeghere and V. Van Speybroeck. 2014. Quasi-1D physics in metal-organic frameworks: MIL-47(V) from first principles. Beilstein J. Nanotechnol. 5: 1738–1748.

Vellingiri, K., D.C.W. Tsang, K.H. Kim, A. Deep, T. Dutta and D.W. Boukhvalov. 2018. The utilization of zinc recovered from alkaline battery waste as metal precursor in the synthesis of metal-organic framework. J. Clean. Prod. 199: 995–1006.

Volkringer, C., M. Meddouri, T. Loiseau, N. Guillou, J. Marrot, G. Ferey, et al. 2008. The Kagome topology of the gallium and indium metal-organic framework types with a MIL-68 structure: Synthesis, XRD, solid-state NMR characterizations, and hydrogen adsorption. Inorg. Chem. 47: 11892–11901.

Wan, H., J. Wang, X. Sheng, J. Yan, W. Zhang and Y. Xu. 2022. Removal of polystyrene microplastics from aqueous solution using the metal–organic framework material of ZIF-67. Toxics. 10: 70.

Wang, Y., H. Wang, H. Chen and H. Liu. 2022a. Towards recycling purpose: Converting PET plastic waste back to terephthalic acid using pH-responsive phase transfer catalyst. Chinese J. Chem. Eng. 51: 53–60.

Wang, Y., H. Wang, S. Li and S. Sun. 2022b. Waste PET plastic-derived CoNi-based metal–organic framework as an anode for Lithium-Ion batteries. ACS Omega. 7: 35180–35190.

Waribam, P., T.R. Katugampalage, P. Opaprakasit, C. Ratanatawanate, W. Chooaksorn, L.P. Wang, et al. 2023. Upcycling plastic waste: Rapid aqueous depolymerization of PET and simultaneous growth of highly defective UiO-66 metal-organic framework with enhanced CO_2 capture via one-pot synthesis. Chem. Eng. J. 473: 145349.

Wu, Y., X. Wang, K.O. Kirlikovali, X. Gong, A. Atilgan, K. Ma, et al. 2022. Catalytic degradation of polyethylene terephthalate using a phase-transitional zirconium-based metal–organic framework. Angew. Chem. Int. Ed. Engl. 61: e202117528.

Wuana, R.A. and F.E. Okieimen. 2011. Heavy Metals in Contaminated Soils: A Review of sources, chemistry, risks and best available strategies for remediation. Int. Sch. Res. Notices. 2011: 1–20.

Xu, J., B. Wang, W.H. Zhang, F.J. Zhang, Y.D. Deng, Y. Wang, et al. 2021. Biodegradation of p-nitrophenol by engineered strain. AMB Express. 11: 124.

Yao, Y., M. Zhu, Z. Zhao, B. Tong, Y. Fan and Z. Hua. 2018. Hydrometallurgical processes for recycling spent Lithium-Ion batteries: A critical review. ACS Sustain. Chem. Eng. 6: 13611–13627.

Yarahmadi, H., S.K. Salamah and M. Kheimi. 2023. Synthesis of an efficient MOF catalyst for the degradation of OPDs using TPA derived from PET waste bottles. Sci. Rep. 13: 19136.

Yusuf, V.F., N.I. Malek and S.K. Kailasa. 2022. Review on metal–organic framework classification, synthetic approaches, and influencing factors: Applications in energy, drug delivery, and wastewater treatment. ACS Omega. 7: 44507–44531.

Zhan, G., W.C. Ng, W.Y. Lin, S.N. Koh and C.H. Wang. 2018. Effective recovery of vanadium from oil refinery waste into vanadium-based metal–organic frameworks. Environ. Sci. Technol. 52: 3008–3015.

Zhang, X., L. Yang, Y. Li, H. Li, W. Wang and B. Ye. 2012. Impacts of lead/zinc mining and smelting on the environment and human health in China. Environ Monit Assess. 184: 2261–2273.

Zhang, S., M. Jian, Q. Zhang, R. Xu, J. Qu, X. Luo, et al. 2020. Recyclable printed circuit boards and alkali reduction wastewater: Approach to a sustainable copper-based metal–organic framework. ACS Sustain. Chem. Eng. 8: 1371–1379.

Zhang, H., X. Hu, T. Li, Y. Zhang, H. Xu, Y. Sun, et al. 2022. MIL series of metal organic frameworks (MOFs) as novel adsorbents for heavy metals in water: A review. J. Hazard. Mater. 429: 128271.

Zhao, W., J. Huang, Y. Fu, T. Yue and W. Hu. 2025. Comparison study of three universal methods for the recovery and purification of valuable heavy metals from electroplating sludge. Sep. Purif. Technol. 352: 128009.

Zheng, X., Z. Zhu, X. Lin, Y. Zhang, Y. He, H. Cao, et al. 2018. A mini-review on metal recycling from spent lithium ion batteries. Engineering. 4: 361–370.

Zope, V.S. and S. Mishra. 2008. Kinetics of neutral hydrolytic depolymerization of PET (polyethylene terephthalate) waste at higher temperature and autogenious pressures. J. Appl. Polym. Sci. 110: 2179–2183.

Zou, Y., X. Wang, A. Khan, P. Wang, Y. Liu, A. Alsaedi, et al. 2016. Environmental remediation and application of nanoscale zero-valent iron and its composites for the removal of heavy metal ions: A review. Environ. Sci. Technol. 50: 7290–7304.

Chapter **11**

Plastics in Health Care Waste-Pathways for Sustainable Strategies in the Indian Scenario

Kavita Gidwani Suneja*[1] and Harsangeet Kaur[2]
[1]Assistant Professor, Energy Centre, MANIT, Bhopal, India
[2]Associate Professor, Department of Civil Engineering, MANIT, Bhopal, India

11.1 INTRODUCTION

Plastic is a material of versatile application used in domestic, commercial, and industrial activities. Production of plastic has increased immensely from the 1970s onwards (Ritchie et al. 2023). An OECD press release in February 2022 raised concern that global plastic waste generated worldwide has increased 100% in 20 years, from 2000, and has reached 353 million tonnes (OECD 2022)

A report by Statistica has put forth that 400 million tonnes of waste plastics are generated annually globally, with an alarming growth rate of 4.6% per year (Alves 2024). Further, almost 98% of single-use plastic products come from fossil fuels, and 8% of global oil production goes into plastic manufacturing. The disposal data of plastics is also worrying, as only about 9% of plastic waste is recycled. About 19% is incinerated, which can be a source of harmful emissions,

*For Correspondence: kavitagidwanisuneja@gmail.com

DOI: 10.1201/9781003531463-11

50% end up in landfills, and 22% evade regulated management systems and find their way into dumpsites (Rizan et al. 2020). It is then either burnt or ends up in aquatic or terrestrial ecosystems, especially in third-world countries. The aftereffects of plastic use are felt for a long time, even after plastic consumption has completely stopped.

11.2 PLASTICS IN HEALTH CARE

Although most nations, including India, support a ban on single-use plastics, they still seem indispensable for sectors like health care. Plastics are prominent in healthcare as they possess an excellent strength-to-weight ratio, toughness, and versatility, and have substantially replaced glass in products like syringes, glucose, and saline bottles. Voluminous generations of single-use plastics have increased significantly worldwide in recent times, particularly in medical equipment and as safety requirements, such as personal protection equipment (PPE) kits like masks, shields, gloves, gowns, etc., which are essential in fighting against various viruses like COVID-19, hepatitis, etc.

11.2.1 Plastics in Health Care: Global Scenario

The global medical plastic market is around 2% of the total plastic production, growing at the rate of 6.1% per annum (Rizan et al. 2020), and this market is only going to increase further because both the developed countries and the developing countries are expanding their health care facilities. The USA is the biggest consumer of medical plastic devices, whereas health care facilities are also expanding tremendously in the developing countries, mainly in Brazil, India, and China. Several factors have contributed to the growth of healthcare facilities, including increased longevity, a larger geriatric population, heightened health awareness, advancements in medical technology, and increased government and private sector involvement, as well as the global pandemic. Consequently, this has led to an increase in health care and related medical waste.

The World Health Organization (WHO) defines medical waste as "The waste generated in the diagnosis, treatment, or immunization of human beings or animals" (Su et al. 2021). Main components of medical waste are plastics, textiles, papers, glass, rubber, metals, etc. Out of this, only about 10–25% of BMW is hazardous, and the remaining 75–95% is non-hazardous (Datta et al. 2018). Medical devices have been an important part of human development and increasing longevity, and various materials have played a major role in medical devices. Initially, metal and ceramics were the main materials being used to prepare medical devices and bone and dental implants. When glass manufacturing became prevalent, various equipment, such as syringes, bottles, tubing, lab equipment, etc., was made of glass. However, these materials suffered from several inherent issues, including heavy weight, high manufacturing costs, and low biocompatibility, as well as rigidity and the fact that glass is easily breakable. As experimentation with plastic started, it emerged as an ideal material for use in the medical sector. Plastics became most preferred because of the ease of sterilization, lightweight, flexibility, and biocompatibility.

A brief compilation of various kinds of plastics used in the medical sector is listed in Table 11.1, and major types of plastics used and their percentage, respectively, are shown in Figure 11.1.

Table 11.1 Listing Commodity Plastics, Engineering Thermoplastics, and High-Temperature Engineering Thermoplastic and Other Polymers in healthcare

Plastic Type	Medical Application
Commodity Plastics	
• Polyethylene	• Tubing Films and packaging
• Polypropylene	• Connectors and labware
• Polystyrene	• IV bags and catheters
• Polyvinyl chloride	• Face masks and drug-delivery components
	• Luers and membranes
	• Sutures
Engineering Thermoplastics	
• Polyamides and nylons	• Luers
• Acetals	• Balloons
• Polyurethanes	• Blood set components
• Acrylics	• Surgical instruments.
• Polyesters	• Syringes
• Poly-carbonates	• Moving parts and components
	• Catheters
	• Blood bowls
	• Blood oxygenators
High-Temperature Engineering Thermoplastic and Other Polymers	• Surgical instruments
• Polyimides	• Syringes,
• Biopolymers	• Dental implant
• Liquid crystalline polymers	• High precision parts
• Polyetherimides	• Luers
• Polyphenylene sulfide	• Bioresorbable sutures
• Polysulfones	• Surgical trays
• Polyether ether ketone	• Bone implants
• Fluoropolymers	• Moving parts and components
• Adhesives and Thermosets.	• Electronic components

Ever since the advent of civilization, various healthcare facilities have been developed, and associated healthcare waste has been generated. This waste was treated as a part of regular waste, and it was only in recent years that it has been given a special status and special provisions laid down to deal with it. The issue of biomedical waste was first discussed by the World Health Organization zonal meeting in 1983 in Norway. Problems related to healthcare waste garnered global concern following beach wash-ups in the USA in 1988. In response, the Bio Medical Waste Management and Handling Rules were introduced under the Environment Protection Act (EPA) of 1986, and the Medical Waste Tracking Act was initiated in 1988 (Meher Wan 2021).

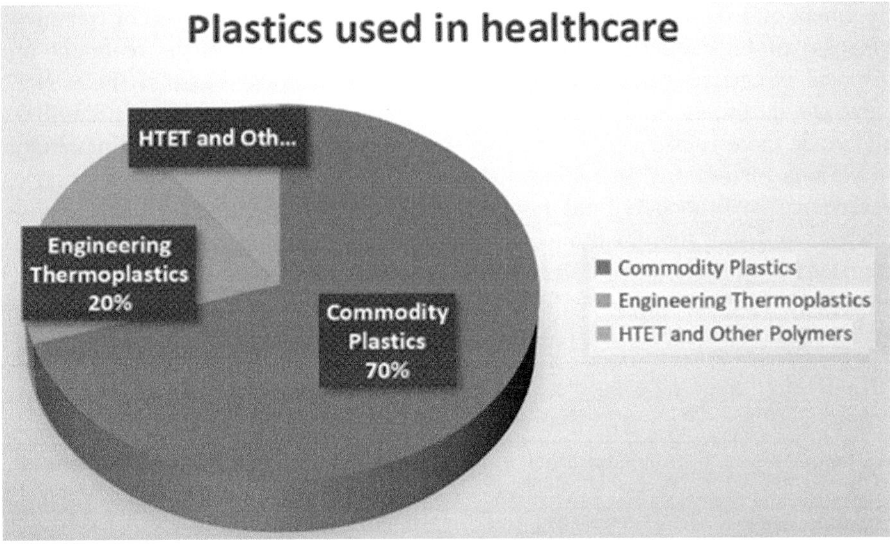

Figure 11.1 Showing a pie chart of plastics used in healthcare: 70% Commodity Plastics, 20% Engineering Thermoplastics, and rest HTET/Others.

11.2.2 Indian Scenario: Policy and Generation

In tune with global concern, in India, the issue was brought to notice by the Honourable Supreme Court of India and the biomedical waste management rules were formulated for the first time in 1998 which had clear provisions for collection, treatment and disposal of biomedical waste. The Ministry of Environment, Forest and Climate Change further modified the rules in 2016 by increasing the scope and coverage, simplifying the categorization, pinning the authorization and responsibility, while recommending improvement in segregation, storage, transportation, and the disposal methods to mitigate environmental pollution. Additional provisions were made in 2020 to deal with the monumental problem of COVID related waste.

India is today the world's most populous country, with a population of more than 1.44 billion (Worldometer Information 2024) and a bustling, growing economy. With expanding healthcare, Indian medical plastics are expected to expand at a compound annual growth rate of 15.2% in the decade from 2020 to 2030 (India Medical Plastics Market Size, Share & Trends Analysis 2019). Increasing government support to improve healthcare, the emergence of the private sector, and public awareness have positively influenced the consumption of plastics in line with societal values. Biomedical waste generation ranges from 0.11 to 0.87 kg per day per patient. India produces 774 metric tonnes of biomedical waste every day (Anurag et al. 2022).

Biomedical waste, (BMW) Rules 2016, defined biomedical waste as "Any waste, which is generated during the diagnosis, treatment or immunization of human beings or animals or research activities pertaining thereto or in the production or testing of biological or in health camps" and "bio-medical waste treatment and disposal facility (BMWTF) means any facility wherein treatment, disposal of bio-medical waste or processes incidental to such treatment and disposal is carried out and includes common bio-medical waste" (CPCB 2022, New Delhi, India). Biological waste management rules have been streamlined and made more user-friendly. The government extended the concept of generators to various institutions, including hospitals, clinics, nursing homes, dispensaries, veterinary institutes, animal houses, pathological laboratories, blood banks, research and education institutes, vaccination camps, medical and surgical camps, etc. BMW categories have decreased from ten (in 1998) to four (in 2016) (Capoor et al. 2017). Waste categories are red, yellow, white, and blue. The segregation and treatment procedure to be followed is given in Table 11.2.

Table 11.2 Showing biomedical waste categories and disposal methods for Yellow, Red, White, and Blue bins

Categories	Bag/Container Used	Type of Waste	Treatment/Disposal Options
	Non-chlorinated plastic bags	**Human/animal** Anatomical waste, expired or discarded medicines, chemical waste, and microbiological waste.	Incineration, plasma pyrolysis, or deep burial
	Non-chlorinated plastic bags or containers	Contaminated waste (recyclable) tubing, bottles, intravenous tubes and sets, catheters, urine bags, syringes, and gloves.	Autoclaving/ microwaving/ hydroclaving and then sent for recycling, not to be sent to landfill.
	(Translucent) puncture, leak, and tamper-proof containers	Waste sharps, including metals	Auto or dry heat sterilization followed by shredding, mutilation, or encapsulation.
	Cardboard boxes with blue markings	Glassware	Disinfection or autoclaving, microwaving, hydroclaving, and then sent for recycling.

Source: Based on annual reports on biomedical waste, CPCB (Ministry of Environment, Forest & Climate Change) Parivesh Bhawan, East Arjun Nagar, Delhi 110032. https://cpcb.nic.in/uploads/Projects/Bio-Medical-Waste/AR_BMWM_2017pdf

Salient points of the BMW management rules (2016) in India are summarized:

1. The polluter pays principle was emphasized.
2. It would be the duty of the common bio-medical waste treatment facility (CBWTF) to collect waste and transport it from the premises of the generator to the treatment facility.
3. The untreated human anatomical waste shall not be stored beyond a period of 48 hours.
4. The urban local bodies would collect biomedical waste generated from the houses in separate bags or containers and then send it to the common biomedical waste treatment facilities.
5. Suitable land has to be allotted by the local bodies for the setting up of CBWTF.
6. The state pollution control boards (SPCB) would be responsible for the inventory and compilation of data sent by generators and CBWTF in the form of an annual report to be submitted to the CPCB. It would be the duty of the state pollution control boards to monitor compliance, conduct third-party audits, and also conduct training programs for healthcare staff.

The Government of India has also initiated the plastic waste management rules, and there are provisions related to the use of plastic in medical and healthcare. In 2023, CPCB issued an update on plastic packaging and other products for manufacturers and importers of medical devices, which included fees for handling plastic waste. The manufacturer or importer must obtain an extended producer's responsibility (EPR) registration to import any medical product using plastic or plastic packaging. The manufacturer needs to quantify their plastic consumption and the waste being produced and report it to the state pollution control board. The Central Pollution Control Board had also set targets for the minimum level of recycling of plastic packaging to be achieved in the plastic waste (Verma 2023).

Table 11.3 Showing 2017-2022 data on healthcare facilities, beds, and waste generation in India

HCF Data	2017	2021	2022	Percentage Increase in 5 Years
No of HCFs	238,259	375,256	393,239	65
No of bedded HCFs	87,281	121,396	121,259	38.93
No of beds	2,094,858	2,561,295	2,465,063	17.67
No of CBWTF	198	215	218	10.1
Quantity of BMW generated (tonnes/day)	559	764	705	26.11
Quantity of BMW treated	558	721	645	15.59

Source: CPCB 2023. Annual Report on Biomedical Waste Management for the year 2023 Accessed on 30th October, 2025.

11.3 BIOMEDICAL WASTE MANAGEMENT PROCESS

The complete biomedical waste management setup consists of activities comprising generation and collection at the health care facilities, transportation and treatment at CBWTF, and subsequent safe disposal, as shown in Figure 11.2.

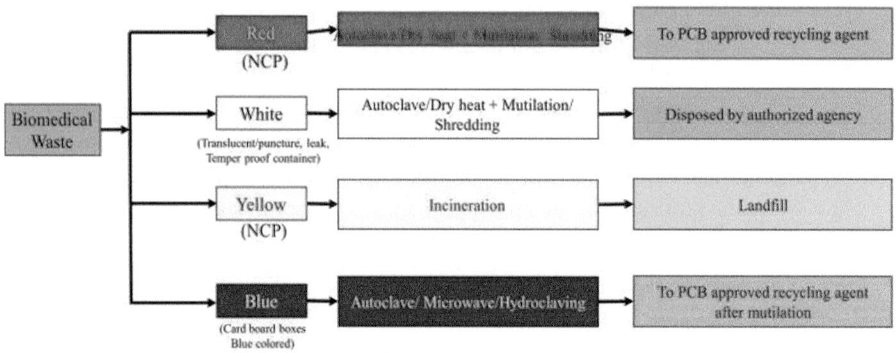

Figure 11.2 Showing a flowchart of treatment and disposal paths for Red, White, Yellow, and Blue biomedical waste categories.

11.3.1 Plastic Waste Management

The plastic waste is mainly collected in red coloured bins. Studies have shown that almost 30% of total waste consists of plastic waste (Rizan et al. 2020). There are two types of biomedical waste treatment facilities: the captive facility, where the hospital has its own biomedical waste treatment facility on its campus, and the common biomedical waste treatment facility (CBWTF), which would be a nodal point and collect waste from different waste generation centers. CBWTFs have adequate storage space, waste treatment facilities depending upon the nature of waste, and subsequent transportation of ash to identified landfills. Most of the generation points are now coordinated with CBWTF, and GPS-enabled vehicles collect this waste from the generators. The plastic waste or the red-coloured waste is sterilized in an autoclave and then shredded to prevent misuse and further disposed of through authorised. Biomedical Waste Management rules 2016 have specified that the serialization should be done for a minimum time cycle of twelve minutes at a pressure of 121 psi.

The process is monitored by the regulatory bodies, and annual reports must be submitted by the generators and CBWTF managers to the respective SPCBs and finally to the central body. Medical plastic waste management is a well-established practice in India, especially in urban areas, and there are 208 operational CBMWTF, having their own fleet of collection vehicles, adequate storage space, and incinerators for yellow waste. These facilities include an autoclave for disinfecting and shredders for shredding the plastic waste. However, the plastic medical waste is an ever-increasing waste that can have monstrous consequences for both the biotic and abiotic environments. Analysis of the current scenario indicates that proactive measures need to be taken in all elements of waste management, which can lead to significant pollution mitigation.

11.4 SUSTAINABLE STRATEGIES FOR PLASTIC WASTE MANAGEMENT

Plastic is a versatile material with multiple applications used in almost all sectors. However, due to its non-biodegradable nature, it is very difficult to manage the huge waste produced. The problem becomes more acute in the healthcare sector because the dangers linked with the proliferation of infected plastic waste are enormous. It is important to evaluate the various activities and explore pathways in tune with the goal of sustainable development.

11.4.1 Generation

It is evident that the healthcare sector is one where it is mandatory to use plastics for medical devices, and even single-use plastics cannot be avoided; hence, steps should be taken to limit the use of plastic in other sectors where plastics can be more conveniently reduced and replaced by other suitable materials. Awareness and training programs conducted for healthcare professionals can prove fruitful, and the use of nonessential plastics can be minimised.

Behera (2021) determined that plastic formed about 22 to 25% of PPE being used and disposed of through incineration, causing substantial pollution in the form of dioxins and toxic metals. It is therefore important to explore avenues of safe reuse after sterilization so that the load is reduced. Laboratory coats and patient gowns can be reused following the disinfection standards, and N95 and other types of masks can be disinfected using 11% hydrogen peroxide vapours, while 75% ethanol and 0.5% sodium hypochlorite solution can be used to decontaminate face shields and goggles (Kuppusamy et al. 2022).

11.4.2 Segregation

Segregation is the keystone for effective waste management. If segregation is not done properly at the generator site, general non-hazardous waste can also get mixed with hazardous waste. Secondly, it is difficult to recycle mixed plastic waste as different types of plastics have different properties. According to current norms, different kinds of plastic waste are collected in the same red-coloured bin; hence, any red-coloured bin would still contain mixed kinds of plastics. It would be an added advantage if segregation and separation of individual kinds of plastics could be done at the source, and from there, only they can be sent to the recycling centre directly or through a transfer station. This would also lead to lower transportation costs in addition to the ease of recycling. However, this would require more awareness and monitoring at the generation point.

11.4.3 Transformation and Waste-to-Energy Alternative

Thermochemical conversion processes, such as pyrolysis (Sharuddin et al. 2018; Su et al. 2021) and gasification (Siwal et al. 2020), have shown promising results in recovering energy from medical waste plastics. Park et al. (2021) fed face masks and food waste into the pyrolysis unit. After varying different operational parameters, they concluded that feeding masks in pyrolytic chambers resulted in the production of CH_4, C_2H_6, C_3H_8, C_2H_4, CO, CO_2, and H_2 as non-condensable hydrocarbons, which have good calorific value. Nema et al. (2002) designed and performed a technoeconomic evaluation of energy recovery from plasma pyrolysis of cotton waste and plastic waste. Blessy et al. (2021) suggested that due to increased use of plastics in the health and packaging sectors, it is imperative to encourage research and methods for polymer recycling, and various conversion treatment methods can be adapted. Other researchers have suggested options like glycolysis (Lalhmangaihzuala et al. 2021; Thiounn and Smith 2020) and aminolysis (Shukla and Harad 2006; Sadeghi et al. 2011) as suitable methods of polymer recycling.

11.4.4 Material Utilisation

Medical plastics sterilized and shredded can subsequently be used in other sectors in a cost-effective manner. Precautions should, however, be taken to minimize the spread of microplastics in the ecosystem. Utilisation of waste plastics in the construction sector is a viable alternative and efficient method to utilize plastic waste. Wahab and Amadi (2020) mixed plastic waste in different proportions with granite dust, sand, and clay to produce sample paver blocks. Akinyele et al. (2020) investigated the use of Polyethylene terephthalate (PET) material as an addition to clay in fired bricks (PET) was mixed with lateritic clay at 0, 5, 10, 15, and 20%. The samples were then tested for compression, flexural strength, firing shrinkage, density, and water absorption. This work concluded that less than 5% PET can be used in fired bricks under controlled conditions (Olawale and Oyelakin 2021).

Dong et al. (2022) studied the characteristics of concrete with various admixtures, including recycled plastics, and found that the addition of plastic waste improved the workability of the concrete. Similarly, Basha et al. (2020) investigated recycled plastic in concrete and reported that it had better thermal insulation and was lighter than typical concrete. Lynch et al. (2021) proposed methods for reducing pandemic-related waste by reusing discarded face masks among other waste materials in civil works. In this study, shredded face masks were blended with recycled concrete aggregate (RCA) for road base and sub-base applications.

11.5 CONCLUSION

The global template today is to ban or minimize disposable plastics, and stringent and comprehensive steps need to be taken for plastic waste in the medical sector.

The UN pollution treaty (Chartres et al. 2025) highlighted the perils caused by healthcare plastics, especially to women and people working in the informal sector, and has identified steps for developing fact sheets for healthcare waste (UNEP, United Nations Environmental Program). Increased opportunities for safe reuse and recycling, encouraging biodegradable plastics, can bring about the desired change. Recovering energy or material can make the process more cost-effective and environmentally friendly. Measures for a circular economy and international collaboration can also go a long way in decreasing carbon and water footprints. It is imperative to provide good health care to human beings, but in a sustainable manner, for achieving the Sustainable Development Goals.

REFERENCES

Akinyele, J.O., U.T. Igba and B.G. Adigun. 2020. Effect of waste PET on the structural properties of burnt bricks. Scientific African. Volume 7.

Anurag, D., V. Sharma and D. Kumar. 2022 Life cycle assessment of biomedical waste management for reduced environmental impacts. J. Cleaner Prod. 349.

Basha, S.I, M.R. Ali, S.U. Al-Dulaijan and M. Maslehuddin. 2020. Mechanical and thermal properties of lightweight recycled plastic aggregate concrete. J. Build. Eng. 32: 101710. https://doi.org/10.1016/j.jobe.2020.101710.

Behera, B.C. 2021. Challenges in handling COVID-19 waste and its management mechanism. A rview. Environ. Nanotechnol. Monit. Manag. 15: 100432. doi:10.1016/j.enmm.2021.100432. Epub 2021 Feb 11. PMID: 33614423; PMCID: PMC7877813.

Blessy, J., J. Jemy, K. Nandakumar and T. Sabu. 2021. Recycling of medical plastics. Adv. Ind. Eng. Polym. Res. 4: 199–208.

Capoor, M.R. and Kumar Tapas Bhowmik. 2017. Current Perspectives on Biomedical Waste Management: Rules, Conventions and Treatment Technologies. Indian J. Med. Microbiol. 35(2): 157–164. Elsevier BV.

Chartres, N., Q. Grundy, F.A. Miller, B. Beeler and T.J. Woodruff. 2025. Negotiating the plastics treaty to protect health and the environment. Bull. World Health Organ. 103(7): 414–414A. doi: 10.2471/BLT.25.294144. PMID: 40630155; PMCID: PMC12231067

CPCB (Central Pollution Control Board). 2022. Annual Report on Biomedical Waste Management for the year 2022 (Ministry of Environment, Forest & Climate Change) Parivesh Bhawan, East Arjun Nagar, Delhi. https://cpcb.nic.in/uploads/Projects/Bio-Medical-Waste/AR_BMWM2_2022.pdf.

CPCB (Central Pollution Control Board). 2023. Annual Report on Biomedical Waste Management for the year 2023 accessed on 30/10/25. (Ministry of Environment, Forest & Climate Change) Parivesh Bhawan, East Arjun Nagar, Delhi. https://cpcb.nic.in/uploads/Projects/Bio-Medical-Waste/AR_BMWM_2023pdf.

Datta, P., G.K. Mohi and J. Chander. 2018. Biomedical waste management in India: Critical appraisal. J. Lab Physicians. 10: 6–14. DOI: 10.4103/JLP.JLP_89_17

Dong, C., Q. Zhang and C. Chen. 2022. Fresh and hardened properties of recycled plastic fiber reinforced self-compacting concrete made with recycled concrete aggregate and fly ash, slag, silica fume. J. Build. Eng. 62: 105384. https://doi.org/10.1016/j.jobe.2022.105384

Kuppusamy, P.P., A. Bhatia and A. Verma. 2022. Accumulation of biomedical waste during the COVID-19 pandemic: Concerns and strategies for effective treatment. Environ. Sci. Pollut. Res. 29: 55528–55540. https://doi.org/10.1007/s11356-022-21086-0

Lalhmangaihzuala, S., Z. Laldinpuii, C. Lalmuanpuia and K. Vanlaldinpuia. 2021. Glycolysis of Poly(Ethylene Terephthalate) Using Biomass-Waste Derived Recyclable Heterogeneous Catalyst. Polymers. 13(1): 37.

Lynch, S.K., M. Saberian, J. Li, R. Roychand and G. Zhang. 2021. Preliminary evaluation of the feasibility of using polypropylene fibres from COVID-19 single-use face masks to improve the mechanical properties of concrete. Constr. Build. Mater. 337(2022): 127576.

Nema, S.K. and K.S. Ganeshprasad. 2002. Plasma pyrolysis of medical waste. Current Science. 83(3): 271–278.

OECD. 2022. Global Plastics Outlook: Policy Scenarios to 2060. OECD Publishing, Paris. https://doi.org/10.1787/aa1edf33-en.

Olawale, A. and S. Oyelakin. 2021. Assessment of the properties of bricks made from stone dust and molten plastic for building and pedestrian pavement. Int. J. Pavement Res. Technol. VL-14, 2021/01/19 DO - 10.1007/s42947-020-1268-5

Wahab, O. and A.N. Amadi. 2020. Production of pavement blocks from plastic waste. Pac. J. Sci. Technol. 21(2): 36–43.

Park, C., H. Choi, K.A. Lin, E.E. Kwon and J. Lee. 2021. COVID-19 mask waste to energy via thermochemical pathway: Effect of Co-Feeding food waste, Energy. Volume 230. https://doi.org/10.1016/j.energy.2021.120876.

Ritchie, H., V. Samborska and M. Roser. 2023. Plastic Pollution. Published online at Our WorldinData.org. Retrieved from: https://ourworldindata.org/plastic-pollution [Online Resource].

Rizan, C., F. Mortimer, R. Stancliffe and M.F. Bhutta. 2020. Plastics in healthcare: Time for a re-evaluation. J. R. Soc. Med. 2020. 113(2): 49–53. DOI: 10.1177/0141076819890554

Thiounn, T. and R.C. Smith. 2020. Advances and approaches for chemical recycling of plastic waste. J. Polym. Sci. 58(10): 1347–1364. https://doi.org/10.1002/ pol.20190261.

Sadeghi, G.M.M., R. Shamsi and M. Sayaf. 2011. From aminolysis product of PET waste to novel biodegradable polyurethanes. J. Polym. Environ. 19(2): 522–534. https://doi.org/10.1007/s10924-011-0283-7.

Sharuddin, S.D.A., F. Abnisa, W.M.A. Daud and M.K. Aroua. 2018. Pyrolysis of plastic waste for liquid fuel production as prospective energy resource. IOP Conf. Ser. Mater. Sci. Eng. 334: 012001. https://doi.org/10.1088/1757-899x/334/1/012001-9.

Shukla, S.R. and A.M. Harad. 2006. Aminolysis of polyethylene terephthalate waste. Polym. Degrad. Stab. 91(8): 1850–1854. https://doi.org/10.1016/j.polymdegradstab.2005.11.005.

Su, G., H.C. Ong, S. Ibrahim, I.M.R. Fattah, M. Mofijur and C.T. Chong. 2021. Valorisation of medical waste through pyrolysis for a cleaner environment: Progress and challenges. Environ. Pollut. 2021(279): 116934–116942.

Siwal, S.S., C. Sun, S. Thakur, V. Kumar, V. Gupta and T. Kumar. 2020. Energy production from steam gasification processes and parameters that contemplate in biomass gasifier—A review, Bioresour. Technol. 297: 122481–122488. https://doi.org/10.1016/j.biortech.2019.122481.

ONLINE RESOURCES CITATIONS

Alves, B. 2024. Plastic waste worldwide - statistics & facts. https://www.statista.com/topics /5401/global-plastic-waste/#topicOverview

https://www.grandviewresearch.com/industry-analysis/india-medical-plastics-market

https://www.unep.org/interactives/beat-plastic-pollution/#:~:text=Around%20the%20 world%2C%20one%20million,are%20used%20worldwide%20every%20year.

India Medical Plastics Market Size, Share & Trends Analysis. 2019. [Report by Product (Rigid Plastics, Soft Plastics), By Application (Medical & Monitoring Equipment Housings, Wearables), and Segment Forecasts. 2025–2030.

Meher Wan. 2021. Tackling a Pandemic of Biomedical Waste.

Verma, G. 2023. How New Plastic Waste Management (PWM) Rules in India Are Affecting Medical Device Manufacturers.

Worldometer Information. 2024. https://www.worldometers.info/world-population/population-by-country/#google_vignette

Chapter **12**

AI-Powered Solutions for Combating Plastic Pollution

Mudita Nagpal*, Parth Malhotra,
Harshita Bangia and Nidhi Sharma
Department of Applied Sciences, Vivekananda Institute of Professional Studies-Technical Campus, New Delhi-110034, India

12.1 INTRODUCTION

The exceptional material properties of various plastics (often used interchangeably with polymers) have enabled numerous innovations in lightweight packaging, agricultural productivity, cosmetics, detergents, and advanced technological applications, among others (Cubas et al. 2022; Glenn et al. 2014). Plastic waste is on the rise and is projected to keep increasing, even under some of the most hopeful scenarios for reducing plastic waste (Figure 12.1). Global estimates of plastic waste emissions in rivers, oceans, and lakes range from 9 to 23 million metric tons annually. Similarly, the terrestrial environment saw emissions of 13 to 25 million metric tons per year as observed in 2016 (MacLeod et al. 2021). The ongoing rise in waste generation is linked to both population growth and a higher amount of waste produced per individual, particularly among urban residents. This trend underscores the need to establish an efficient process chain for waste disposal.

Many traditional plastic waste disposal systems have been used for a long time. A key conventional method includes organizing and executing waste collection and cleanup campaigns. These initiatives frequently engage local communities,

*For Correspondence: mudita.nagpal@vips.edu

DOI: 10.1201/9781003531463-12

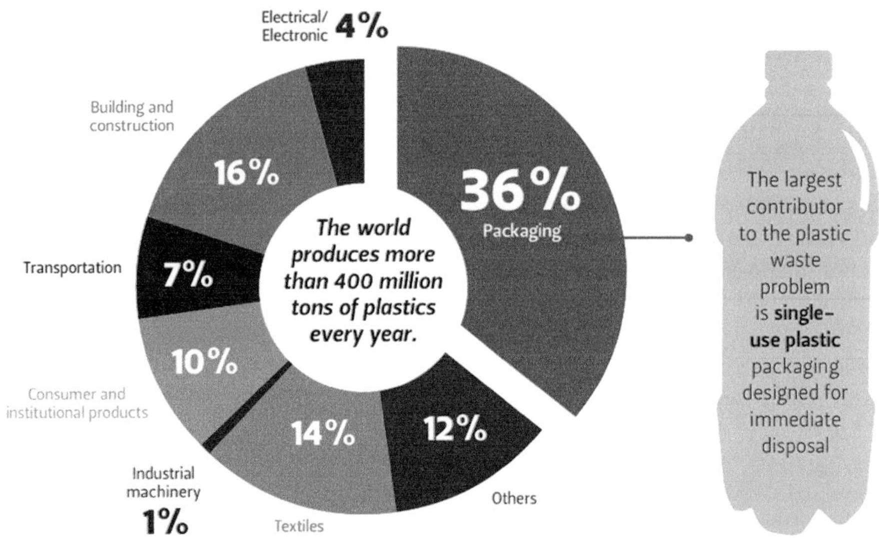

Figure 12.1 Plastic is produced globally from various industrial sectors (Meidl 2018).

volunteers, and environmental organizations to manually gather plastic debris from beaches, shorelines, and other impacted areas. Although this approach aids in the immediate removal of waste, it may not offer a long-term solution as it fails to tackle the underlying causes of plastic pollution (Prata et al. 2019). Traditionally, disposing of plastic waste in landfills has been a common practice. However, this approach is now seen as problematic due to the limited capacity of landfills, the risk of harmful substances leaching into soil and water, and the slow decomposition rate of plastics (Wan et al. 2022). Incineration, which involves burning plastic waste, has also been a conventional method. Although it can decrease waste volume and produce energy, it presents environmental challenges, including air pollution and the emission of toxic substances. Additionally, burning plastics can result in greenhouse gas emissions (Shen et al. 2020).

The integration of AI (Artificial Intelligence) and technology offers promising avenues for mitigating plastic pollution and promoting sustainable practices for a healthier planet. This paper highlights promising technological advancements designed to improve waste management practices. It delves into the development of waste sorting systems utilizing robotics and AI, emphasizing their potential to transform practices for waste management. Additionally, the paper examines the use of AI in 3D printing, materials modeling, and design, which fosters sustainable material choices and innovative waste reduction strategies. Furthermore, the article explores studies for the use of AI in accurately identifying sources of plastic waste pollution, which can help researchers better trace pathways and implement targeted strategies. In addition, the advantages and disadvantages of the discussed AI solutions have been highlighted.

12.2 TECHNOLOGICAL SOLUTIONS FOR PLASTIC WASTE MANAGEMENT

12.2.1 Robotic Waste Sorting Systems

Waste materials can be categorized based on their reusability, which includes: (a) waste like plastic, glass, metal, paper, and cardboard, which can be recycled and reused; (b) organic waste, which can be composted for agricultural use; and (c) non-disposable waste, which cannot be reused and must be buried due to its environmental harm (Diya et al. 2018). Effective categorization and management of these waste types are crucial for better resource utilization, cost-effective production and cultivation, and environmental sustainability. Waste sorting aids in recycling and conserving energy (Rousta et al. 2017).

Robotic waste sorting systems depend on advanced imaging technology, leveraging recent developments in AI to detect and classify waste. Presently, highly renowned commercially used robots employed in the waste management field are produced by Sadako, SamurAI, and ZenRobotics (manages construction waste) and AMP Robotics (Koskinopoulou et al. 2021). Paulraj and group introduced an algorithm for detecting waste materials from thermal camera images (Paulraj et al. 2016). Na et al. (2022) and Patrizi et al. (2021) have worked on expanding datasets for waste recognition. Additionally, Bircanoğlu et al. (2018) proposed a deep learning architecture called RecycleNet for classifying recyclables.

Seredkin and group introduced a method for object detection and classification in a sorting unit where waste is fed sparsely onto a conveyor belt to minimize object overlap. A machine vision system identifies and classifies individual objects while assessing their position on the belt. A height profile scanner measures the height profile of objects on the conveyor belt. The method uses a laser sheet directed vertically downwards and a camera positioned above the belt to capture the shape of the laser sheet on the object surface. Coordinates, height, and results of classification of objects are sent to a module for collection, which uses a handling device to move objects to designated collection bins (Seredkin et al. 2019).

Machine learning algorithms have become a primary approach for addressing classification challenges (Witten et al. 2005). Applying these algorithms to waste sorting has been reported in various studies testing neural network models (Bircanoğlu et al. 2018; Yang and Thung 2016). Typically, such studies focus on two main components: the model itself that is evaluated, and the dataset employed for testing and training the network. A recent development in neural networks is the region proposal convolutional neural network (R-CNN). The R-CNN model integrates various modules for detecting objects in images and subsequently classifying those objects. This approach allows for the recognition of multiple objects of different types within a single image (Huynh et al. 2020). In the study by Seredkin and group, waste collected from separate bins in the Novosibirsk region was utilized to create a database. The waste underwent careful pre-sorting to ensure accurate labeling of objects. Images of municipal solid waste (MSW) items were captured on a stationary conveyor belt to simulate real-world sorting scenarios in a sorting facility. Images of MSW objects obtained were over 1400. The database includes three main types of objects: polyethylene terephthalate (PET) bottles, high-density polyethylene (HDPE) bottles, and aluminum cans.

Additionally, a dataset labeled as "other" was created to encompass various types of MSW not classified under the main categories. Each individual object and groups of objects were categorized based on their material composition. Moreover, PET bottles were classified according to their respective colors (Figure 12.2). The highest average precision was observed for the HDPE class (Seredkin et al. 2019).

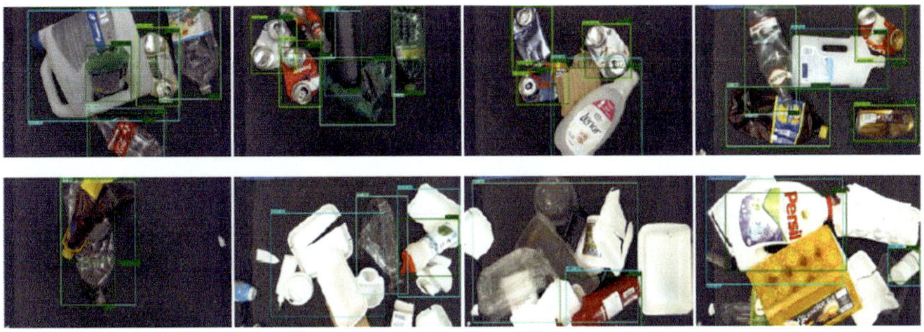

Figure 12.2 Correct detection of regions is shown in the first row, and incorrect detection of regions is shown in the second row (reproduced from Seredkin et al. 2019. This article is published under CC BY licence).

12.2.2 Application of AI for Detecting the Source of Plastic Waste

With the sources of plastics in the environment growing increasingly intricate, accurately identifying these sources helps researchers better trace pathways and implement targeted strategies (Figure 12.3). This knowledge is crucial for informed policy-making aimed at curbing environmental pollution. Traditional methods for tracing plastic pollution sources typically estimate their abundance based on environmental factors, but this approach is complex, lacks scientific precision, and can lead to errors. The use of AI technologies to trace and predict environmental pollutants can reveal pollution patterns and correlations, helping to understand the sources, transmission routes, and hazardous effects of plastic waste, while also improving the effectiveness of conventional tracing techniques (Jin et al. 2024).

Currently, common methods for statistically analyzing plastic sources include cluster analysis (CA), principal component analysis (PCA), absolute principal component score-multiple linear regression (APCS-MLR) approach, and redundancy analysis (RDA). Although these techniques have shown effectiveness in tracing sources, they require data to meet exact assumptions and distributions for optimal outcomes (Mohan and Lakshmanan 2023). Recent studies indicate that machine learning (ML) can be used to identify potential sources of plastic pollution. Various ML classifiers have been utilized to investigate pollution sources, using PCA to determine the key factors influencing the fate of plastics. A combination of decision trees (DT), random forests (RF), and support vector machines (SVM) proved highly accurate in identifying different MP sources (Wu et al. 2023).

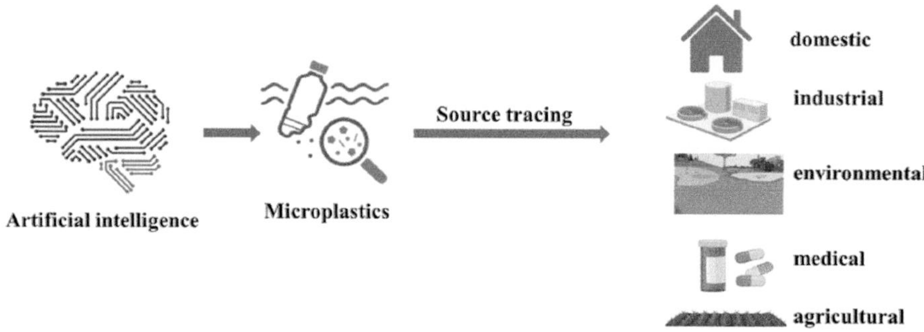

Figure 12.3 Use of AI technology for tracing the source of microplastic waste (reproduced with permission from Jin et al. 2024).

Additionally, artificial neural networks (ANNs) have been used to study the sources of MP pollutants. Kida and group (Kida et al. 2024) developed an ANN model that excels at predicting plastic pollutant migration. The model's simulation data closely correlated with experimental results, though it currently only accounts for ideal lab situations, without including many environmental variables. Advanced and complex processes that consider these intricate environmental variables could significantly enhance the accuracy and dependability of plastic source detection. Using ML in conjunction with multivariate procedures to determine plastic waste levels is critical for evaluating risks and impacts on human health and the environment. SVM and RF methods have been effective in predicting the presence of MPs in specific study areas, enabling the classification of pollution severity in multiple levels (Chakraborty et al. 2024). Moreover, ML has also been found effective in forecasting regional MP pollution levels. Three ML models—long short-term memory (LSTM), RF, and least squares support vector machine (LS-SVM)—have been employed to determine the levels of plastic in samples, showing strong precision (Tran et al. 2023). However, data of high quality with long-term results are needed to ensure the stability of these models in assessing pollution levels across diverse environments.

12.2.3 Application of AI in 3D Printing of Plastic Waste

3D printing has garnered substantial attention and transformed the industrial sector. This cutting-edge technology enables the production of three-dimensional objects by layering material, usually plastic or metal, according to a digital blueprint (Liu et al. 2021). Traditional subtractive manufacturing techniques employ cutting, drilling, sawing, and transforming materials to produce structures, which require eliminating portions of raw materials to achieve the final product of desired dimensions. Consequently, these subtractive processes often generate waste that cannot be recycled further (Hogan 2023). In contrast, 3D printing facilitates the creation of complex, customized objects with ease. By using recycled products as raw materials for 3D printing, we can explore a variety of material properties and characteristics (Oladapo et al. 2023). Plastic waste can be converted to filaments, which can be printed, allowing for the production of objects with different ranges of rigidity, flexibility, or even transparency (Figure 12.4). This adaptability leads to design innovation, which can result in the production of products that can be tailored to match specific requirements. Utilizing 3D printing with recycled materials is also in accordance with the circular economy principles, as it allows for the transformation of discarded waste to feedstock which can be printed and allows for creation of new products (Oladapo et al. 2021).

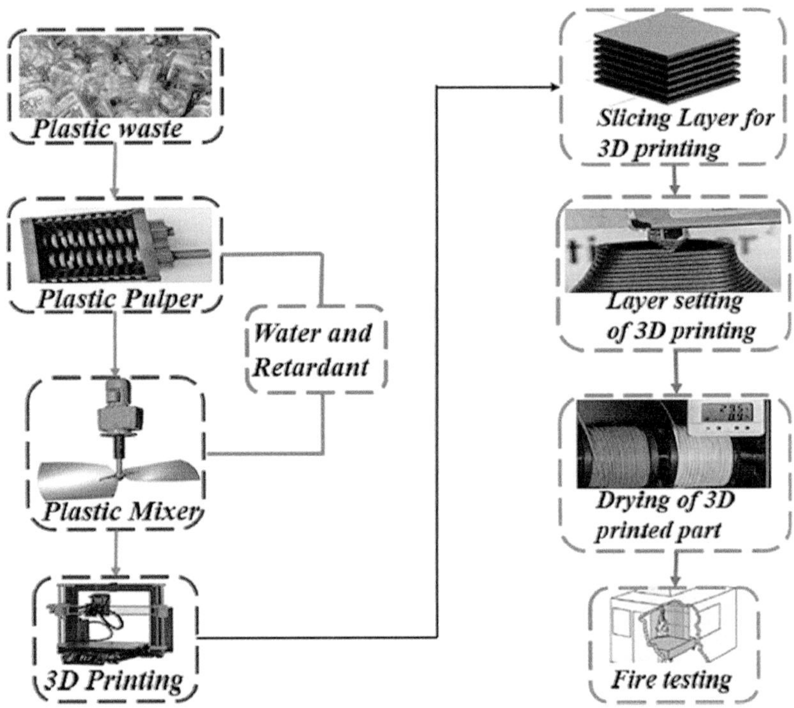

Figure 12.4 Flowchart for production of specimen using plastic waste (reproduced from Olawumi et al. 2023. This article is published under CC BY licence).

Operating a 3D printer involves a complex procedure, but incorporating AI can significantly enhance its efficiency and performance (Talaat and Hassan 2021). Lu introduced the printability checker (PC) scheme for assessing the material suitability for 3D printing, aiming to streamline product manufacturing. While the PC scheme has proven effective in testing printability, its practical application lacks functionality and reliability when handling diverse features or data. To address this, Lu suggested using machine learning (ML) to adjust laws and parameters automatically, thus enhancing printability checks. An alternative method involves training an estimation model with a support vector machine (SVM) for printability prediction, bypassing predefined rules. This approach, validated by experiments, reduced 3D model feature extraction time without compromising product precision (Lu 2016). In real-world manufacturing, printability or complexity assessments consider multiple factors such as cost, time, model size, geometry, and raw material. Genetic algorithms (GAs) and genetics-based ML methods can optimize these indicators, achieving the best effect proportion and minimal complexity value (Talaat and Hassan 2021).

Jin and the group utilized a deep learning network-based method for real-time anomaly monitoring in the 3D bioprinting process. Closed-loop AI printing is a sophisticated technique of manufacturing which integrates 3D printing with AI to ensure high quality and precision. This method involves real-time monitoring and adjustment of various printing parameters using a feedback loop, which helps in automatically controlling the adaptation to new requirements and correcting errors. The system relies on two main components: sensors and a control algorithm. Sensors collect real-time data on factors like pressure, temperature, material flow, and speed, and the control algorithm analyzes this data to adjust parameters including nozzle temperature, print speed, and material flow. When environmental changes result in distortions in the printed product, the system immediately corrects these errors, maintaining quality and accuracy (Jin et al. 2021). Paraskevoudis and group introduced a remote defect detection technique using computer vision and AI to identify defects like beading in real-time (Paraskevoudis et al. 2020). Wang and group introduced a closed-loop control framework to enhance this process (Wang et al. 2018).

AI applications in the post-processing stage of 3D printing are uncommon because the product cannot undergo further modifications. Consequently, after printing, a quality inspection is necessary to ensure that defective models are not used for surgical training. Sundaram and group developed a deep learning algorithm using a CNN architecture and developed a Smart Quality Inspection (SQI) technique based on AI for manufacturing (Sundaram and Zeid 2023). Lee and group employed techniques for data enhancement for processing images of precast concrete components and used CNNs to train and recognize these processed images for crack detection and quality assessment (Lee et al. 2022).

12.2.4 AI-Assisted Synthesis of Plastic Substitutes

Petrochemical plastics are valued for being lightweight, durable, and cost-effective, leading to their widespread use (Xie 2024). Still, less than 10% of these plastics undergo recycling, and nearly 80% are dumped in landfills or as environmental pollutants, contributing to global plastic pollution (Rosenboom et al. 2022). A promising solution involves developing sustainable, biodegradable plastic alternatives from natural components, which can reduce plastic waste and prevent microplastic pollution (Shen et al. 2020). Nonetheless, finding biodegradable options that follow specific standards, such as fire retardancy, mechanical strength and optical transparency, is challenging (Wang et al. 2021). Current methods often involve trial-and-error experiments and an unfocused parameter exploration (Nigam et al. 2022). As the requirement for replacements of plastic grows, the cost and time to identify appropriate plastic alternatives which are biodegradable, will also rise. Recently, robotics and machine learning are being integrated to forecast and create a range of nanocomposite materials that replicate the properties of different plastics, using natural building blocks as the foundation (Liu 2024).

In a study by Chen and group, four naturally safe components—glycerol, gelatin, montmorillonite (MMT) nanosheets, and cellulose nanofibres (CNFs)—were chosen as the foundational materials for creating various all-natural plastic substitutes. A robot (the OT-2 robot) specialized in automatic pipetting was utilized to synthesize 286 nanocomposites with different ratios of MMT/glycerol CNF/gelatin, and the quality of the films was assessed to instruct a support-vector machine (SVM) for classification. Following this, 14 active learning loops with data augmentation were conducted, resulting in the stage-wise fabrication of 135 all-natural nanocomposites. This approach facilitated the creation of an ANN model that boasts highly accurate prediction throughout the design space. Utilizing the predictive strengths of this model, two design tasks were successfully completed: (1) accurate forecasting of multiple properties of all-natural nanocomposites based on their composition, and (2) automatic recommendation of biodegradable alternatives to plastic according to user-defined specifications. As illustrated in Figure 12.5, by entering specific property requirements, the predictive model was able to pinpoint appropriate all-natural substitutes for various replacements of plastic without necessitating iterative experiments for optimizations. Moreover, several insights rooted in data science were obtained using Shapley Additive explanations (SHAP) for the interpretation of models and further validation using MD simulations. (Chen et al. 2024).

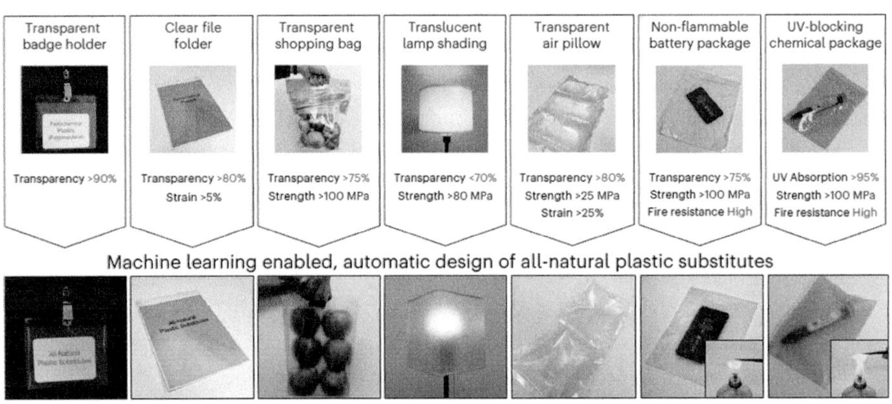

Figure 12.5 Designing plastic substitutes by inputting various parameters. The photos in the top row display non-biodegradable plastic products along with their specifications. The bottom row showcases the all-natural, biodegradable alternatives recommended using an ANN model (reproduced from Chen et al. 2024). This article is published under a CC BY 4.0 licence.

To evaluate whether the all-natural substitutes were biodegradable, two of these materials were placed deep in soil alongside polyethylene films and polystyrene, which served as controls. After 5 weeks, complete decomposition of natural substitutes was observed, while the petrochemical plastics exhibited no signs of degradation. The exceptional biodegradability of the natural substitutes for plastic was verified using many cytotoxicity tests performed on L929 cells. (Chen et al. 2024).

12.2.5 AI-Assisted Monitoring and Disposal of Marine Plastic Waste

The majority of marine pollution stems from land-based sources, primarily as plastic waste, which varies widely and takes an extended period for decomposition (Worm et al. 2017). Over time, plastic waste causes significant harm to marine ecosystems, affecting the habitats of various species in the sea. To combat marine pollution, two primary methods are employed: monitoring and waste disposal (Maximenko et al. 2019). Both methods utilize AI to operate automatically and in real-time.

In waste monitoring, identification of the location of marine waste is crucial for mapping the ocean environment. The primary method involves observing the sea surface and analyzing the variations in sea color. This observation is conducted using satellites or drones, which capture images of the sea surface. Researchers then use these images to identify and mark objects floating in the ocean. These annotated images serve as a data source for machine learning. By integrating this data with AI algorithms, it becomes possible to detect and classify marine debris (Fazri et al. 2023). This detailed classification and quantification of plastic litter offer critical benchmark information necessary for formulating decisions by citizens, policymakers, and various stakeholders in both public and private sectors. Multiple monitoring systems have been developed to support this effort, including NOAA's Marine Debris Tracker, the European Environment Agency's MarineLitterWatch, the Ocean Conservancy's CleanSwell, and the Ocean Networks' Coastbuster (Villarrubia-Gomez et al. 2018).

A new machine learning algorithm utilizing CNNs has been developed for quantification and detection of plastic waste, both floating in water and washed ashore by Wolf and group. This system, named APLASTIC-Q (Aquatic Plastic Litter Detection, Classification, and Quantification System), underwent training using high-resolution geo-spatial images (approximately 5 pixels per cm, or 0.002 meters per pixel) from surveys conducted in aerial mode in Cambodia. Two components are present in APLASTIC-Q, which are based on machine learning: Quantifier (PLQ-CNN) and Plastic Litter Detector (PLD-CNN). The PLD-CNN effectively categorized objects into plastic litter, water, vegetation, and sand, with an accuracy rate of 83% and resulted in a qualitative litter count based on a threshold, indicating low or high amounts. The PLQ-CNN further identified and counted specific types of litter, including Styrofoam, shoes, cartons, cups, water bottles, polystyrene packaging, canisters, bowls, textiles, and large or small carry bags (Wolf et al. 2020).

Several technologies have been introduced to collect plastic waste from the marine systems. A Dutch company, The Ocean Cleanup, has created an automated setup designed to clean the marine ecosystems. This setup consists of a ship equipped with AI technology. The AI-powered ship autonomously navigates the oceans to collect trash (Sil et al. 2021). In addition, many AI-powered drones and robots have been developed to collect macroplastics. Some of these include FRED (Solar-powered floating robot for eliminating debris), Jellyfishbot, Seabin (automated bucket to collect debris from water), BluePhin (Battery-powered robot with zero carbon emissions), and WasteShark (Drone in the shape of a shark to collect floating plastic) (Schmaltz et al. 2020).

12.3 CONCLUSIONS

Artificial intelligence (AI) has been playing a crucial role in combating plastic pollution by offering innovative solutions across various domains, including detection, monitoring, recycling, cleanup, and public engagement. AI-powered technologies are used to detect and monitor plastic pollution in various environments. Equipped with AI-driven image recognition, drones can survey and monitor hard-to-reach areas for plastic debris, providing real-time images. The sea surface images can be analyzed using AI algorithms and used to identify and map areas with high concentrations of plastic waste in oceans, rivers, and coastal regions. This detailed classification and quantification of plastic litter offer critical benchmark information necessary for formulating decisions by citizens, policymakers, and various stakeholders in both public and private sectors. In addition, AI aids in the cleanup of plastic waste from natural environments. Drones and robots equipped with AI can detect and collect plastic debris from water bodies.

AI also enhances the accuracy and efficiency of sorting waste and recycling processes. It can help in accurately identifying the sources of plastic waste which can help the researchers better trace pathways and implement targeted strategies. AI-powered robots and conveyor belts efficiently use computer vision and machine learning to identify and sort different types of plastics, improving recycling rates and reducing contamination. In addition, AI accelerates the development of sustainable materials and alternatives to plastics. Machine learning models can predict the properties of new materials, facilitating the creation of biodegradable or more recyclable plastics. AI optimizes plastic recycling by improving the efficiency and performance of 3D printing, which can be used to transform discarded items or waste into printable feedstock and create new products.

However, there are still some bottlenecks associated with the large-scale use of AI for tackling plastic pollution. AI models require large, high-quality datasets to function effectively. Obtaining and maintaining such datasets can be challenging. In addition, implementing AI solutions on a global scale involves significant logistical and financial challenges. Further, effective use of AI in combating plastic pollution requires collaboration between technologists, environmental scientists, policymakers, and the public. As AI technology continues to evolve, its applications in combating plastic pollution are expected to become more sophisticated, providing powerful tools to address this critical environmental issue.

REFERENCES

Bircanoğlu, C., M. Atay, F. Beşer, Ö. Genç and M.A. Kızrak. 2018. RecycleNet: Intelligent waste sorting using deep neural networks. Innovations in intelligent systems and applications (INISTA). IEEE. 1–7

Chakraborty, T.K., M.S. Rahman, M.S. Nice, B.N. Netema, K.R. Islam, P.C. Debnath, et al. 2024. Application of machine learning and multivariate approaches for assessing microplastic pollution and its associated risks in the urban outdoor environment of Bangladesh. J. Hazard. Mater. 472: 134359.

Chen, T., Z. Pang, S. He, Y. Li, S. Shrestha, J.M. Little, et al. 2024. Machine intelligence-accelerated discovery of all-natural plastic substitutes. Nat. Nanotechnol. 1–10.

Cubas, A.L.V., R.T. Bianchet, I.M.A.S.D. Reis and I.C. Gouveia. 2022. Plastics and microplastic in the cosmetic industry: aggregating sustainable actions aimed at alignment and interaction with UN sustainable development goals. Polymers. 14: 4576.

Diya, S.Z., R.A. Proma, M.N. Islam, T.T. Anannya, A. Al Mamun, R. Arefeen, et al. 2018. Developing an intelligent waste sorting system with robotic arm: A step towards green environment. International Conference on Innovation in Engineering and Technology (ICIET). IEEE. 1–6

Fazri, M.F., L.B. Kusuma, R.B. Rahmawan, H.N. Fauji and C. Camille. 2023. Implementing Artificial Intelligence to Reduce Marine Ecosystem Pollution. ITSDI. 4: 101–108.

Glenn, G.M., W. Orts, S. Imam, B.S. Chiou and D.F. Wood. 2014. Starch plastic packaging and agriculture applications. 421–452. *In*: Halley, P.J. and L. Avérous (eds). Starch Polymers. Elsevier.

Hogan, M. 2023. How Does 3D Printing Reduce Waste? Understanding Sustainable 3D Printing. Nexa 3D, https://nexa3d.com/blog/3d-printing-products-reduce-waste/#:~:text=Material%20recyclability%20in%203D%20printing,need%20for%20new%20material%20inputs. Accessed 14-06-2024

Huynh, M.H., P.T. Pham-Hoai, A.K. Tran and T.D. Nguyen. 2020. Automated waste sorting using convolutional neural network. 7th NAFOSTED Conference on Information and Computer Science (NICS). IEEE. 102–107.

Jin, Z., Z. Zhang, X. Shao and G.X. Gu. 2021. Monitoring anomalies in 3D bioprinting with deep neural networks. ACS Biomater. Sci. Eng. 9: 3945–3952.

Jin, H., F. Kong, X. Li and J. Shen. 2024. Artificial intelligence in microplastic detection and pollution control. Environ. Res. 119812.

Kida, M., K. Pochwat and S. Ziembowicz. 2024. Assessment of machine learning-based methods predictive suitability for migration pollutants from microplastics degradation. J. Hazard. Mater. 461: 132565.

Koskinopoulou, M., F. Raptopoulos, G. Papadopoulos, N. Mavrakis and M. Maniadakis. 2021. Robotic waste sorting technology: Toward a vision-based categorization system for the industrial robotic separation of recyclable waste. IEEE Robot. Autom. Mag. 28: 50–60.

Lee, S., M. Jeong, C.S. Cho, J. Park and S. Kwon. 2022. Deep learning-based PC member crack detection and quality inspection support technology for the precise construction of OSC projects. Applied Sciences. 12: 9810.

Liu, W., H. Huang, L. Zhu and Z. Liu. 2021. Integrating carbon fiber reclamation and additive manufacturing for recycling CFRP waste. Compos. B Eng. 215: 108808.

Liu, Z. 2024. Artificial intelligence assists the synthesis of all-natural plastic substitutes. Nat. Nanotechnol. 1–2.

Lu, T. 2016. Towards a fully automated 3D printability checker. *In*: 2016 IEEE International Conference on Industrial Technology (ICIT). IEEE. 922–927.

MacLeod, M., H.P.H. Arp, M.B. Tekman and A. Jahnke. 2021. The global threat from plastic pollution. Science. 373: 61–65.

Maximenko, N., P. Corradi, K.L. Law, E. Van Sebille, S.P. Garaba, R.S. Lampitt, et al. 2019. Toward the integrated marine debris observing system. Front. Mar. Sci. 6: 447.

Meidl, R.A. 2018. Plastic Waste Management: Are We on the Right Path to Sustainability? Issue brief no. 11. 2018 Rice University's Baker Institute for Public Policy, Houston, Texas.

Mohan, K. and V.R. Lakshmanan. 2023. A critical review of the recent trends in source tracing of microplastics in the environment. Environ. Res. 117394.

Na, S., S. Heo, S. Han, Y. Shin and M. Lee. 2022. Development of an artificial intelligence model to recognise construction waste by applying image data augmentation and transfer learning. Buildings. 12: 175.

Nigam, S., A.K. Das, F. Matkawala and M.K. Patidar. 2022. An insight overview of bioplastics produced from cellulose extracted from plant material, its applications and degradation. Environ. Sustain. 5: 423–441.

Oladapo, B.I., S.A. Zahedi, V.A. Balogun, S.O. Ismail and Y.A. Samad. 2021. Overview of additive manufacturing biopolymer composites. 915–928. *In*: Dermot Brabazon (ed.). Encyclopedia of Materials: Composites. Elsevier.

Oladapo, B., A. Zahedi, S. Ismail, W. Fernando and O. Ikumapayi. 2023. 3D-printed biomimetic bone implant polymeric composite scaffolds. Int. J. Adv. Manuf. Tech. 126: 4259–4267.

Olawumi, M.A., B.I. Oladapo, O.M. Ikumapayi and J.O. Akinyoola. 2023. Waste to wonder to explore possibilities with recycled materials in 3D printing. Sci. Total Environ. 905: 167109.

Paraskevoudis, K., P. Karayannis and E.P. Koumoulos. 2020. Real-time 3D printing remote defect detection (stringing) with computer vision and artificial intelligence. Processes. 8: 1464.

Patrizi, A., G. Gambosi and F.M. Zanzotto. 2021. Data augmentation using background replacement for automated sorting of littered waste. J. Imaging. 7: 144.

Paulraj, S.G., S. Hait and A. Thakur. 2016. Automated municipal solid waste sorting for recycling using a mobile manipulator. International design engineering technical conferences and computers and information in engineering conference. American Society of Mechanical Engineers. 50152: V05AT07A045

Prata, J.C., A.L.P. Silva, J.P. Da Costa, C. Mouneyrac, T.R. Walker, A.C. Duarte, et al. 2019. Solutions and integrated strategies for the control and mitigation of plastic and microplastic pollution. Int. J. Environ. Res. Public Health. 16: 2411.

Rosenboom, J.G., R. Langer and G. Traverso. 2022. Bioplastics for a circular economy. Nat. Rev. Mater. 7: 117–137.

Rousta, K., I. Ordoñez, K. Bolton and L. Dahlén. 2017. Support for designing waste sorting systems: A mini review. Waste Manag. Res. 35: 1099–1111.

Schmaltz, E., E.C. Melvin, Z. Diana, E.F. Gunady, D. Rittschof, J.A. Somarelli, et al. 2020. Plastic pollution solutions: emerging technologies to prevent and collect marine plastic pollution. Environ. Int. 144: 106067.

Seredkin, A.V., M.P. Tokarev, I.A. Plohih, O.A. Gobyzov and D.M. Markovich. 2019. Development of a method of detection and classification of waste objects on a conveyor for a robotic sorting system. J. Phys. Conf. Ser. IOP Publishing. 1359: 012127.

Shen, M., B. Song, G. Zeng, Y. Zhang, W. Huang, X. Wen, et al. 2020. Are biodegradable plastics a promising solution to solve the global plastic pollution? Environ. Pollut. 263: 114469.

Shen, M., W. Huang, M. Chen, B. Song, G. Zeng and Y. Zhang. 2020. (Micro) plastic crisis: un-ignorable contribution to global greenhouse gas emissions and climate change. J. Clean. Prod. 254: 120138.

Sil, R., A. Das, F. Shamim, A. Chowdhury, R. Mukherjee, S. Ali, et al. 2021. DAMONA: A multi-robot system for collection of waste in ocean and sea. *In*: International Conference on Micro-Electronics and Telecommunication Engineering. Singapore: Springer Nature Singapore. 149–157.

Sundaram, S. and A. Zeid. 2023. Artificial intelligence-based smart quality inspection for manufacturing. Micromachines, 14: 570.

Talaat, F.M. and E. Hassan. 2021. Artificial Intelligence in 3D Printing. Enabling Machine Learning Applications in Data Science: Proceedings of Arab Conference for Emerging Technologies. Springer Singapore. 77–88.

Tran, H.T., M. Hadi, T.T.H. Nguyen, H.G. Hoang, M.K. Nguyen, K.N. Nguyen, et al. 2023. Machine learning approaches for predicting microplastic pollution in peatland areas. Mar. Pollut. Bull. 194: 115417.

Villarrubia-Gómez, P., S.E. Cornell and J. Fabres. 2018. Marine plastic pollution as a planetary boundary threat–the drifting piece in the sustainability puzzle. Mar. Policy. 96: 213–220.

Wan, Y., X. Chen, Q. Liu, H. Hu, C. Wu and Q. Xue. 2022. Informal landfill contributes to the pollution of microplastics in the surrounding environment. Environ. Pollut. 293: 118586.

Wang, T., T.H. Kwok, C. Zhou and S. Vader. 2018. In-situ droplet inspection and closed-loop control system using machine learning for liquid metal jet printing. J. Manuf. Syst. 47: 83–92.

Wang, Z., Y. Xu, R. Liu and X. Zhu. 2021. Planstic: biodegradable plastic with high-entropy fibers made from waste plastic and plant leaves. ACS Appl. Polym. Mater. 3: 5355–5360.

Witten, I.H., E. Frank, M.A. Hall, C.J. Pal and M. Data. 2005. Practical machine learning tools and techniques. Data mining. Amsterdam. The Netherlands: Elsevier. 2: 403–413.

Wolf, M., K. van den Berg, S.P. Garaba, N. Gnann, K. Sattler, F. Stahl, et al. 2020. Machine learning for aquatic plastic litter detection, classification and quantification (APLASTIC-Q). Environ. Res. Lett. 15: 114042.

Worm, B., H.K. Lotze, I. Jubinville, C. Wilcox and J. Jambeck. 2017. Plastic as a persistent marine pollutant. Annu. Rev. Environ. Resour. 42: 1–26.

Wu, P., B. Wang, Y. Lu, G. Cao, P. Xie, W. Wang, et al. 2023. Machine Learning-Assisted Insights into Sources and Fate of Microplastics in Wastewater Treatment Plants. ACS ES&T Water. 4: 1107–1118.

Xie, Y. 2024. The Intersection of Plastic Electronics and Medical Innovation. Highl. Sci. Eng. Technol. 96: 191–197.

Yang, M. and G. Thung. 2016. Classification of trash for recyclability status. CS229 project report. 2016: 3.

Chapter **13**

Plastic Waste Amendments to the Basel Convention: A Sri Lankan Stakeholder Perspective

Thenuwara Arachchige Omila Kasun Meetiyagoda[1],
Induka Werellagama[2],
Premakumara Jagath Dickella Gamaralalage[3]
and Amila Abeynayaka*[3,4]

[1]Saitama University, Saitama 338-8570, Japan

[2]The Open Polytechnic, Waterloo, 5040, New Zealand

[3]Institute for Global Environmental Strategies (IGES), Japan

[4]Quantitative Sustainability Assessment,
Department of Environmental and Resource Engineering,
Technical University of Denmark (DTU), Kongens Lyngby 2800, Denmark

13.1 INTRODUCTION

13.1.1 Global Context of Plastic Waste Transboundary Movement

The global issue of plastic waste transboundary movement has gained significant international attention due to its extensive environmental, social, and economic

*For Correspondence: amiab@dtu.dk

DOI: 10.1201/9781003531463-13

implications (Rucevska et al. 2015). This matter has been exacerbated by the unregulated export of plastic waste from developed to developing nations (Man et al. 2013; Rucevska et al. 2015), often masked as recycling, resulting in environmental pollution, health risks, and strained diplomatic relations (Uhm 2021). Consequently, countries worldwide are increasingly turning to international conventions and domestic policies as a means to restrict, manage, and regulate the transboundary movement of plastic waste. Among these initiatives, the Basel Convention's recent amendments regarding plastic waste play a central role.

The Basel Convention is a multinational environmental treaty officially adopted in March 1989 and entered into force in May 1992 (Man et al. 2013). Presently, it has a membership of 191 Parties (as of 4 September 2023) (United Nations 1989) and has achieved near-universal participation. The United States has refrained from ratifying the convention, citing insufficient domestic legal provisions to execute its requirements fully. The primary objectives of the Basel Convention revolve around regulating the transboundary movements of hazardous wastes and their disposal, as well as promoting environmentally sound management (ESM) of such wastes. These activities are undertaken with the principal goal of protecting human health and mitigating environmental hazards, reducing improper handling of hazardous and other wastes. The Basel Convention stands as one of the most comprehensive global environmental agreements focusing on the management of hazardous wastes. The convention exercises authority over a range of waste categories, encompassing poisonous, toxic, corrosive, explosive, flammable, infectious, and ecotoxic wastes. Signatory Parties bear the responsibility to curtail the volume of waste transport, treat and dispose of waste in proximity to its generation, and proactively minimize or prevent waste generation at its source. This collective effort, facilitated by the Basel Convention, serves to ensure the proper management and disposal of hazardous and other wastes, thereby contributing to the global endeavour of environmental protection (BRS 2011).

In recent decades, China has held a prominent position as a major destination for the world's plastic waste, processing a substantial portion of global plastic scrap (Man et al. 2013). However, the adverse environmental consequences associated with this influx of waste, coupled with domestic concerns regarding pollution and resource efficiency, prompted China to enact a comprehensive policy shift. In 2018, China imposed a significant ban on the import of 24 categories of solid waste, including various types of plastic waste (Man et al. 2013; Wen et al. 2021). The repercussions of China's waste import ban were profound and reverberated throughout the global waste management landscape (Man et al. 2013). The consequences of China's ban on global waste trade dynamics were particularly pronounced, as waste-exporting countries were compelled to seek alternative destinations for their plastic waste, leading to a redirection of waste to other Southeast Asian nations. This shift emphasized the urgent need for a coordinated global response. Moreover, the ban significantly impacted the global recycling industry, which had come to rely heavily on China as a primary processing hub. Recycling industries in waste-exporting countries face substantial challenges in finding new markets, reconfiguring their supply chains, and ensuring the quality of recycled materials (Wen et al. 2021).

While the ban was intended to address environmental concerns within China, it also raised concerns about the potential diversion of waste to countries with less stringent environmental regulations, potentially exacerbating pollution problems elsewhere. Furthermore, waste-exporting countries were compelled to reevaluate their own waste management practices in response to China's ban. This re-evaluation has led to a renewed focus on reducing plastic consumption, improving domestic recycling infrastructure, and implementing more sustainable waste management policies. China's waste import ban has prompted other nations to reexamine their strategies for managing plastic waste transboundary movement. Several notable examples highlight how countries have responded to the challenges posed by this complex issue (Wen et al. 2021; Liu and Liu 2023). After China's waste import ban, Malaysia and the Philippines saw a surge in waste imports. Still, concerns over improper disposal and environmental damage led both countries to implement stricter regulations and repatriate contaminated shipments. In Greenpeace Malaysia (2019), returned 4,120 tons of plastic waste to 13 countries and shut down 200 illegal recycling centers. Tensions also rose between the Philippines and Canada over 2,700 tons of mislabelled Canadian waste. President Duterte recalled diplomats after Canada missed a deadline to return the waste, prompting environmentalists to claim the delay violated the Basel Convention (CBC 2019). Additionally, Thailand banned electronic waste imports in 2018 and aimed to end plastic waste imports by 2021, while Vietnam plans to ban all plastic scrap imports by 2025 (Jain 2020). The European Union (EU) has embraced a circular economy approach, committed to minimizing waste generation and promoting recycling. Initiatives such as the Single-Use Plastics Directive and the Circular Economy Action Plan set ambitious targets for reducing plastic waste and enhancing recycling efforts (EU 2019, 2020). As discussed earlier, the Basel Convention is another avenue for addressing plastic waste transboundary movement.

Until 2019, plastic waste was not clearly defined or categorized within the original classification of hazardous substances outlined in the Basel Convention. This convention, which was adopted in 1989, aimed to regulate the transboundary movement and disposal of hazardous wastes. Recognizing this deficiency, the 14th Conference of the Parties (COP) acted in 2019 to amend the Basel Convention, incorporating plastic waste provisions into a legally binding framework. These amendments were made to Annexes II, VIII, and IX of the Basel Convention (Figure 13.1a), with the intention of strengthening control over the international movement of plastic waste while also prioritizing human and environmental safety in its management (BRS 2019).

The convention establishes a "notice and consent" regime, known as Prior Informed Consent (PIC), for exporting hazardous and other specific waste to importing countries (Figure 13.1b). The PIC procedure applies to waste that is hazardous or requires special consideration, including the types of plastic waste listed in Annex II and Annex VIII, before the transboundary movement of hazardous wastes and other wastes. It has a mechanism with strict requirements and procedures. It is required to obtain and disseminate the decisions of importing parties regarding whether they wish to receive future shipments (Kojima 2020).

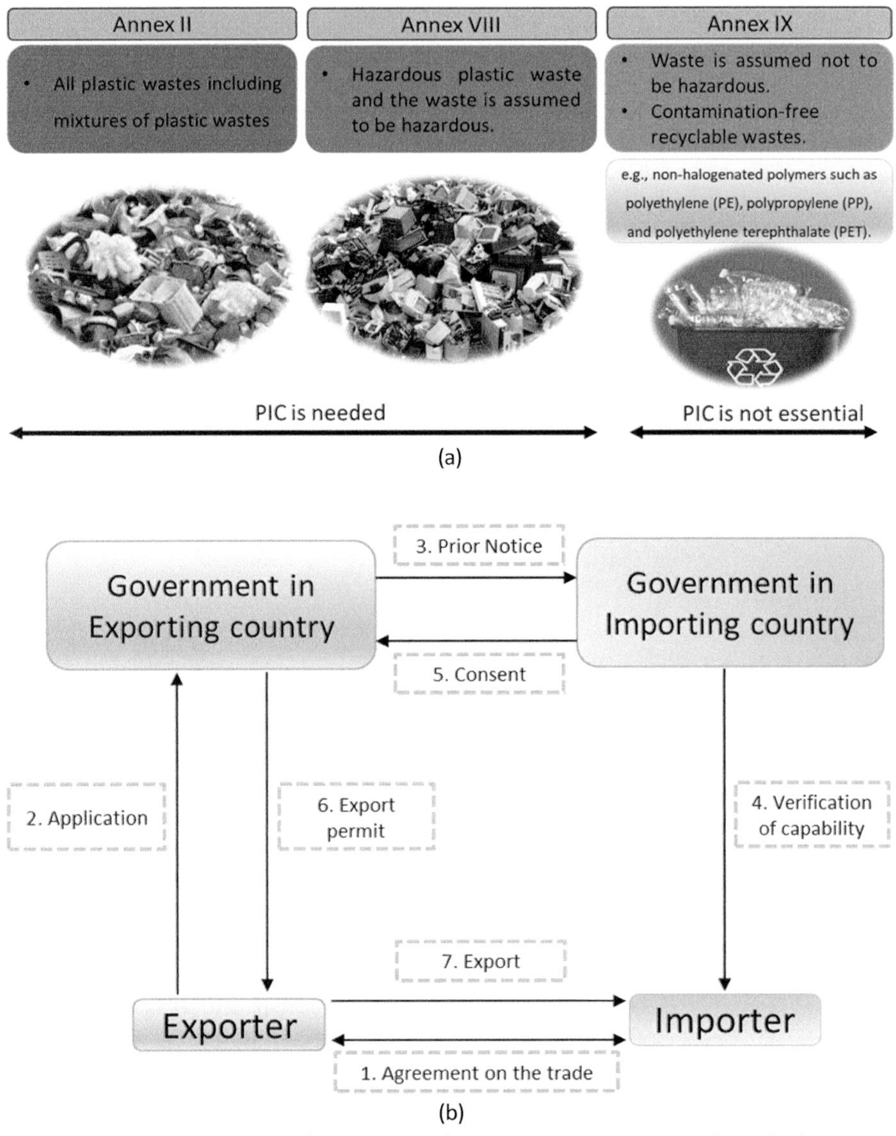

Figure 13.1 Basel convention: (a) amendments to annexes concerning plastic waste (BRS 2019); (b) Prior Informed Consent (PIC) procedure (Kojima 2020).

13.1.2 Stakeholder Perspective Collection

This chapter is based on an original study conducted by the authors, which assesses the current state of implementation of the plastic waste amendments in Sri Lanka. The primary objective was to understand the potential impacts of these amendments on Sri Lanka. To achieve this, a questionnaire was developed to gather input from key stakeholders across various sectors, including policymakers, businesses, academics, civil society representatives, and the general public in Sri Lanka. This survey requested opinions and insights regarding the consequences of the Basel Convention's plastic waste amendments. The ultimate goal of the study is to formulate a set of recommendations based on the findings of this survey. Figure 13.2 illustrates the stakeholder perspective collection procedure.

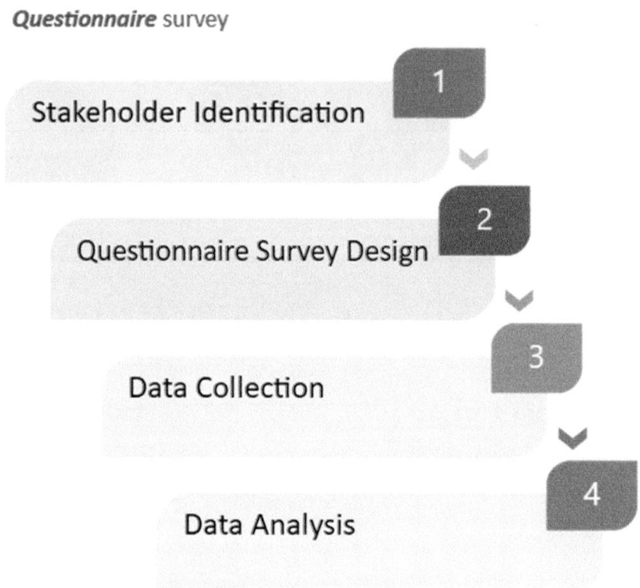

Figure 13.2 Procedure for collecting stakeholder perspectives.

13.1.2.1 Stakeholder Identification

Stakeholders in plastic waste management were categorized into five key groups: government sector, private sector, recycling sector, academia, and other sectors such as international development projects, Non-Governmental Organizations (NGOs), and consultancy services related to waste management and plastic activities (Figure 13.3). These stakeholders were carefully selected based on their specific roles and expertise in the field, such as their involvement in plastic waste management, experience with environmental policies, or participation in related sectors. The diverse perspectives gathered from these groups were integral to understanding the consequences of the Basel Convention's plastic waste amendments and formulating relevant recommendations for Sri Lanka's policy and practice.

13.1.2.2 Questionnaire Survey Design

The questionnaire design process involved several key stages to ensure the development of a comprehensive and practical survey (Figure 13.4). Initially, existing questionnaire-based surveys from different countries were reviewed to understand best practices and gain a deeper understanding of effective survey design in the context of plastic waste management. This provided a solid

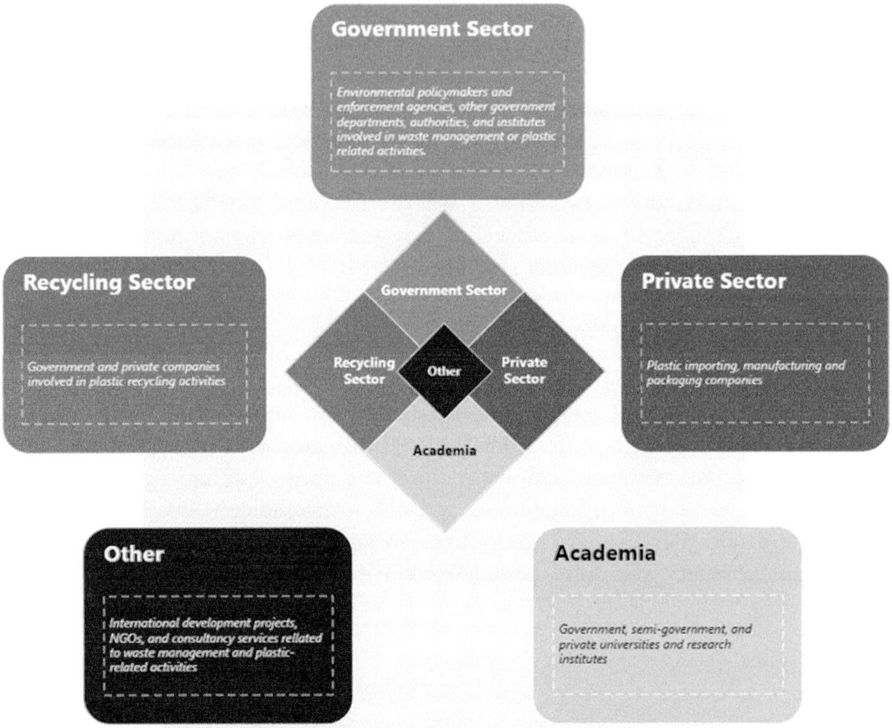

Figure 13.3 Categorization of stakeholders involved in plastic waste management.

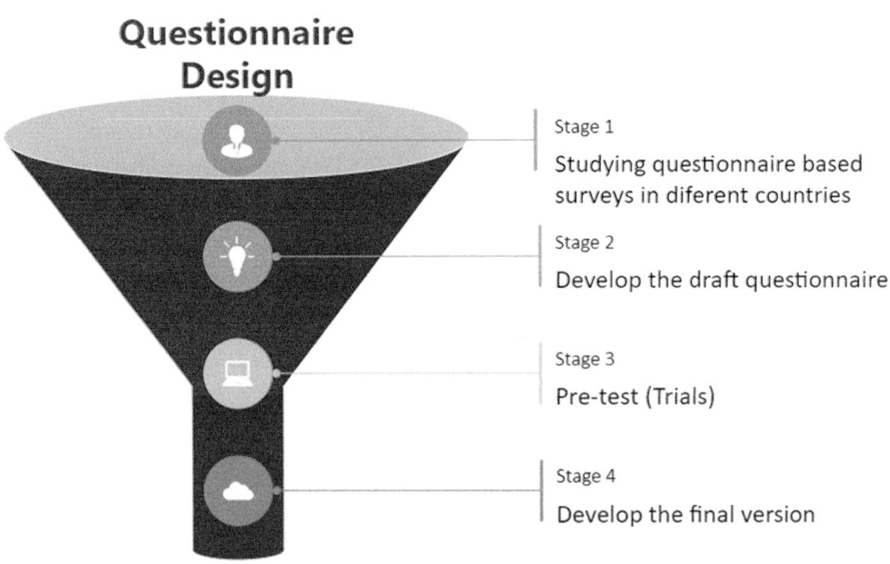

Figure 13.4 Key stages in the questionnaire design process.

foundation for developing a draft questionnaire, which focused on key areas related to the Basel Convention's plastic waste amendments and their potential impact on Sri Lanka. The draft questionnaire was then pre-tested through trials with a small group of stakeholders to identify any issues with clarity, question wording, or survey flow. Feedback from these trials led to necessary adjustments and refinements to improve the questionnaire's effectiveness. Finally, based on the pre-test results, the questionnaire was revised and finalized, incorporating all necessary changes to ensure it effectively gathered relevant and reliable data from stakeholders across various sectors.

13.1.2.3 Data Collection

The questionnaire, available in both online and offline formats, along with a summary of the Basel Convention's plastic waste amendments, was emailed to all 250 selected participants from the various categories outlined previously. These participants were chosen from diverse sectors, including government, private industry, recycling, academia, and other stakeholders involved in waste management and plastic-related activities. To supplement the survey responses, key informant interviews were conducted via Zoom or WhatsApp, allowing for

Figure 13.5 Data collection methodology.

in-depth discussions on the plastic waste amendments and their implications. Participants for the interviews were randomly selected from each category to ensure a representative sample and capture a wide range of perspectives. Figure 13.5 illustrates the process of collecting stakeholder insights through both the survey and interviews.

13.1.2.4 Data Analysis

The response data from the email-based questionnaire survey were collected through a Google Form, while responses from key informant interviews were recorded on an Excel sheet. Subsequently, all data were consolidated and subjected to descriptive analysis using Microsoft Excel 2013. Table 13.1 shows the number of respondents and survey response rate in each category.

Table 13.1 Survey response rate of participants in each category

Category	No. of Selected Participants	No. of Respondents		Survey Response Rate (%)
		Questionnaire Survey	Key Informant Interviews	
Government sector	31	11	1	39
Private sector	20	1	3	20
Recycling sector	40	5	3	20
Academia	56	10	1	20
Other	68	7	1	11
Total	215	34	9	20

13.2 SRI LANKAN CONTEXT: RELEVANT POLICIES AND AMENDMENTS

13.2.1 Constitutional Provisions

The Constitution of Sri Lanka, established in 1978 and amended up to 2021, includes provisions aimed at protecting, preserving, and improving the environment for the benefit of the community. It further emphasizes that it is the fundamental duty of every individual in Sri Lanka to protect nature and conserve its resources (GOSL 2022).

13.2.2 Criminal Procedure and Penal Codes

The Code of Criminal Procedure Act No. 15 of 1979 and the Penal Code Ordinance No. 2 of 1883 address the removal or abatement of public nuisances, including the illegal dumping of waste. These legal instruments provide the framework for addressing waste-related offences (GOSL 1979).

13.2.3 National Environmental Act (NEA)

The National Environmental Act broadly defines "waste" as any matter that alters the environment negatively. It includes mechanisms such as:

- Environmental Protection Licensing (EPL)
- Environmental Impact Assessment (EIA)
- Initial Environmental Examination (IEE)

The Act also established the Central Environmental Authority (CEA) to oversee these processes (CEA 2000).

13.2.4 National Environmental Policy (2003)

This policy focuses on preventing the pollution of surface and groundwater caused by human and domestic wastes. It promotes improved sanitation and effective waste disposal practices to safeguard water resources (GOSL 2003).

13.2.5 Cleaner Production Policy (2004)

Aiming to reduce excessive exploitation and waste, this policy encourages changes in consumption patterns to minimize energy consumption and overall waste production (GOSL 2005).

13.2.6 National Policy on Solid Waste Management (2007)

This policy aims to establish a robust and integrated system for managing waste, focusing on maximizing resource recovery to achieve a significant reduction in the volume of waste sent for disposal. It emphasizes environmental accountability and social responsibility and includes key principles such as:

- 3R Principle (Reduce, Reuse, and Recycle)
- Polluter Pays Principle
- Resource Maximization

It also encourages cost-effective methods for waste collection, transportation, storage, and treatment of solid waste while ensuring stakeholder participation in waste management (van der Wel and Post 2007).

13.2.7 Marine Pollution Prevention Act No. 2 (2008)

The Marine Pollution Prevention Act focuses on preventing, reducing, and controlling marine pollution within Sri Lanka's territorial waters. The Marine Environmental Prevention Authority (MEPA) is tasked with enforcing this Act (GOSL 2008).

13.2.8 Environment Conservation Levy Act No. 26 (2008)

This Act imposes levies on specified consumer items, imports, manufactured goods, and services that are likely to harm the environment. It aims to mitigate the environmental impact of these products and services (Chathurika 2012; Kamalakkannan et al. 2022).

13.2.9 Plastic Ban (2017)

In 2017, Sri Lanka implemented a ban on the manufacture, sale, and free distribution of polythene products under 20 µm. The ban also includes prohibiting HDPE grocery bags and the open burning of plastics (CEA 2017).

13.2.10 National Policy on Waste Management (2020)

This policy aims to establish a comprehensive and integrated waste management system by providing clear, detailed guidelines for policymakers and implementers at all levels. It focuses on aligning various stakeholders across administrative and management structures to work together within a unified framework. The policy outlines specific actions and strategies for efficiently managing different types of waste, including municipal solid waste, hazardous waste, construction and demolition waste, and specialized waste streams such as healthcare and e-waste. By offering well-defined directions, the policy ensures that both vertical and horizontal coordination within administrative and management levels is achieved, promoting effective implementation and long-term sustainability of waste management practices throughout the country (MOE-SL 2020).

13.2.11 National Environmental Regulations (2021)

These regulations require clearly marking manufactured plastic items according to Plastic Material Identification Standards. They also ban the use of PET (polyethylene terephthalate) and PVC (polyvinyl chloride) for agrochemical packaging, certain sachets (volume less than 20 mL), inflatable toys, and plastic-stemmed cotton buds (GOSL 2021).

13.2.12 National Action Plan on Plastic Waste Management (2021–2030)

The National Action Plan outlines 16 goals to be achieved from 2021 to 2030, including (MOE-SL 2021):

- Developing a plastic waste inventory
- Implementing preventive 3R strategies
- Addressing crosscutting issues related to education, health, safety, and community participation in plastic waste management.

13.3 PIC PROCEDURE AND KEY STAKEHOLDERS' PERCEPTIONS

As shown in Figure 13.6a, 20 out of 43 participants are familiar with the PIC procedure, while the remaining participants are not. Notably, all participants involved in policymaking and enforcement agencies possess a thorough understanding of the PIC procedure. Conversely, a significant portion of respondents from specific government departments and institutes, constituting 42% of the total government sector respondents, need more awareness of the PIC procedure. Additionally, the survey revealed that 63% of academic professionals are not aware of PIC. The findings further indicate that all of the participants from the private sector, specifically those involved in manufacturing plastic products, need to be made aware of either the plastic waste amendments or the PIC procedures.

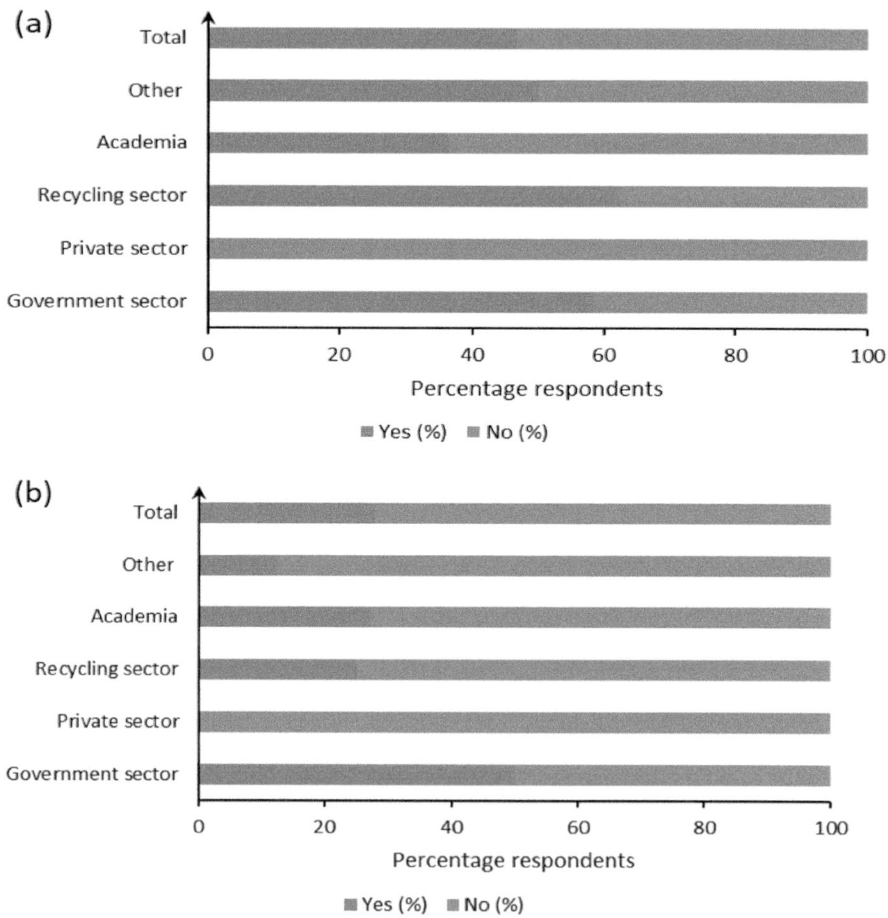

Figure 13.6 Awareness levels: (a) respondents' understanding of the PIC procedure and (b) respondents' knowledge of the waste importing procedure in Sri Lanka.

As shown in Figure 13.6b, most participants need to be made aware of the waste-importing procedures in Sri Lanka. However, 50% of participants from the government sector are aware of these procedures, as they are directly involved in handling the import process. According to the CEA, importers must obtain prior approval from the CEA before importing any waste or recycled material into Sri Lanka. Since Sri Lanka lacks disposal facilities, the importation of List A (Annex VIII) waste under the Basel Convention is prohibited. For importing List B (Annex IX) waste, importers must submit a proposal to the Basel Technical Expert Committee at the CEA. Figure 13.7 illustrates the procedure for importing recycling materials into Sri Lanka. The Technical Expert Committee (TEC) conducts the proposal evaluation, which includes various responsible stakeholder authorities. The National Coordinating Committee makes the final decision on implementing the Basel Convention under the Ministry of Environment. This decision is then communicated to the Ports Authority and Sri Lanka Customs (CEA 2023).

The following information is required in order to grant permission for importing waste:

- Authenticity of the country of origin and source of materials.
- Valid Environmental Protection License (EPL).
- Details of the production process in the country of origin, the composition of the compound, and the total quantity of raw material received from the sources available in the country of origin.
- An analytical report obtained from an accredited laboratory to ensure the non-toxicity and cleanliness of the material.
- Sample/photograph of raw material.
- List of main manufactured products and capacities.
- List of by-products.
- Details of waste arising from the process and disposal mechanism.
- An agreement with the country of exporting B waste (Annex IX) is allowed.

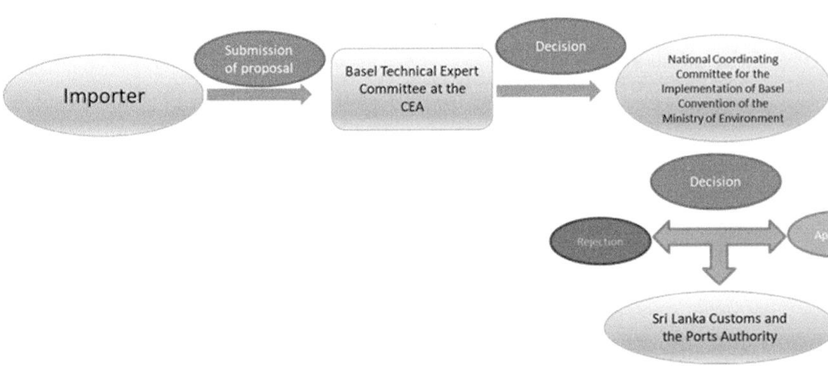

Figure 13.7 Recycling waste importing procedure in Sri Lanka (CEA 2023).

Importing activities should be carried out in accordance with Basel Convention procedures. However, according to the Ministry of Environment, Sri Lanka 2023, there are records related to illegal imports. Furthermore, post-consumer waste is not imported, and it is banned under the national policy on waste management, and the establishment of new recycling companies based on imported waste is prohibited in Sri Lanka.

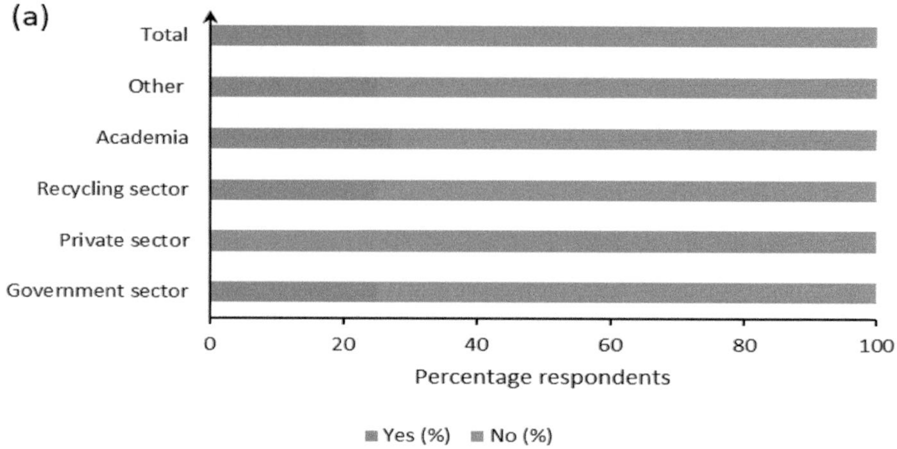

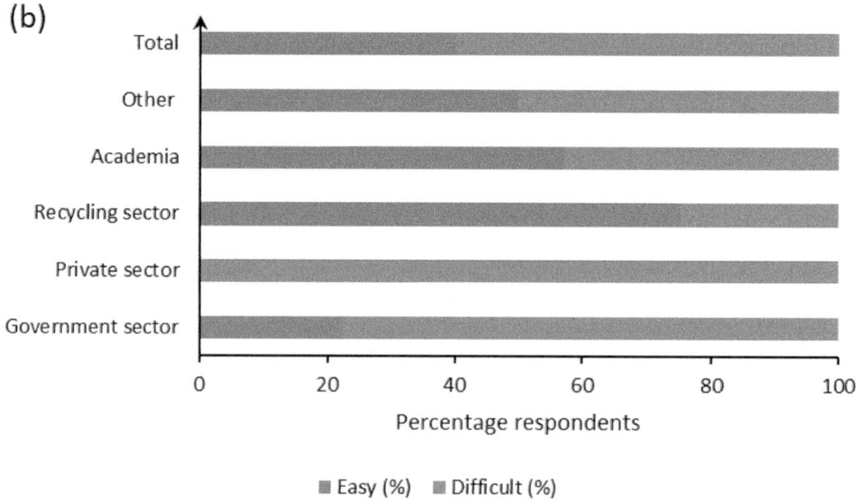

Figure 13.8 Participants' awareness: (a) imported waste handling procedure, (b) perception of implementing and enforcing plastic waste amendments.

According to the survey, 77% of participants need to gain knowledge of the procedures for handling imported waste in Sri Lanka (Figure 13.8a). Only 23% are aware of these procedures, primarily those involved in government policymaking, enforcement agencies, and academia. These knowledgeable participants indicated that industries must obtain approval from the Ministry of Environment, the CEA, the Waste Management Authority, and Sri Lanka Customs to import waste.

In a notable example from 2019, Sri Lanka Customs intercepted illegal waste imports from the United Kingdom, which included plastic waste, clinical waste, used cushions and mattresses, plant parts, and other hazardous materials. Following a court order in June 2020, the waste was sent back to the UK (CEJ 2020).

More than 60% of those familiar with Sri Lanka's plastic waste amendments expressed concerns about the challenges of implementing and enforcing these changes (Figure 13.8b). Despite these concerns, the government has taken significant steps to address plastic waste issues, such as banning PET and PVC for agrochemical packaging, introducing plastic material identification standards, and prohibiting polythene less than 20 microns thick. Additionally, 3R plastic manufacturers and recyclers must obtain an EPL from the CEA before starting operations. The government has also promoted the use of environmentally friendly degradable covers for products like peritoneal dialysis fluid, as highlighted in gazette notifications on January 19 and January 21, 2021, demonstrating a solid commitment to plastic waste management.

On January 19, 2021, the government enacted prohibitions on PET and PVC materials for agrochemical packaging. It banned certain single-use plastic items, including small sachets and inflatable toys, with exceptions for specific uses. Plastic cotton buds with plastic stems were also restricted, except for medical or clinical purposes. On January 21, 2021, plastic material identification standards were introduced, requiring clear marking on all manufactured plastic items to ensure compliance.

Survey respondents from academia noted increased research and development activities related to plastic waste management, which can provide recommendations to authorities. Policymakers are encouraged to develop a conceptual framework for proper directives, and awareness campaigns can promote plastic waste separation. Participants from NGOs and waste management consultancy services categorized under "other" sectors emphasized the need to reduce usage, promote the 3R concept, decrease PET consumption, avoid plastic waste contamination, and develop infrastructure to strengthen the plastic value chain and local strategies for managing plastic waste.

13.4 PROPOSED MEASURES TO ENSURE THE ENFORCEMENT OF PLASTIC WASTE AMENDMENTS

Table 13.2 lists the measures proposed by participants in different sectors to ensure the implementing and enforcement of plastic waste amendments in Sri Lanka.

Table 13.2 Proposed measures by different stakeholder sectors for the enforcement of plastic waste amendments

Government Sector	Recycling Sector	Academia	Other
Legal enforcement with awareness	Use 3R concept	Awareness	Reduce the usage.
Prepare guidelines	–	Close monitoring	Reduce the PET consumption.
Continuous monitoring	–	Coordination among authorities	Avoid contamination of plastic waste.
Identification of plastic wastes and their categories clearly	–	Create task forces in each district to manage the plastic waste.	Develop strategies to manage plastic waste locally.
–	–	–	Develop infrastructure to strengthen the plastic value chain.

13.4.1 Impacts of Plastic Waste Amendments

According to Figure 13.9, more than 48% of participants are familiar with the PIC procedure, and there is no discernible positive or negative impact on business and daily operations related to the import and export of plastic waste. However, professionals from universities and research institutes within the academic sector identified positive impacts of the PIC procedure on plastic waste management. They indicated that the procedure aids in controlling the influx of plastic waste into Sri Lanka, allowing the Ministry to mitigate the adverse effects of hazardous plastic waste.

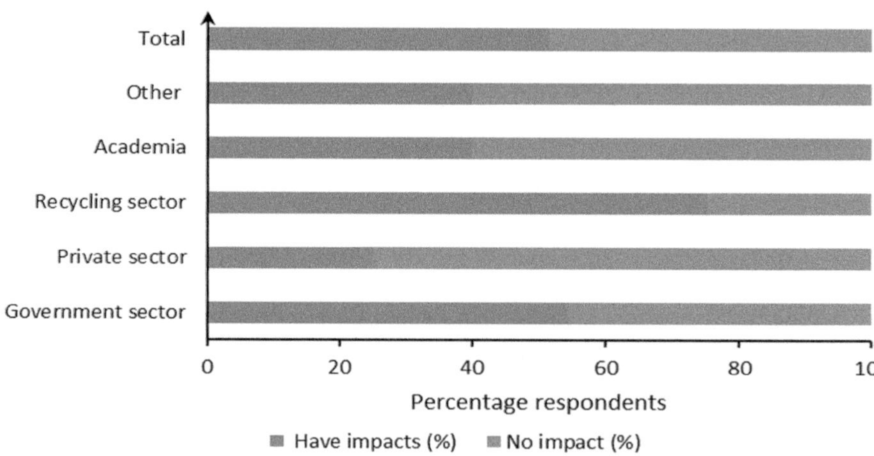

Figure 13.9 Knowledge of the PIC procedure and its impact on business and daily operations.

In scenarios where imported plastic wastes are hazardous, the PIC procedure ensures that confirmation is obtained from the receiving country, thereby minimizing environmental and health risks. Furthermore, all participants, except one from the government sector, agreed that the impact of implementing the PIC procedure for plastic waste imports and exports would be long-term, extending beyond three years.

13.4.2 Impact of Amendments on Reducing the Plastic Waste Volume Handled

As illustrated in Figure 13.10, the majority of participants from the government and academic sectors predicted that plastic amendments would lead to a reduction in waste volume handling over the medium to long term. In contrast, representatives from private small-scale plastic manufacturing companies anticipated that such amendments would result in a short-term reduction in plastic volume.

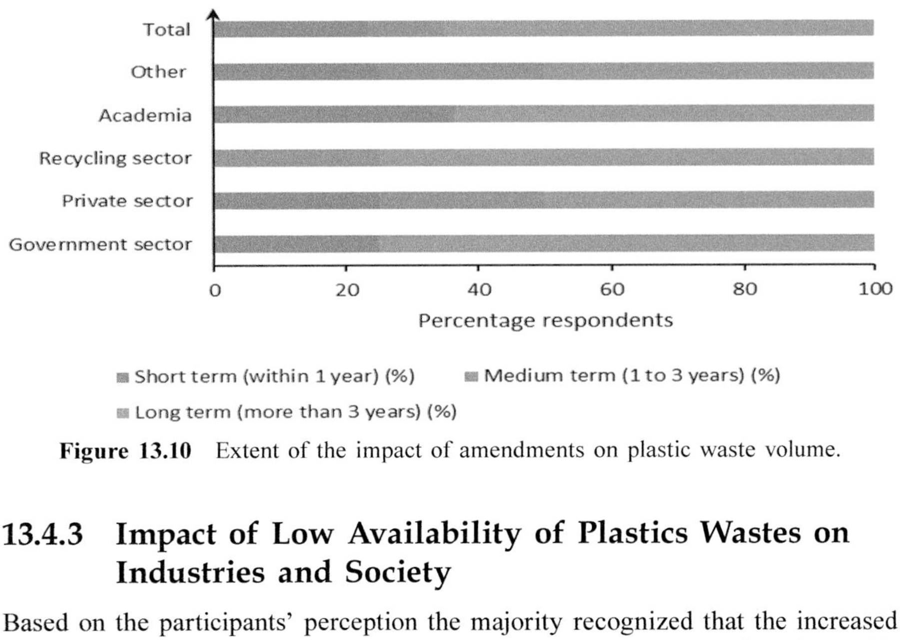

Figure 13.10 Extent of the impact of amendments on plastic waste volume.

13.4.3 Impact of Low Availability of Plastics Wastes on Industries and Society

Based on the participants' perception the majority recognized that the increased availability of plastic could positively impact the environment and society. This increase can reduce the burden on waste management activities, subsequently improving water and air quality. Such changes would benefit local and global ecosystems, enhance agricultural health, and contribute to overall national hygiene, ultimately reducing the dependency on plastics in daily life. This progression could lead to a plastic-free country and a more pleasant environment.

However, some participants from publicly owned waste management companies indicated that while this shift is beneficial for society, it may adversely affect the operation of waste-to-energy projects in Sri Lanka. From the perspective of small-scale plastic manufacturing and recycling companies, the availability of plastic waste in Sri Lanka has remained the same. Figure 13.11 illustrates the typical plastic waste recycling procedure in the country, involving numerous small-scale plastic collection and recycling companies.

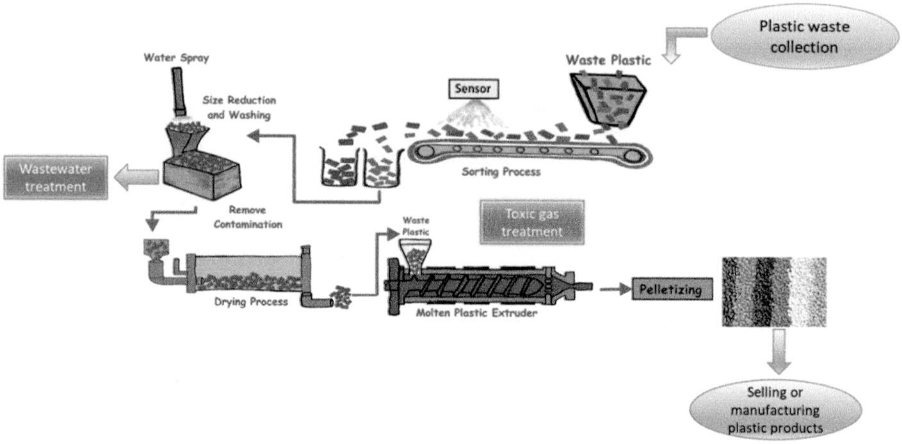

Figure 13.11 Plastic recycling procedure in Sri Lanka.

Plastic waste is collected from various sources, ranging from households to industries. The collected plastics are sorted according to resin type and pelletized following a series of cleaning processes. These plastic pellets, categorized by resin, can be utilized to manufacture various plastic products. However, the segregation process of plastic waste at the domestic level requires significant improvement. Enhanced segregation practices would facilitate easier collection and recycling of plastic waste, reduce negative environmental impacts, and provide livelihood opportunities for local communities.

13.4.4 Impacts of Plastic Waste Amendments on Different Sector

As shown in Figure 13.12, 80% of academic professionals reported no impact of plastic waste amendments on their sector. In contrast, the majority of participants from the government, private, and recycling sectors acknowledged a positive impact of plastic waste amendments on the environment and society. They noted that unauthorized importation could be regulated, and reducing public usage of plastics would promote environmentally friendly products. This, in turn, would lead to a decrease in waste collection at sources and an improvement in local production.

Additionally, due to the lack of proper waste management in the country, imported plastics pose a risk of contaminating water resources. However, a few respondents indicated negative impacts, such as material shortages. They also noted that, amidst the current economic crisis, legislators might view plastic imports as an opportunity to generate foreign currency.

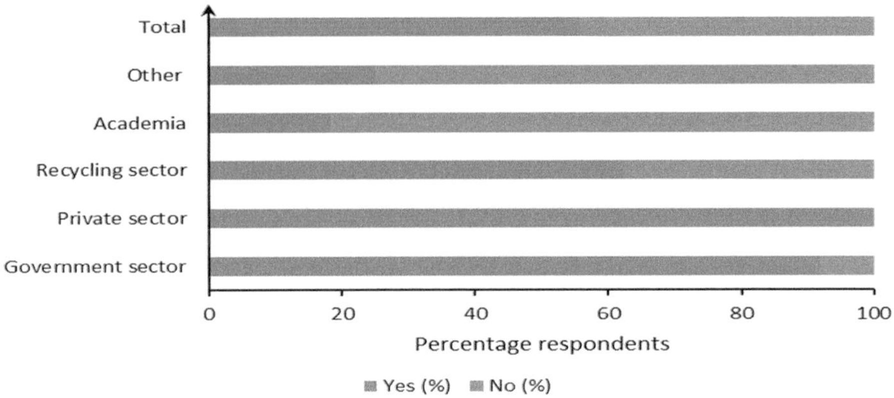

Figure 13.12 Impacts of plastic waste amendments in different sectors.

13.4.5 Financial Impacts of Plastic Waste Amendments on Different Sector

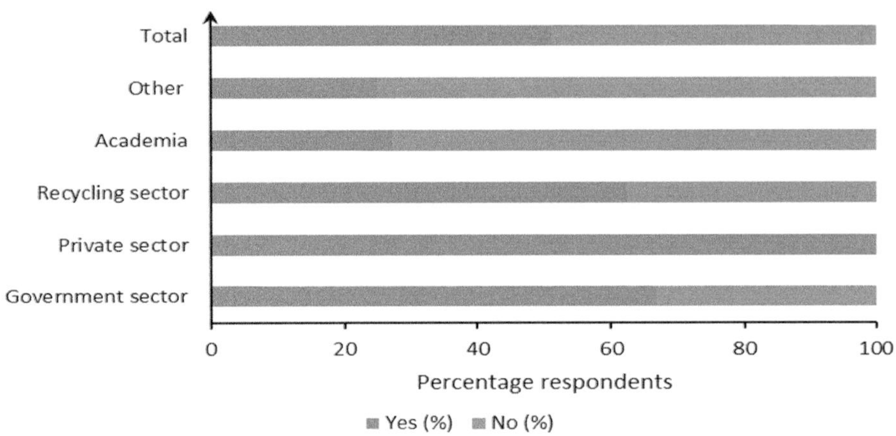

Figure 13.13 Financial impacts of plastic waste amendments.

Approximately 72% of respondents from the academia category believe that the financial impact on their institutions due to plastic waste amendments is minimal (Figure 13.13). However, there has been a recent increase in research activities on microplastics at universities and research institutes. In contrast, about 67% of respondents from the government sector reported significant financial impacts of plastic waste amendments. Government officers face considerable work regarding policies, rules, and regulations. Governments can introduce charges on consignment clearances. The accumulation of plastics in reservoirs can affect reservoir operations, and contaminated water can impact drinking water supply and irrigation production. Controlling transboundary plastic waste can reduce the treatment cost of microplastics in wastewater by decreasing the need for purification chemicals. Furthermore, managing transboundary movements of plastics can reduce production and operation costs associated with handling plastics and minimize the disposal costs of imported waste.

The majority of respondents from the private and recycling sectors identified a high impact on their companies due to plastic waste amendments. Environmental consulting service providers suggested that controlling the transboundary movement of plastics would reduce air and water toxicity, simplifying wastewater treatment for environmental engineering companies and reducing the complexity of the issues they need to address.

13.4.6 Impacts of Plastic Waste Amendments on Overall Economy

Figure 13.14 illustrates the responses of various sectors regarding their perception of the impact of plastic waste management on the overall economy of Sri Lanka. The majority of respondents anticipate that private plastic manufacturing and recycling companies will report a positive effect on the economy due to controlling plastic waste. All government sector respondents believe that while there will be a positive impact, it will manifest over the long term. Reducing plastic waste will decrease costs associated with environmental remediation. A cleaner environment will lead to fewer health hazards and foster the development of local businesses that produce environmentally friendly products, creating new job opportunities and boosting tourism. Health issues such as respiratory problems from open burning and diseases like Dengue will diminish, saving government spending on these health concerns. This will ultimately reduce societal stress and enhance overall well-being. From an academic perspective, an environment free from plastic pollution will attract numerous opportunities. However, it is crucial to implement this policy while considering adverse environmental impacts and addressing economic concerns simultaneously.

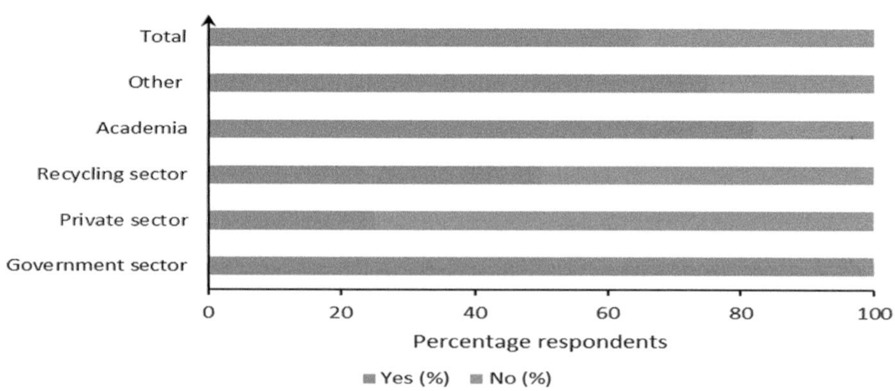

Figure 13.14 Impact of plastic waste amendments on the overall economy.

13.4.7 Environmental Impacts of Plastic Waste Amendments

A total of 88% of respondents indicated that plastic waste amendments have a significant impact on the environment (Figure 13.15), primarily through reducing pollution caused by plastic waste. The government sector highlights that these amendments will aid in controlling unauthorized waste imports, including post-consumer plastic waste, into Sri Lanka. Reducing single-use plastics through these amendments is also critical in mitigating environmental pollution. Effective plastic waste management will lead to fewer microplastics in water bodies, thereby decreasing the adverse effects on human health via the food chain and protecting aquatic ecosystems and wildlife.

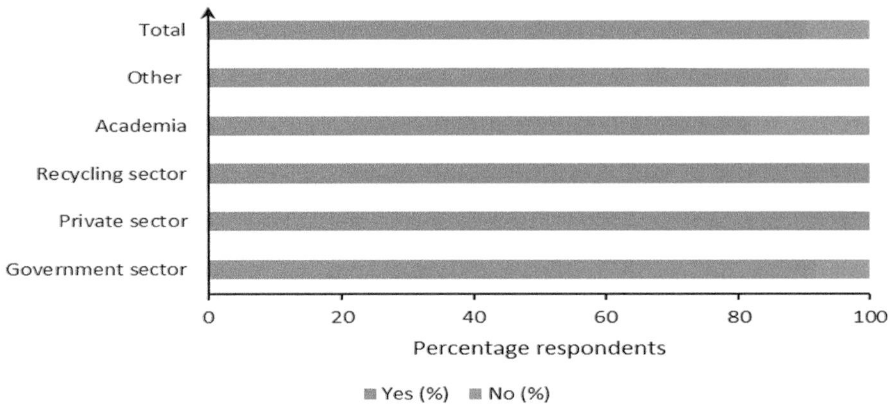

Figure 13.15 Environmental impacts of plastic waste amendments.

From the perspective of academic professionals, these amendments are essential in curbing the improper disposal of hazardous plastic waste, thereby safeguarding human health and the environment. They advocate for the use of natural and biodegradable materials as sustainable alternatives to conventional plastics.

13.4.8 Impacts of Plastic Waste Amendments Administrative Process

As shown in Figure 13.16, the majority of respondents from academic and other sectors indicated that plastic waste amendments have no significant impact on the administrative processes of their institutions or businesses. However, a minority of academic professionals believe that these amendments do affect the administrative process, as they necessitate adjustments and changes to routine activities, which can be inconvenient for personnel.

Despite the introduction, amendment, and improvement of existing administrative procedures in response to recent plastic waste regulations, the government sector has highlighted the need for comprehensive training for officials. This training is essential for effectively implementing the changes, modifying procurement procedures, and enhancing the control and monitoring processes.

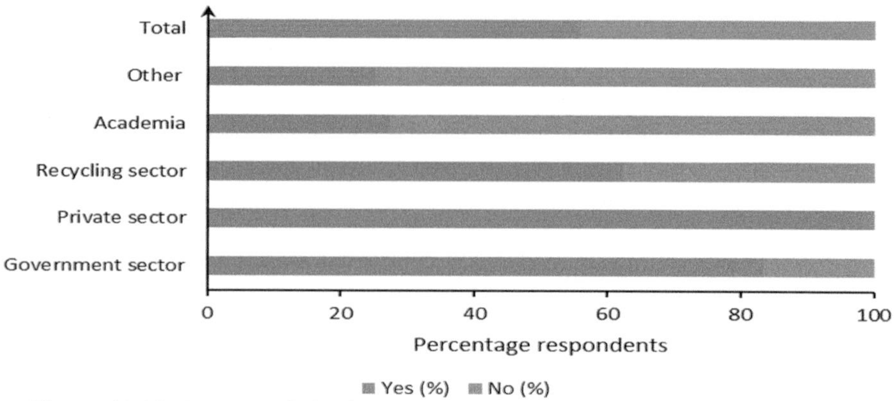

Figure 13.16 Impacts of plastic waste amendments on the administrative process.

13.5 CONCLUSION AND RECOMMENDATIONS

In summary, this study aimed to assess the current implementation status of plastic waste amendments in Sri Lanka and evaluate their potential impacts. Through a questionnaire targeting key stakeholders across various sectors and supplemented by email-based surveys and key informant interviews, the research revealed that while approximately half of the respondents lacked awareness of the plastic waste amendments or the PIC procedure, policymakers and enforcement agencies were well-informed due to recent administrative adjustments. Notably, Sri Lanka prohibits the importation of List A waste under the Basel Convention, while List B waste undergoes a rigorous screening process. Stakeholders, particularly government and academic professionals, highlighted numerous positive impacts of the amendments, including improved waste management leading to a cleaner environment, reduced health hazards, and economic benefits such as job creation and enhanced tourism. However, challenges still need to be addressed, particularly in implementing waste-to-energy projects. Recommendations include improving awareness among academics and government officials through targeted workshops, enhancing monitoring of plastic waste activities, developing coordination among government authorities, and educating the general public through various outreach initiatives. Overall, there is a call for careful consideration of the Basel Convention's plastic waste amendments to ensure they support the country's economy while safeguarding the environment.

REFERENCES

BRS. 2011. Basel Convention—Overview. Basel, Rotterdam and Stockholm Conventions Secretariat, UNEP. https://www.basel.int/TheConvention/Overview/tabid/1271/Default.aspx. Accessed 10 Oct 2023.

BRS. 2019. Basel Convention on the Control of Transboundary Movements of Hazardous Wastes and Their Disposal. Plastic Waste Amendments FAQs. Basel, Rotterdam and Stockholm Conventions Secretariat, UNEP. https://www.basel.int/Implementation/Plasticwaste/PlasticWasteAmendments/FAQs/tabid/8427/Default.aspx. Accessed 3 Sep 2023.

CBC. 2019. The Philippines recalling its diplomats from Canada over trash shipments. CBC News. https://www.cbc.ca/news/world/philippines-recalls-envoys-canada-trash-1.5138105. Accessed 28 Sep 2024.

CEA. 2000. National Environmental (Amendment) Act, No. 53 of 2000. Central Environmental Authority. http://www.cea.lk/web/images/pdf/acts/act53-2000.pdf. Accessed 18 Sep 2023.

CEA. 2017. Regulations on Polythene and Plastic Management. Central Environmental Authority. http://www.cea.lk/web/?option=com_content&view=article&layout=edit&id=1080. Accessed 10 Sep 2023.

CEA. 2023. Granting Consent for Transboundary movement of Hazardous Waste (Import/Export/Transit). Central Environmental Authority. https://www.cea.lk/web/granting-consent-for-transboundary-movement-of-hazardous-waste-import-export-transit. Accessed 10 Sep 2023.

CEJ 2020. Case Update- Case against Imported Waste Containers CA/Writ/303/19 https://ejustice.lk/case-update-case-against-imported-waste-containers-ca-writ-303-19/. Accessed 17 Sep 2023.

Chathurika, G.D.P.N. 2012. A Critical analysis of environmental taxation for Sri Lanka in a sustainomic framework. Master's theses, University of Moratuwa. Institutional Repository University of Moratuwa. http://dl.lib.mrt.ac.lk/handle/123/11371

EU. 2019. Single Use Plastics Directive. European Commission. https://environment.ec.europa.eu/topics/plastics/single-use-plastics_en. Accessed 12 Oct 2023.

EU. 2020. Circular Economy Action Plan. European Commission. https://environment.ec.europa.eu/strategy/circular-economy-action-plan_en. Accessed 10 Jul 2023.

GOSL. 1979. Code of Criminal Procedure. Government of Sri Lanka. https://www.lawnet.gov.lk/code-of-criminal-procedure-2/. Accessed 18 Sep 2023.

GOSL. 2003. National Environment Policy and Strategies. Government of Sri Lanka. https://env.gov.lk/web/images/pdf/policies/National_Environment_Policy_-_English.pdf. Accessed 18 Sep 2023.

GOSL. 2005. National Policy and the Strategy on Cleaner Production. Government of Sri Lanka. https://env.gov.lk/web/images/downloads/environment_planning/publications/sustainable_consumption/clener_production_policy/2005_cp_eng.pdf. Accessed 18 Sep 2023.

GOSL. 2008. Marine Pollution Prevention Act. No 35 of 2008. Parliament of the Democratic Socialist Republic of Sri Lanka. https://env.gov.lk/web/images/downloads/acts/marine_pollution_prevention_act_no_35_of_2008.pdf

GOSL. 2021. The Extraordinary Gazette under the Section 23W (1) (a) of the National Environmental Act, No. 47 of 1980. Government of Sri Lanka. https://www.cea.lk/web/images/pdf/2021/regulations/2211-51_E.pdf

GOSL. 2022. The Constitution of the Democratic Socialist Republic of Sri Lanka. Published by the Parliament Secretariat, Sri Lanka. https://www.parliament.lk/files/pdf/constitution.pdf

Greenpeace Malaysia. 2019. Southeast Asia's Struggle Against the Plastic Waste Trade. Greenpeace. https://www.greenpeace.org/malaysia/publication/1905/southeast-asias-struggle-against-the-plastic-waste-trade/ (accessed Sep 11, 2023).

Jain, A. 2020. Trash Trade Wars: Southeast Asia's Problem with the World's Waste. https://www.cfr.org/in-brief/trash-trade-wars-southeast-asias-problem-worlds-waste. Accessed 10 Oct 2023.

Kamalakkannan, S., A. Abeynayaka, A.K. Kulatunga, R.K. Singh, M. Tatsuno and P.J.D. Gamaralalage. 2022. Life cycle assessment of selected single-use plastic products towards evidence-based policy recommendations in Sri Lanka. Sustainability. 14(21): 14170.

Kojima M. 2020. The impact of recyclable waste trade restrictions on producer recycling activities. Int. J. Autom. Technol. 14: 873–881. https://doi.org/10.20965/ijat.2020.p0873

Liu, C. and C. Liu. 2023. Exploring plastic-management policy in China: Status, challenges and policy insights. Sustainability (Switzerland) 15. https://doi.org/10.3390/su15119087

Man, M., R. Naidu and M.H. Wong. 2013. Persistent toxic substances released from uncontrolled e-waste recycling and actions for the future. Sci. Total Environ. 463–464: 1133–1137. https://doi.org/10.1016/j.scitotenv.2012.07.017

Ministry of Environment, Sri Lanka. 2023. Environment Pollution Control and Chemical Management Division. Ministry of Environment. https://www.env.gov.lk/web/index.php/en/international-day-of-forests/19-environment-pollution. Accessed 10 Sep 2023.

MOE-SL (Ministry of Environment). 2021. National Action Plan on Plastic Waste Management. Ministry of Environment, Sri Lanka. https://resolutions.unep.org/resolutions/uploads/national_action_plan_on_plastic_waste_management.pdf

MOE-SL. 2020. National Policy on Waste Management. Environment Pollution Control and Chemical Management Division Ministry of Environment. https://www.env.gov.lk/web/images/pdf/policies/National_Policy_on_Waste_Management_English.pdf

Rucevska, I., C. Nellemann, N. Isarin, W. Yang, N. Liu, K. Yu, et al. 2015. Waste crime – waste risks: Gaps in meeting the global waste challenge. A UNEP Rapid Response Assessment. United Nations Environment Programme and GRID-Arendal, Nairobi and Arendal. https://www.unep.org/resources/report/waste-crime-waste-risks-gaps-meeting-global-waste-challenge-rapid-response

Uhm, Y. 2021 Plastic waste trade in Southeast Asia after China's import ban: Implications of the new Basel Convention amendment and recommendations for the future. California Western Law Review. 57(1): 9. Available at: https://scholarlycommons.law.cwsl.edu/cwlr/vol57/iss1/9

United Nations. 1989. Basel Convention on the Control of Transboundary Movements of Hazardous Wastes and Their Disposal. United Nations. https://treaties.un.org/pages/viewdetails.aspx?src=treaty&mtdsg_no=xxvii-3&chapter=27&clang=_en. Accessed 4 Sep 2023

van der Wel, A. and V. Post. 2007. Solid Waste Management in Sri Lanka: Policy and Strategy. https://v1.cepa.lk/content_images/publications/documents/173-S-WASTE-SWM%20in%20SL.pdf. Accessed 10 Jan 2023

Wen, Z., Y. Xie, M. Chen and C.D. Dinga. 2021. China's plastic import ban increases prospects of environmental impact mitigation of plastic waste trade flow worldwide. Nat. Commun. 12(1): 425. https://doi.org/10.1038/s41467-020-20741-9

Chapter 14

Plastic Pollution and Global Warming: Integrated Approach to Develop Policy-Based Solution

Jaber Bin Abdul Bari*[1], Md. Sohel Parvez[1, 2] and Syed Masiur Rahman[3]

[1]Department of Oceanography, Noakhali Science and Technology University, Noakhali 3814, Bangladesh

[2]Department of Hydrobiology, University of Debrecen, 4032 Debrecen, Hungary

[3]King Fahd University of Petroleum & Minerals (KFUPM), Dhahran 31261, Saudi Arabia

14.1 INTRODUCTION

Plastics are indispensable in modern times due to their affordability, widespread availability, and versatility in various fields, including but not limited to healthcare, technology, building, clothing, and footwear (Horton 2022). At present, production and consumption have reached such high levels that they are falsely considered a "crucial asset" in the global market and a "bargaining tool" for numerous countries (Williams and Rangel-Buitrago 2019). Conventionally, plastics are synthesized from fossil fuels, although renewable biomass sources can also be

*For Correspondence: jaber.abdulbari@gmail.com

DOI: 10.1201/9781003531463-14

used for plastic production. Plastics have exceptional corrosion resistance and possess the advantageous property of flexibility, allowing them to be shaped into many forms. From the chemical perspective, plastics are essentially polymers, formed by molecules of various monomers. The two main types of plastics that are often utilized are thermoset and thermoplastic. Thermoset plastics usually harden when heat is applied, while thermoplastics soften when exposed to heat (Alhazmi et al. 2021). Additionally, plastics are further classified based on the source of raw materials and their biodegradable properties. Plastics possess a variety of distinctive characteristics: they are cost-effective, lightweight, robust, long-lasting, resilient to corrosion, and exhibit excellent thermal and electrical insulation capabilities (Wabnitz and Nichols 2010).

Nevertheless, the environmental concerns associated with plastic are exacerbated by its inherent characteristics, which render it advantageous. Upon entering the environment, its exceptional endurance enables it to persist for extended periods ranging from hundreds to thousands of years, while its low weight facilitates its transportation over large distances through wind and air currents. Plastic debris has been observed in various environments, from the highest mountain (Napper et al. 2020) to the greatest depths of the oceans (Jamieson et al. 2017), rivers (Tibbetts et al. 2018), and the atmosphere (O'Brien et al. 2023). Plastic contamination can also occur in locations as diverse as the poles, the equator, and even the most distant islands all around the globe. Due to their persistence in the environment, the rates of natural degradation occur over decades to centuries. Therefore, the levels of plastic are anticipated to rise due to the combination of these ongoing sources of emissions and the exceptionally sluggish rates of degradation.

However, plastic pollution is a long-standing global problem with significant influence on both living and non-living aspects of the ecosystem (Iroegbu et al. 2021). Several studies have conclusively shown that every stage of the plastic life cycle significantly contributes to global warming, climate change, and biodiversity loss (Shen et al. 2020; Vishwakarma 2020; Lee et al. 2021; Ford et al. 2022; Parvez et al. 2024). Due to the swift growth of plastic manufacturing, the plastic industry has emerged as the primary and fast-increasing contributor to industrial greenhouse gas (GHG) emissions. GHG emissions arise not just from the manufacturing and production process, but also from the extraction and transportation of raw materials used in production, as well as the management of plastic waste and the release of plastics into the environment (Hamilton and Feit 2019). As the concentration of GHG emissions increases, it intensifies the greenhouse effect in the atmosphere. This, in turn, alters the climate of the globe, resulting in changes in climatic patterns: higher average temperatures and an increase in severe weather events (Zandalinas et al. 2021). However, global warming has the potential to worsen plastic pollution, and in turn, plastic pollution could exacerbate global warming in a feedback loop (Tang 2023). Global warming leads to the melting of sea ice and glaciers, thus increasing the sea level. Rising sea levels could accelerate coastal erosion and floods, resulting in an increasing influx of plastics from land to ocean (Oberbeckmann et al. 2018). Furthermore, the distribution of plastic in the environment is also influenced by the intensity or frequency of extreme weather events resulting from global warming. Increased soil erosion and discharge volume can result from extreme rainfall events and inundation, which in turn transports a greater amount of terrestrial plastic from urban areas, agricultural fields, and waste disposal sites to rivers and the ocean (Haque and Fan 2023). Biota, nutrients, contaminants, and pathogens may also be influenced by global warming, which can also impact the interactions of plastics with other environmental factors (Tang 2023).

Previous studies strongly suggest that the climate impacts of GHG emissions from plastics require urgent attention (Shen et al. 2020; Ford et al. 2022; Haque and Fan 2023; Tang 2023), although the precise impact of plastics on global warming is mostly unknown (Parvez et al. 2024). Nevertheless, solutions that strictly concentrate on the reduction of GHG emissions or that only address plastic contamination will be demonstrably insufficient in the coming days. Therefore, the present chapter suggests a framework for integrated policy-based solutions that address both issues simultaneously.

14.2 INFLUENCES OF PLASTIC POLLUTION ON GLOBAL WARMING

14.2.1 Indirect Contribution by Plastic Production, Transportation, and Processing

Petroleum-based synthetic plastics are usually manufactured by a sequential process, beginning with the distillation of crude oil in an oil refinery. This process involves the separation and fractionation of heavy crude oil into distinct groupings of lighter components, known as segments. Every segment consists of a blend of polymeric hydrocarbon chains that vary in size and structure. Naphtha, one of these fractions, is an essential component required to produce monomers like ethylene, propylene, and styrene, which are used in the manufacturing of plastics. These individual molecules combine to create polymers via the processes of polyaddition and/or polycondensation, with the assistance of specific catalysts (Saleh and Gupta 2016; Atiwesh et al. 2021). To produce functional plastic products, polymers and other materials must undergo processing. Plastic processing involves a set of sequential operations to generate the specific portion needed for the ultimate application (Shrivastava 2011). Research evidence has effectively demonstrated that GHG emissions continue throughout the whole life cycle of plastic, encompassing the extraction, transportation, manufacturing, processing, waste management, and the release of plastic into the environment (Shen and Worrell 2014; Zheng and Suh 2019; Benavides et al. 2020; Nielsen et al. 2020; Parvez et al. 2024). However, GHG emissions mostly occur during the manufacturing of plastic, which involves sophisticated processes for obtaining and transferring fossil fuels. Although the emissions throughout the manufacturing process primarily rest with the manufacturing facilities, which often depend on the equipment's efficiency, configuration, and lifespan. In 2019, the initial manufacturing of plastic released over 1.8 billion tonnes of carbon dioxide equivalent (CO_2eq), which is equivalent to more than 3.3% of the total global emissions from fossil fuels (Ritchie 2023). The plastic refining industry is also one of the most GHG-intensive sectors in manufacturing. The refining of fuels and the production of plastics often include chemical processes that result in significant emissions during polymerization, plasticization, and several other stages (Masnadi et al. 2018). These chemical processes alone emitted a global total of 184.3–213.0 million metric tons CO_2eq in 2015. Nevertheless, the current manufacture of plastics is anticipated to generate 850 million tons of greenhouse gases. By 2050, these emissions are projected to rise to 2.8 gigatons of carbon dioxide per year, which is similar to the output of 615 coal-fired power plants (Hamilton et al. 2019).

14.2.2 Direct Contribution by Plastic Disposal, Waste Management, and Plastic Degradation

The end-of-life (EOL) of plastic typically involves disposal, recycling, landfill, and incineration, which might directly contribute to emissions. Plastic waste management processes, including landfilling and incineration, generate direct GHG emissions. Sanitary landfills remain the primary method of disposal for plastic packaging waste. The emissions associated with landfill plastic waste arise from the sorting and processing of waste before landfilling, as well as the utilization of fossil fuels for transporting waste from collection points to landfills (Zhang et al. 2019). The inadequate handling of plastic garbage worldwide has both direct and indirect consequences for the global climate situation.

Incineration is presently regarded as a straightforward remedy for widespread plastic pollution (Ágnes and Rajmund 2016). It can efficiently control plastic pollution and also generate energy and heat for a variety of purposes. However, incineration is widely acknowledged to have the most significant climate impact (Eriksson and Finnveden 2009; Sharma et al. 2023). A substantial quantity of CO_2 is released during the incineration of plastics. Similar processes, such as pyrolysis, also contribute to CO_2 emissions as a result of incomplete combustion. On average, every metric ton of plastic packaging waste contains around 79% flammable carbon. This carbon, when burned, will emit roughly 790 kilograms of carbon into the atmosphere, which is equivalent to about 2.9 metric tons of carbon dioxide (CO_2) (Hamilton and Feit 2019).

Environmental plastic degradation is a significant contributor to GHG emissions. The plastic litter that is not recycled transforms, and plastic deterioration initiates upon exposure to different environmental conditions. Plastic undergoes degradation and fragmentation, gradually breaking down into ever-tinier pieces, ultimately resulting in the formation of microplastics (MPs) and nanoplastics (NPs) (Napper and Thompson 2020). The decomposition of extended carbon chains in plastic items naturally releases harmful gases, primarily methane and ethylene, which possess a significant warming impact. Polyethylene, which represents 36% of all plastic varieties, is the most abundant source of methane and ethylene emissions among the degraded polymers (Geyer et al. 2017). However, the degraded small plastic pieces possess not only higher surface-to-volume ratios but also longer edge lengths compared to bigger plastic objects (Royer et al. 2018). This implies that plastic particles will emit a greater quantity of greenhouse gases per unit mass, as they degrade and shrink in the environment.

14.3 CO-OCCURRENCES AND IMPACTS OF GLOBAL WARMING AND PLASTIC POLLUTION ON THE ENVIRONMENT

14.3.1 Combined Impacts on Terrestrial Ecosystem

The plastics industry has witnessed significant growth in both size and variety over the years, with an expanded range of materials and increased global production. Additionally, the emissions from plastic are also on the rise and are projected to continue growing, even in the most hopeful scenarios for reduction (Borrelle et al. 2020; Soares et al. 2021). Hence, it is widely recognized that plastic waste can accumulate in terrestrial systems, as plastics are made, used, and disposed of in large quantities (Chae and An 2018; de Souza Machado et al. 2018; Zeb et al. 2023). Plastic can lead to adverse effects on the terrestrial ecosystem by modifying the fundamental characteristics of soil. Furthermore, the presence of plastic pieces induces a significant change in soil temperature and modifies soil decomposition rate, potentially affecting the soil fertility (Sharma et al. 2023). Fragmentation is also a crucial process that significantly impacts the terrestrial environment. Plastic waste, released into the environment, undergoes fragmentation, weathering, and other processes, resulting in the formation of MPs (Rai et al. 2023). In terrestrial environments, MPs can absorb and convey contaminants, including pesticides, heavy metals, and persistent organic pollutants, resulting in bioaccumulation and toxicity in soil organisms (Cai et al. 2021).

Nevertheless, plastics in the terrestrial system contribute to global warming in addition to serving as a significant source of contamination. An important consequence of global warming is the alteration or amplification of existing natural variability. The rise in global mean surface temperature (GMST), which rose by 0.87 degrees Celsius between 2006 and 2015 compared to the period from 1850 to 1900, has led to more frequent and severe consequences (Masson-Delmotte et al. 2019). This provides more substantial evidence of the potential effects that could occur if the temperature increases further, impacting both natural and human systems. Anthropogenic global warming has already resulted in numerous documented alterations in the climate system. The changes encompass rising temperatures in both terrestrial and marine environments, along with an enhanced occurrence of heatwaves in the majority of land areas (Perkins-Kirkpatrick and Gibson 2017). Global warming also has a direct impact on precipitation; increased heating results in increased evaporation, which results in a significantly drier surface. Thus, the degree and frequency of droughts in certain regions increase (Kumar and Pooja 2020). Tropical ecosystems are expected to be particularly vulnerable to global warming due to the limited temperature tolerance range of many tropical species (Deutsch et al. 2008; Tewksbury et al. 2008). Forest fires, extreme weather events, and the dissemination of invasive species, parasites, and diseases are all potential risks associated with biodiversity that might result from increased temperature. Thus, the co-occurrence of global warming and plastic pollution poses a significant threat to terrestrial biodiversity; direct impact through bioaccumulation and global warming disturbs natural ecosystems and changes species distributions, endangering wildlife health.

14.3.2 Combined Impacts on Marine Ecosystem

The marine environment has been proposed as the ultimate destination for plastic that is released into the terrestrial system (Villarrubia-Gómez et al. 2018). Once introduced into the marine environment, plastics also undergo degradation, resulting in the formation of MPs and NPs (Ter Halle et al. 2016). Plastic pollution affects marine species in several ways, including entanglement, ingestion, bioaccumulation, and disruption of ecosystems. Macroplastic debris is the primary cause of entanglement, although several marine organisms consume both micro- and macroplastics (Vegter et al. 2014). However, micro- and macro plastic ingestion has been observed across various levels of the marine food chain, including zooplankton (Khan et al. 2024), fish (Collard et al. 2019), seabirds (Provencher et al. 2018), sea turtles (White et al. 2018), and marine mammals (Zantis et al. 2021). The ingestion of plastic can also result in a variety of adverse effects, including internal and external lesions, impaired nutrition capacity, obstruction of the digestive tract, satiation and starvation, and general debilitation that frequently leads to mortality (Franzellitti et al. 2019). Furthermore, plastics ingestion by invertebrates has the potential to transmit noxious substances to higher trophic levels (van Raamsdonk et al. 2020).

Similar to the terrestrial environment, plastic also contributes to GHG emissions in the marine environment. The elevated temperature, which is a consequence of global warming, leads to extreme weather events, including the thawing of glaciers and the rise in ocean temperature, eventually resulting in ocean acidification and a rise in sea level. It is anticipated that changes in water temperatures will prompt the relocation of certain species (e.g., plankton, fish) to higher latitudes and the formation of new ecosystems (Alfonso et al. 2021). However, other ecosystems (e.g., coral reefs and kelp forests) are relatively less mobile and are expected to encounter high rates of mortality. Coral reefs are regarded as particularly susceptible to temperature fluctuation, as numerous coral species reside within a 1–2°C range of their critical thermal limits (Frieler et al. 2013). Although the loss of living coral habitat would directly impact the distribution and abundance of numerous reef-associated organisms, these organisms will also experience the direct effects of rising temperatures. Thus, the combined impacts of global warming and plastic pollution might result in a "double whammy" for vulnerable species, bringing them closer to extinction.

14.4 INTEGRATED APPROACHES TO MITIGATION

14.4.1 Bio-Based and Biodegradable Plastics as a Sustainable Alternative

Plastic disposal has become a long-lasting and significant environmental issue globally. Efficiently handling plastic garbage is crucial for the development of sustainable and habitable cities. Nevertheless, in several nations, effective waste management continues to be a daunting task, leading to inadequate and/or unregulated disposal approaches (Hahladakis et al. 2020). In addition, plastics can be extremely slow to decompose, resulting in their prolonged presence in the environment. This can result in social and economic impacts, as well as damage to biodiversity and ecosystem services. Therefore, substitutes for fossil-based plastics can be investigated to solve these problems permanently. Biopolymers derived from renewable sources and plastics that can naturally decompose have been proposed as a viable substitute for conventional plastics (Table 14.1) (Altman 2021). These biodegradable plastics, whether derived from biological sources or petroleum, break down when exposed to naturally occurring microbes.

Similarly, bio-based plastics, which are either derived entirely from biomass or partially from biomass, provide a potential option to address both plastic pollution and the resulting emissions (Moshood et al. 2022; Rajvanshi et al. 2023). Bio-based plastics often exhibit lower GHG emissions throughout their lifecycle compared to their non-bio-based counterparts (Zheng and Suh 2019). For instance, the production of polylactic acid conserves two-thirds of the energy required to manufacture conventional plastics. In addition, it has been demonstrated that the breakdown of PLA bioplastics does not result in an aggregate rise in carbon dioxide gas (Elsawy et al. 2017). Significantly, PLA emits 70% fewer greenhouse gases during its degradation in landfills (Iwata 2015). However, Bio-based plastic production includes numerous concerns, such as GHG emissions, negative effects on land use, and the substantial consumption of water and land (Atiwesh et al. 2021). However, the benefits of using bioplastics and biodegradable plastics seem to outweigh those of traditional plastics, especially as more environmentally friendly plastics are being produced. Thus, bio-based and biodegradable plastics are potential alternatives for achieving sustainable growth in the plastic industry.

14.4.2 Switching to Low-Carbon Energy

The integration of low-carbon energy into industrial processes throughout the life cycle of plastics could potentially reduce their greenhouse gas emissions. At different stages of their life cycles, all plastics necessitate process-related electricity and heat (on-site resources). By substituting renewable energy sources (e.g., biogas and wind power) for utility electricity and direct fossil fuel combustion, emissions can be reduced without altering the fundamental processes used to manufacture plastics (Posen et al. 2017). GHG emissions can be reduced by 50% by selecting zero-carbon or low-carbon energy sources during the plastic

Table 14.1 Sustainable alternatives to conventional plastics

Origin	Types	Composition	Advantage	Circularity Aspect	Suitable Alternative	Reference
Bio-based	Biodegradable	Starch-based polymers, (Polylactic acid, and polyhydroxyalkanoat)	Perfect for biodegradation and a viable end-of-life solution	Renewable; Recyclable	Yes	(Van den Oever et al. 2017; Kumari et al. 2022)
	Non-Biodegradable	Biobased polyethylene (bio-PE)	Minimal emission of carbon is accompanied by less energy requirement for recycling.	Renewable; Recyclable	Yes	(Robertson 2005; Siracusa and Blanco 2020)
Petroleum-Based	Biodegradable	Polycaprolactone (PCL), polyamide (PA), poly(propylene carbonate) (PPC), and poly(vinyl alcohol) (PVOH)	It has a lower carbon footprint and is biodegradable under suitable conditions.	Non-renewable; Recyclable	Yes	(Halley and Smith 2005, and Li et al. 2022)
	Non-Biodegradable	Polyethylene (PE), Polyvinyl chloride (PVC), Polystyrene (PS)	Readily available, affordable, and adaptable for a vast array of applications	Non-renewable; Recyclable	No	(Marsh and Bugusu 2007, Peng et al. 2023)

polymer manufacturing process (Bauman 2019). However, the transition to 100% renewable energy would result in a 62% reduction in life-cycle GHG emissions from plastics by 2050, provided that demand continues to grow at a rate of 4% per year (Zheng and Suh 2019).

Additionally, the plastic industry can acquire raw materials from renewable energy sources, whenever feasible, to mitigate GHG emissions. The ultimate resin produced remains unaffected by energy substitution. The new, "greener" polymer can be used as a substitute in the market without requiring any modifications to downstream production methods or product functionality (Álvarez-Chávez et al. 2012).

However, the adverse effect of plastics on global climate change is not fully addressed by the application of renewable energy to plastics. Various chemical processes are responsible for a significant portion of GHG emissions from plastic production, and the use of renewable energy does not influence their reduction (Posen et al. 2017). Nevertheless, this energy alteration can be accomplished without making any fundamental changes to current manufacturing processes, existing capital structures in the chemical sector, or requiring further testing to guarantee the continuity of product quality, as would be necessary when transitioning to bio-based plastics. Therefore, energy substitution offers the greatest promise for reducing GHG emissions until sophisticated bio-plastic production methods are developed.

14.4.3 Circular Economy Approach

Change in industrial manufacture and proper disposal at the conclusion of a product's life cycle are the most effective methods for reducing the negative effects of plastics and the resulting emissions (Wiesinger et al. 2021). Given the imperative to reduce GHG emissions, the circular economy is increasingly recognized as a viable solution to address this problem (Liu et al. 2018). The rationale for this is that plastic products are derived from fossil fuels, which are a highly carbon-intensive resource. Hence, the reuse and recovery of plastic waste will lead to a decrease in the consumption of fossil fuels, thereby reducing the corresponding GHG emissions (Wang and Chang 2014). Simultaneously, this approach also helps in mitigating post-consumer plastic waste pollution.

Therefore, the sector adopts a circular economy by focusing on resource efficiency, which in turn mitigates the problem of plastics disposed of in landfills or burned. Furthermore, the circular economy method can effectively mitigate the environmental impact of plastic by implementing strategies such as material reuse and recycling (Figure 14.1). Recent studies on individual or partially integrated circular technologies demonstrate significant forecasted drops in greenhouse gas emissions on a broad scale (Kätelhön et al. 2019; Schwarz et al. 2021).

The idea of a circular economy has been implemented worldwide, however, with differing degrees of success. A couple of countries have implemented particular policies to aid in the shift to a circular economy (Liu et al. 2018). The National Plastics Plan 2021 of Australia, for example, highlights plans to phase

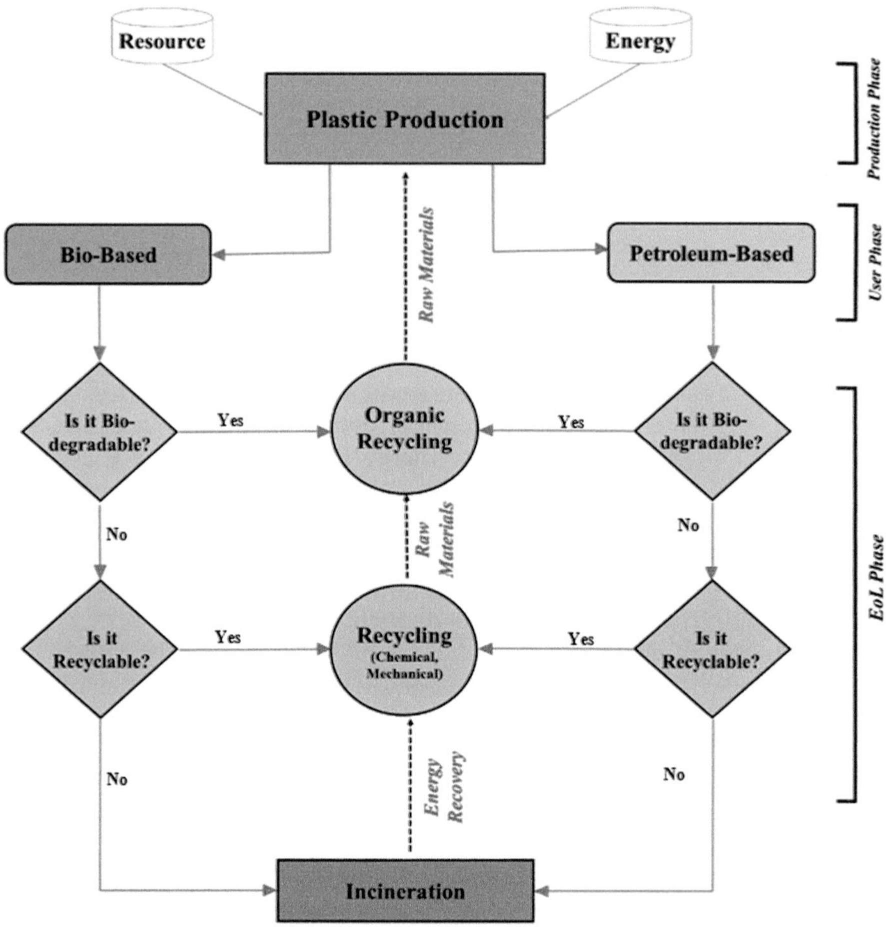

Figure 14.1 Plastic management through a circulatory approach [redrawn from (Payne et al. 2019; Hahladakis et al. 2020)].

out hazardous and obsolete plastics by 2025, especially single-use plastics like plastic straws and polystyrene containers (Hossain et al. 2022). Similarly, Canada's Zero Plastic Waste Policy by 2030, coupled with its wider Net-Zero Emissions Strategy, highlights the need to eliminate plastic waste to cut greenhouse gas emissions. Canada is using a lifecycle strategy to manage plastic resources by prohibiting single-use plastics, requiring recycled content in plastic goods, and establishing extended producer responsibility (EPR) programs (Aldag 2019). In the implementation of circular economy policies, the European Union (EU) countries currently hold a leadership position. By 2030, the Circular Economy Action Plan (2020) requires that all plastic packaging in the European Union be either recyclable or reusable (Mazur-Wierzbicka 2021). Policies pertaining to the circular economy in Asia (e.g., Korea and Japan) seek to make consumers more accountable for their plastic use and waste. In North America, corporations have used the concept of circular economy to improve their programs for recycling, reusing, and reducing waste, as well as to carry out life cycle assessments at the product level (Winans et al. 2017).

Nevertheless, the problem of plastic pollution cannot be effectively addressed solely by enhancing the recycling and reutilization of products. Hence, it is imperative to prioritize efforts in both lowering excessive dependence on plastics and enhancing the capacity of these materials to be recycled within the system. This approach is crucial for avoiding the adverse environmental, economic, technical, and social consequences linked to plastic waste (Hahladakis et al. 2020).

14.4.4 Blue Carbon Ecosystem as Plastic and Carbon Sink

Blue carbon ecosystems, which include salt marshes, seagrasses, and mangroves, are distinguished by their enormous organic carbon storage, which resides mainly within layers over the course of millennia (Duarte et al. 2005). The soils and sediments of these ecosystems accumulate carbon, and they have high carbon burial rates on a per unit area basis. Hence, the blue carbon ecosystem has garnered global attention for its potential to reduce emissions (Gao et al. 2016). However, its impact is restricted to coastlines, which limits its global reach.

Interestingly, they offer numerous non-climatic advantages. For instance, the high sediment accumulation rates that are supported by blue carbon coastal habitats are also capable of burying and trapping plastics. Additionally, they sequester substantial quantities of carbon in their sediments (Martin et al. 2020). High rates of sedimentation in vegetated coastal ecosystems, such as mangroves, tidal marshes, and seagrass beds, have often been linked to significant MPs stocks. Even more importantly, the mangrove ecosystem's distinctive characteristics were recognized as significant sinks for a variety of contaminants originating from both marine and terrestrial activities (Bandh et al. 2023). The complex structure of mangrove forests is facilitated by the unique root systems, which effectively capture plastics, increasing plastic accumulation (Wang et al. 2023). In contrast to mangroves, MPs can accumulate in greater quantities in salt marshes, which is correlated with the sediments of the salt marsh (Jiang et al. 2022). Remarkably, seagrass meadows have the capacity to capture particulate matter, too, suggesting that they may also serve as a significant habitat to intercept MPs (Boström et al. 2010).

14.5 SYNERGIES IN POLICY: ADDRESSING PLASTIC POLLUTION AND GLOBAL WARMING

14.5.1 Plastic Policies Mitigating Global Warming

Plastic pollution has drawn minimal attention in the international community, unlike carbon emissions and other global pollutants, such as chlorofluorocarbons (CFCs) and Persistent Organic Pollutants (POPs), which are addressed by the Paris Agreement, Montreal Protocol, Stockholm Convention, etc (Lallas 2001; Velders et al. 2012; Borrelle et al. 2017). At the global scale, before 2000, five binding international policies were established that pertain to the issue, but only pollution originating from marine sources was specifically addressed. Since 2000, there have been 28 international policies that attempt to address the plastic issue (Virdin et al. 2020). These policies are usually nonbinding agreements, which are considered "soft law." They mostly focus on the sources of pollution that originate from land. As global production of plastic and its related pollution continues to rise, the array of national and international agreements to curb plastic pollution is also on its way up. (Lau et al. 2020). At the national level, the increase in policies mainly results from the implementation of fresh policies specifically targeting pollution caused by plastic bag manufacture, import, transportation, and usage (Zen et al. 2013; Nielsen et al. 2019; Thomas et al. 2019). The most commonly employed measure in these policy approaches was a regulatory prohibition on plastic at various stages of its life cycle. According to a report by the United Nations Environment Programme (UNEP), nearly 127 countries had implemented legislation that restricted the use of plastic bags to a minimum in 2018. As part of these bills and regulations, the manufacturing and usage of single-use plastic products were also restricted (Cho 2020). The amendments to the Basel Convention have also implemented a legally enforceable structure to enhance the ability of 180 nations to avoid, reduce, and effectively handle plastic pollution (Avery-Gomm et al. 2019; Wingfield and Lim 2022). This includes regulating the export and import of plastic waste, with the ultimate goal of protecting the environment and human health. Nevertheless, the United Nations Environment Program established a global plastics treaty to reduce plastic pollution, which encompasses all phases of the plastic life cycle, including in the marine environment. At the fifth UN Environment Assembly (UNEA-5.2) in Nairobi, in March 2022, 175 UN Member States voted to form an Intergovernmental Negotiating Committee (INC) to establish a legally binding instrument on plastic pollution by 2024 (UNEP 2022). There were four negotiating sessions (INC 1–4) of the UN Plastics Treaty conducted between 2022 and 2024 to keep the international negotiations on track (Syberg et al. 2024). The last planned session, INC–5, is scheduled for Busan, Republic of Korea, and will be a pivotal moment in delivering a robust global plastic treaty.

An envisioned scenario of the global ambition to achieve a significant reduction in plastic usage by the year 2040 also entails a corresponding decrease in GHG emissions. The drop in plastic lifecycle emissions is mostly attributed to the reduction in plastic use, which is also a result of policies (such as plastic taxes) that restrict plastic usage, as well as the transition to secondary plastics (Walker et al. 2020). The reduction of mismanaged waste is anticipated as an outcome of these policies that are designed to enhance plastic waste management, while an increase in the incineration, landfilling, and recycling of waste can be expected too (Wichai-Utcha and Chavalparit 2019; Vanapalli et al. 2021). However, enhancing recycling infrastructure and implementing effective recycling programs can divert plastic waste from landfills and incineration (Dai et al. 2022). This reduces methane emissions from landfills and decreases the energy requirements of producing new plastics from virgin materials, thus lowering GHG emissions. Extended Producer Responsibility (EPR) policies can make producers financially responsible for the end-of-life disposal of their products, incentivizing them to design plastic products that are more easily recyclable and less damaging to the environment, ultimately leading to reduced overall plastic production and associated GHG emissions (Tumu et al. 2023). Therefore, the progression and enforcement of the international legally binding instrument on plastic pollution offer a distinctive opportunity to further global efforts to resolve both issues permanently and effectively.

14.5.2 Global Warming Mitigation Policies: Reducing Plastic-Related Emissions

Plastic lifecycle emissions are substantially affected by the GHG mitigation scenario. Thus, strengthening efforts to reduce overall GHG emissions from the plastic life cycle is feasible under the UNFCCC and the Paris Agreement (Falkner 2016, Hermwille et al. 2017). The primary pathway via which the GHG mitigation scenario impacts the emissions of plastic comprises a transition in the energy sources used in plastics-related processes, such as manufacturing and conversion. This transition involves a move from high carbon-intensive sources to minimal carbon-intensive sources, such as electricity and gas (Kuusela et al. 2021). The impact on emissions at the end-of-life stage is also present, although to a lesser degree. Primary plastics manufacturing, which requires more energy, has a more significant effect compared to secondary production (Klotz et al. 2022). However, the total effect on the usage of plastics is minimal. Therefore, the most crucial chance for achieving these emission reductions is through the decarburization of plastics and the transition of energy during the manufacturing process (Bauer et al. 2022). Under the Paris Agreement, the parties have been trying to achieve a significant milestone in the fight against climate change and the strengthening of the global response by ensuring that the worldwide temperature rise remains at or below 2°C above pre-industrial levels and by making further efforts to restrict the temperature increase to 1.5°C (Falkner 2016). However, the lack of formal mention of plastics as a material and the plastics sector as a significant emission sector is a crucial oversight in the UNFCCC and the Paris Agreement. Nevertheless, the UNFCCC and the Paris Agreement have introduced new advancements that are driving a more focused effort toward addressing the issue of the plastics cycle (GRID-Arendal 2024). The COP 28 declaration, which advocates for the "transition from fossil fuels in energy systems to achieve net zero by 2050," is a critical step in the decarbonization of plastic life (Ivanova 2016; Jiang et al. 2024).

While it is crucial to focus on enhancing the accessibility of plastic alternatives through policy-making, it is equally necessary to acknowledge that existing policies subsidize oil exploration and production (Kojima 2016). Consequently, these policies contribute to the availability of inexpensive and plentiful raw materials for plastic manufacturing. Therefore, given the existing framework, implementing a carbon pricing system would only have a modest effect on reducing emissions from plastic production (Meys et al. 2021). In order to achieve a significant reduction in emissions and enhance the circularity potential, it is crucial to include the whole life cycle of plastics within the framework of carbon pricing. This can indirectly influence industries involved in plastic production and use to adopt cleaner technologies and practices, thereby reducing their overall environmental impact. Setting and enforcing energy efficiency standards for buildings, vehicles, and appliances can reduce energy consumption and associated emissions (Dixon et al. 2010). This lowers the demand for energy-intensive processes used in plastic production and recycling, indirectly reducing the environmental footprint of plastics. Supporting afforestation and preventing deforestation efforts globally can indirectly contribute to reducing plastic pollution by enabling people to use more wooden or bio-based products and reducing the dependency on plastic-based materials, for example, in household appliances, building materials, furniture, etc., that otherwise would have been made of plastic materials and produce more plastic waste (Sun et al. 2022). International cooperation can lead to shared strategies for reducing plastic waste and promoting recycling on a global scale (Simon and Schulte 2017; Barrowclough and Birkbeck 2022). Therefore, implementing generally acknowledged policy instruments to enhance the accessibility of plastic alternatives (such as bio-based and biodegradable plastics) and offering financial incentives can accelerate the progress towards achieving net-zero emission plastics (MacArthur et al. 2016).

14.5.3 Synergistic Policies to Tackle Plastic Pollution and Global Warming

GHG emission policies and plastic policies have distinct impacts on the emissions generated throughout the lifetime of plastics. GHG emission policies primarily affect the carbon footprint of plastic manufacturing, whereas plastic policies largely influence the reduction in plastic usage (Liu et al. 2021). Meanwhile, individual policies may serve their own interests but can also give rise to complications. One significant concern is that a specific strategy might potentially redirect the movement of plastics along a pathway with a greater risk of emission. A particular example of this concept is the waste unit pricing scheme. Imposing taxes or charges on household waste might lead to an upward trend in illicit disposal (Linderhof et al. 2001).

Given the multitude of policy objectives, including the reduction of plastic waste, a boost in recycling, and a decrease in illegal disposal, it is highly unlikely, if not impossible, to look for a single policy instrument that can effectively address both issues (Walls 2011). Instead, a comprehensive approach using several instruments will be required, such as implementing a carbon tax that accounts for all environmental consequences, providing incentives for recyclable designs, and establishing a deposit-refund mechanism (Alpizar et al. 2020). In addition, plastic is responsible for just 4% of the total GHG emissions. Thus, merely decreasing emissions connected to plastic will not be enough to accomplish ambitious goals for minimizing overall emissions. Simultaneously, it is imperative to resolve the emission sources in order to accomplish ambitious climate goals, including net-zero emissions. This indicates that the two policies have the potential to optimize the reduction of emissions throughout the lifecycle of plastics (Figure 14.2). The most effective way to promote global environmental action and optimize climate benefits is to establish a more cohesive relationship between the global plastic and GHG emissions reduction agreements (GRID-Arendal 2024). The synergistic policy should be consistent with the Paris Agreement's global objective to limit the global temperature increase to 1.5°C in comparison to pre-industrial levels. It should also encourage actions that tackle plastic pollution and support efforts to mitigate climate change.

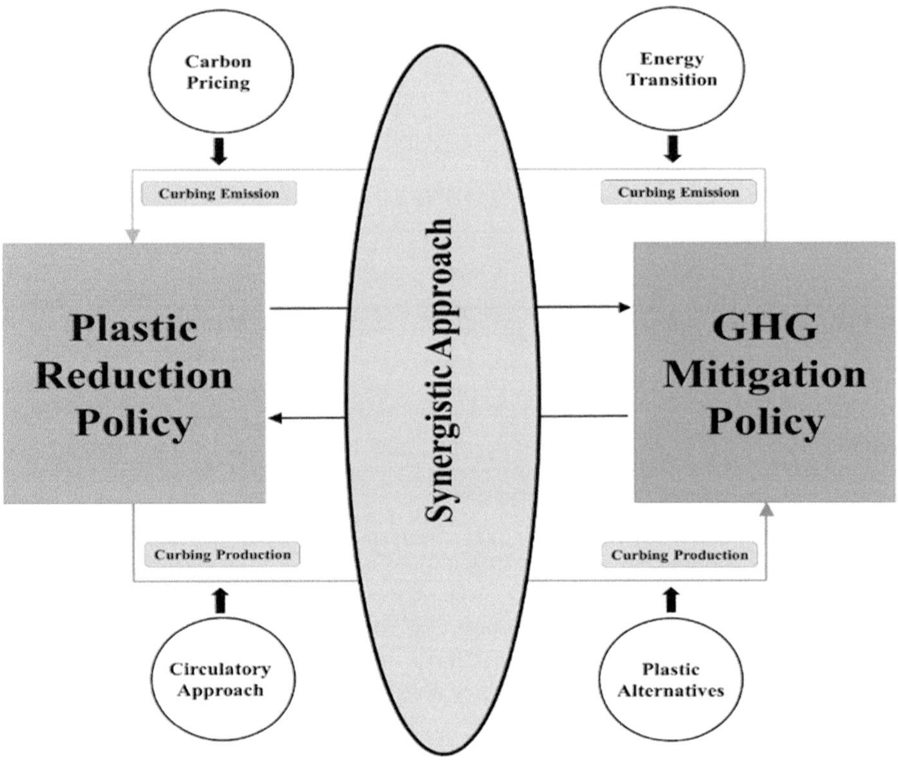

Figure 14.2 Addressing GHG emissions and plastic pollution through synergistic approaches.

14.6 CONCLUSIONS

While often addressed and analyzed independently, the seemingly disparate environmental concerns of plastic pollution and global warming have garnered international recognition. To address the difficulties posed by these issues, a complete global solution might be achieved by the use of multiple instruments within an internationally binding framework. Nevertheless, this chapter illustrates how the synergistic approach in policy design may effectively guide the regime toward either success or failure. When implemented in tandem, policies aimed at reducing plastics and climate policies interact with each other to provide synergistic effects in decreasing the GHG emissions from plastics. There is no single policy concept that can ensure success since different situations necessitate different responses. Hence, policies solely aimed at reducing GHG emissions cannot be relied upon as a replacement for policies targeting plastic waste reduction. Similarly, policies addressing plastic waste cannot serve as an alternative for specific actions focused on mitigating GHG emissions. Only by aggressively implementing two policies—renewable energy in production and widespread plastic recycling, for example—along with multilayered methods aimed at reducing plastic-related use, might help to curb emissions as well as global warming. It is also evident that demand control, enhanced recycling capabilities, the use of bio-based plastics, and the decarbonization of the energy infrastructure are all necessary for a complete drop in the GHG emissions associated with the plastics life cycle. Therefore, exceptional efforts are needed from the international community to ensure sustainable plastic use as well as to minimize plastic waste and the resulting emissions.

REFERENCES

Aldag, J. 2019. The last straw: Turning the tide on plastic pollution in Canada. Report of the Standing Committee on Environment and Sustainable Development.

Ágnes, N. and K. Rajmund. 2016. The environmental impact of plastic waste incineration. AARMS–Academic and Applied Research in Military and Public Management Science. 15: 231–237.

Alfonso, S., M. Gesto and B. Sadoul. 2021. Temperature increase and its effects on fish stress physiology in the context of global warming. J. Fish Biol. 98: 1496–1508.

Alhazmi, H., F.H. Almansour and Z. Aldhafeeri. 2021. Plastic Waste Management: A Review of Existing Life Cycle Assessment Studies. Sustainability. 13: 5340.

Alpizar, F., F. Carlsson, G. Lanza, B. Carney, R.C. Daniels, M. Jaime, et al. 2020. A framework for selecting and designing policies to reduce marine plastic pollution in developing countries. Environ. Sci. Policy. 109: 25–35.

Altman, R. 2021. The myth of historical bio-based plastics. Science. 373: 47–49.

Álvarez-Chávez, C.R., S. Edwards, R. Moure-Eraso and K. Geiser. 2012. Sustainability of bio-based plastics: general comparative analysis and recommendations for improvement. J. Cleaner Prod. 23: 47–56.

Atiwesh, G., A. Mikhael, C.C. Parrish, J. Banoub and T.-A.T Le. 2021. Environmental impact of bioplastic use: A review. Heliyon. 7(9).

Avery-Gomm, S., T.R. Walker, M.L. Mallory and J.F. Provencher. 2019. There is nothing convenient about plastic pollution. Rejoinder to Stafford and Jones "Viewpoint–Ocean plastic pollution: A convenient but distracting truth?" Marine Policy. 106: 103552.

Bandh, S.A., F.A. Malla, I. Qayoom, H. Mohi-Ud-Din, A.K. Butt, A. Altaf, et al. 2023. Importance of blue carbon in mitigating climate change and plastic/microplastic pollution and promoting circular economy. Sustainability. 15: 2682.

Barrowclough, D. and C.D. Birkbeck. 2022. Transforming the global plastics economy: the role of economic policies in the global governance of plastic pollution. Social Sciences. 11: 26.

Bauer, F., T.D. Nielsen, L.J. Nilsson, E. Palm, K. Ericsson, A. Fråne, et al. 2022. Plastics and climate change—Breaking carbon lock-ins through three mitigation pathways. One Earth. 5: 361–376.

Bauman, B. 2019. How plastics contribute to climate change. Yale Climate Connections. 20.

Benavides, P.T., U. Lee and O. Zarè-Mehrjerdi. 2020. Life cycle greenhouse gas emissions and energy use of polylactic acid, bio-derived polyethylene and fossil-derived polyethylene. J. Cleaner Prod. 277: 124010.

Borrelle, S.B., J. Ringma, K.L. Law, C.C. Monnahan, L. Lebreton, A. McGivern, et al. 2020. Predicted growth in plastic waste exceeds efforts to mitigate plastic pollution. Science. 369: 1515–1518.

Borrelle, S.B., C.M. Rochman, M. Liboiron, A.L. Bond, A. Lusher, H. Bradshaw, et al. 2017. Why we need an international agreement on marine plastic pollution. Proceedings of the National Academy of Sciences. 114: 9994–9997.

Boström, C., A. Törnroos and E. Bonsdorff. 2010. Invertebrate dispersal and habitat heterogeneity: expression of biological traits in a seagrass landscape. J. Exp. Mar. Biol. Ecol. 390: 106–117.

Cai, Y., Q. Wang, B. Zhou, R. Yuan, F. Wang, Z. Chen, et al. 2021. A review of responses of terrestrial organisms to perfluorinated compounds. Sci. Total Environ. 793: 148565.

Chae, Y. and Y.-J. An. 2018. Current research trends on plastic pollution and ecological impacts on the soil ecosystem: A review. Environ. Pollut. 240: 387–395.

Cho, R. 2020. More plastic on the way: What it means for climate change. State of the Planet-News from the Columbia Climate School.

Collard, F., J. Gasperi, G.W. Gabrielsen and B. Tassin. 2019. Plastic particle ingestion by wild freshwater fish: A critical review. Environ. Sci. Technol. 53: 12974–12988.

Dai, L., N. Zhou, Y. Lv, Y. Cheng, Y. Wang, Y. Liu, et al. 2022. Pyrolysis technology for plastic waste recycling: A state-of-the-art review. Prog. Energ. Combust. Sci. 93: 101021.

de Souza Machado, A.A., W. Kloas, C. Zarfl, S. Hempel and M.C. Rillig. 2018. Microplastics as an emerging threat to terrestrial ecosystems. Glob. Chang. Biol. 24: 1405–1416.

Deutsch, C.A., J.J. Tewksbury, R.B. Huey, K.S. Sheldon, C.K. Ghalambor, D.C. Haak, et al. 2008. Impacts of climate warming on terrestrial ectotherms across latitude. Proceedings of the National Academy of Sciences. 105: 6668–6672.

Dixon, R.K., E. McGowan, G. Onysko and R.M. Scheer. 2010. US energy conservation and efficiency policies: Challenges and opportunities. Energy Policy. 38: 6398–6408.

Duarte, C.M., J.J. Middelburg and N. Caraco. 2005. Major role of marine vegetation on the oceanic carbon cycle. Biogeosciences. 2: 1–8.

Elsawy, M.A., K.-H. Kim, J.-W. Park and A. Deep. 2017. Hydrolytic degradation of polylactic acid (PLA) and its composites. Renew. Sustain. Energy Rev. 79: 1346–1352.

Eriksson, O. and G. Finnveden. 2009. Plastic waste as a fuel-CO_2-neutral or not? Energy Environ. Sci. 2: 907–914.

Falkner, R. 2016. The Paris Agreement and the new logic of international climate politics. Int. Aff. 92: 1107–1125.

Ford, H.V., N.H. Jones, A.J. Davies, B.J. Godley, J.R. Jambeck, I.E. Napper, et al. 2022. The fundamental links between climate change and marine plastic pollution. Sci. Total Environ. 806: 150392.

Franzellitti, S., L. Canesi, M. Auguste, R.H.G.R Wathsala and E. Fabbri. 2019. Microplastic exposure and effects in aquatic organisms: a physiological perspective. Environ. Toxicol. Pharmacol. 683: 7–51.

Frieler, K., M. Meinshausen, A. Golly, M. Mengel, K. Lebek, S.D. Donner, et al. 2013. Limiting global warming to 2°C is unlikely to save most coral reefs. Nat. Clim. Change. 3: 165–170.

Gao, Y., G. Yu, T. Yang, Y. Jia, N. He and J. Zhuang. 2016. New insight into global blue carbon estimation under human activity in land-sea interaction area: A case study of China. Earth Sci. Rev. 1593: 6–46.

Geyer, R., J.R. Jambeck and K.L. Law. 2017. Production, use and fate of all plastics ever made. Sci. Adv. 3: e1700782.

GRID-Arendal. 2024. Climate Impacts of Plastics. Arendal: GRID-Arendal.

Hahladakis, J.N., E. Iacovidou and S. Gerassimidou. 2020. Plastic waste in a circular economy. 481–512. *In*: Trevor M. Letcher (ed.). Plastic Waste and Recycling. Elsevier.

Halley, P.J. and R. Smith. 2005. Biodegradable polymers for industrial applications. Thermoplastic Starch Biodegradable Polymers. 140–162.

Hamilton, L.A., S. Feilt, C. Muffett, M. Kelso, S. Rubright and C. Bernhardt. 2019. The hidden costs of a plastic planet. Center for International Environmental Law. Available at https://www.ciel.org/wp-content/uploads/2019/05/Plastic-and-Climate-FINAL-2019.pdf.

Hamilton, L.A. and S. Feit. 2019. Plastic and climate: The hidden costs of a plastic planet.

Haque, F. and C. Fan. 2023. Fate of microplastics under the influence of climate change. iScience. 26(9): 107649

Hermwille, L., W. Obergassel, H.E. Ott and C. Beuermann. 2017. UNFCCC before and after Paris–what's necessary for an effective climate regime? Clim. Policy. 17: 150–170.

Horton, A.A. 2022. Plastic pollution: When do we know enough? J. Hazard. Mater. 422: 126885.

Hossain, R., M.T. Islam, A. Ghose and V. Sahajwalla. 2022. Full circle: Challenges and prospects for plastic waste management in Australia to achieve circular economy. J. Cleaner Prod. 368: 133127.

Iroegbu, A.O.C., S.S. Ray, V. Mbarane, J.C. Bordado and J.P. Sardinha. 2021. Plastic pollution: a perspective on matters arising: challenges and opportunities. ACS Omega. 6: 19343–19355.

Ivanova, M. 2016. Good COP, bad COP: Climate reality after Paris. Global Policy. 7: 411–419.

Iwata, T. 2015. Biodegradable and bio-based polymers: future prospects of eco-friendly plastics. Angewandte Chemie (International Ed. in English). 54: 3210–3215.

Jamieson, A.J., T. Malkocs, S.B. Piertney, T. Fujii and Z. Zhang. 2017. Bioaccumulation of persistent organic pollutants in the deepest ocean fauna. Nat. Ecol. Evol. 1: 51.

Jiang, Y., F. Yang, S.S.U.H. Kazmi, Y. Zhao, M. Chen and J. Wang. 2022. A review of microplastic pollution in seawater, sediments and organisms of the Chinese coastal and marginal seas. Chemosphere. 286: 131677.

Jiang, T., X. He, B. Su, P.H. Havea, K. Wei, Z.W. Kundzewicz, et al. 2024. COP 28: Challenge of coping with the climate crisis. The Innovation. 5.

Kätelhön, A., R. Meys, S. Deutz, S. Suh and A. Bardow. 2019. Climate change mitigation potential of carbon capture and utilization in the chemical industry. Proceedings of the National Academy of Sciences. 116: 11187–11194.

Khan, N.S., J.B.A. Bari, S.M. Mahatab Uddin, M.S. Rahman, M. Uddin, S. Bhowmik, et al. 2024. A First Record on Microplastic Ingestion by Tropical Estuarine Copepods of Bangladesh. Bull. Environ. Contam. Toxicol. 113: 1.

Klotz, M., M. Haupt and S. Hellweg. 2022. Limited utilization options for secondary plastics may restrict their circularity. Waste Management. 141: 251–270.

Kojima, M. 2016. Fossil fuel subsidy and pricing policies: recent developing country experience. World Bank Policy Research Working Paper. 7531.

Kumar, P.M. and R. Pooja. 2020. Global warming, impacts and mitigation measures: An overview. Disaster Adv. 13: 82–96.

Kumari, S.V.G., K. Pakshirajan and G. Pugazhenthi. 2022. Recent advances and future prospects of cellulose, starch, chitosan, polylactic acid and polyhydroxyalkanoates for sustainable food packaging applications. Int. J. Biol. Macromol. 221: 163–182.

Kuusela, K., V. Uusitalo, J. Ahola and J. Levänen. 2021. The transformation of plastics production from net positive greenhouse gas emissions to net negative: An environmental sustainability assessment of CO_2-based polypropylene. J. CO_2 Utilization. 52: 101672.

Lallas, P.L. 2001. The Stockholm Convention on persistent organic pollutants. Am. J. Int. Law. 95: 692–708.

Lau, W.W.Y., Y. Shiran, R.M. Bailey, E. Cook, M.R. Stuchtey, J. Koskella, et al. 2020. Evaluating scenarios toward zero plastic pollution. Science. 369: 1455–1461.

Lee, J.G., H.G. Chae, S.R. Cho, H.-J. Song, P.J. Kim and S.T. Jeong. 2021. Impact of plastic film mulching on global warming in entire chemical and organic cropping systems: Life cycle assessment. J. Cleaner Prod. 308: 127256.

Li, X., L. Meng, Y. Zhang, Z. Qin, L. Meng, C. Li, et al. 2022. Research and application of polypropylene carbonate composite materials: A review. Polymers. 14: 2159.

Linderhof, V., P. Kooreman, M. Allers and D. Wiersma. 2001. Weight-based pricing in the collection of household waste: The Oostzaan case. Resource and Energy Economics. 23: 359–371.

Liu, Z., M. Adams, R.P. Cote, Q. Chen, R. Wu, Z. Wen, et al. 2018. How does circular economy respond to greenhouse gas emissions reduction: An analysis of Chinese plastic recycling industries. Renew. Sustain. Energy Rev. 91: 1162–1169.

Liu, Z., W. Liu, T.R. Walker, M. Adams and J. Zhao. 2021. How does the global plastic waste trade contribute to environmental benefits: Implication for reductions of greenhouse gas emissions? J. Environ. Manage. 287: 112283.

MacArthur, D.E., D. Waughray and M.R. Stuchtey. 2016. The new plastics economy. Rethinking the Future of Plastics.

Marsh, K. and B. Bugusu. 2007. Food packaging—roles, materials and environmental issues. J. Food Sci. 72: R39–R55.

Martin, C., F. Baalkhuyur, L. Valluzzi, V. Saderne, M. Cusack, H. Almahasheer, et al. 2020. Exponential increase of plastic burial in mangrove sediments as a major plastic sink. Sci. Adv. 6: eaaz5593.

Masnadi, M.S., H.M. El-Houjeiri, D. Schunack, Y. Li, J.G. Englander, A. Badahdah, et al. 2018. Global carbon intensity of crude oil production. Science. 361: 851–853.

Masson-Delmotte, V., P. Zhai, H.-O. Pörtner, D. Roberts, J. Skea, P.R. Shukla, et al. 2019. Global warming of 1.5 C. An IPCC Special Report on the Impacts of Global Warming Of. 1: 93–174.

Mazur-Wierzbicka, E. 2021. Circular economy: advancement of European Union countries. Environ. Sci. Eur. 33: 1–15.

Meys, R., A. Kätelhön, M. Bachmann, B. Winter, C. Zibunas, S. Suh, et al. 2021. Achieving net-zero greenhouse gas emission plastics by a circular carbon economy. Science. 374: 71–76.

Moshood, T.D., G. Nawanir, F. Mahmud, F. Mohamad, M.H. Ahmad and A. AbdulGhani. 2022. Sustainability of biodegradable plastics: New problem or solution to solve the global plastic pollution? Curr. Res. Green Sustainable Chem. 5: 100273.

Napper, I.E., A.C. Barrett and R.C. Thompson. 2020. The efficiency of devices intended to reduce microfibre release during clothes washing. Sci. Total Environ. 738: 140412.

Napper, I.E. and R.C. Thompson. 2020. Plastic debris in the marine environment: history and future challenges. Glob. Chall. 4: 1900081.

Nielsen, T.D., J. Hasselbalch, K. Holmberg and J. Stripple. 2020. Politics and the plastic crisis: A review throughout the plastic life cycle. Wiley Interdisciplinary Reviews: Energy Environ. 9: e360.

Nielsen, T.D., K. Holmberg and J. Stripple. 2019. Need a bag? A review of public policies on plastic carrier bags–Where, how and to what effect? Waste Manage. 87: 428–440.

O'Brien, S., C. Rauert, F. Ribeiro, E.D. Okoffo, S.D. Burrows, J.W. O'Brien, et al. 2023. There's something in the air: a review of sources, prevalence and behaviour of microplastics in the atmosphere. Sci. Total Environ. 874: 162193.

Oberbeckmann, S., B. Kreikemeyer and M. Labrenz. 2018. Environmental factors support the formation of specific bacterial assemblages on microplastics. Front. Microbiol. 8: 2709.

Parvez, M.S., H. Ullah, O. Faruk, E. Simon and H. Czédli. 2024. Role of Microplastics in Global Warming and Climate Change: A Review. Water Air Soil Pollut. 235: 201.

Payne, J., P. McKeown and M.D. Jones. 2019. A circular economy approach to plastic waste. Polym. Degrad. Stab. 165: 170–181.

Peng, B.-Y., Y. Sun, X. Zhang, J. Sun, Y. Xu, S. Xiao, et al. 2023. Unveiling the residual plastics and produced toxicity during biodegradation of polyethylene (PE), polystyrene (PS), and polyvinyl chloride (PVC) microplastics by mealworms (Larvae of Tenebrio molitor). J. Hazard. Mater. 452: 131326.

Perkins-Kirkpatrick, S.E. and P.B. Gibson. 2017. Changes in regional heatwave characteristics as a function of increasing global temperature. Sci. Rep. 7: 12256.

Posen, I.D., P. Jaramillo, A.E. Landis and W.M. Griffin. 2017. Greenhouse gas mitigation for US plastics production: energy first, feedstocks later. Environ. Res. Lett. 12: 34024.

Provencher, J.F., J.C. Vermaire, S. Avery-Gomm, B.M. Braune and M.L. Mallory. 2018. Garbage in guano? Microplastic debris found in faecal precursors of seabirds known to ingest plastics. Sci. Total Environ. 644: 1477–1484.

Rai, M., G. Pant, K. Pant, B.N. Aloo, G. Kumar, H.B. Singh, et al. 2023. Microplastic pollution in terrestrial ecosystems and its interaction with other soil pollutants: a potential threat to soil ecosystem sustainability. Resources. 12: 67.

Rajvanshi, J., M. Sogani, A. Kumar, S. Arora, Z. Syed, K. Sonu, et al. 2023. Perceiving biobased plastics as an alternative and innovative solution to combat plastic pollution for a circular economy. Sci. Total Environ. 874: 162441.

Ritchie, H. 2023. How much of global greenhouse gas emissions come from plastics? Published online at OurWorldInData.org. Retrieved from: "https://ourworldindata.org/ghg-emissions-plastics" [Online Resource].

Robertson, G.L. 2005. Food Packaging: Principles and Practice. CRC Press.

Royer, S.-J., S. Ferrón, S.T. Wilson and D.M. Karl. 2018. Production of methane and ethylene from plastic in the environment. PloS One. 13: e0200574.

Saleh, T.A. and V.K. Gupta. 2016. Nanomaterial and Polymer Membranes: Synthesis, Characterization and Applications. Elsevier.

Schwarz, A.E., T.N. Ligthart, D.G. Bizarro, P. De Wild, B. Vreugdenhil and T.V. Harmelen. 2021. Plastic recycling in a circular economy; determining environmental performance through an LCA matrix model approach. Waste Manage. 121: 331–342.

Sharma, S., V. Sharma and S. Chatterjee. 2023. Contribution of plastic and microplastic to global climate change and their conjoining impacts on the environment-A review. Sci. Total Environ. 875: 162627.

Shen, L. and E. Worrell. 2014. Plastic Recycling. Handbook of Recycling: State-of-the-Art for Practitioners, Analysts and Scientists. Elsevier Inc.

Shen, M., W. Huang, M. Chen, B. Song, G. Zeng and Y. Zhang. 2020. (Micro) plastic crisis: un-ignorable contribution to global greenhouse gas emissions and climate change. J. Cleaner Prod. 254: 120138.

Shrivastava, A. 2011. Plastics Processing. *In*: Applied Plastics Engineering Handbook: Processing and Materials.

Simon, N. and M.L. Schulte. 2017. Stopping Global Plastic Pollution: The Case for an International Convention: Simon, Nils| USchulte, Maro Luisa.

Siracusa, V. and I. Blanco. 2020. Bio-polyethylene (Bio-PE), Bio-polypropylene (Bio-PP) and Bio-poly (ethylene terephthalate)(Bio-PET): Recent developments in bio-based polymers analogous to petroleum-derived ones for packaging and engineering applications. Polymers. 12: 1641.

Soares, J., I. Miguel, C. Venâncio, I. Lopes and M. Oliveira. 2021. Public views on plastic pollution: Knowledge, perceived impacts and pro-environmental behaviours. J. Hazard. Mater. 412: 125227.

Sun, X., M. Xie, L. Mai and E.Y. Zeng. 2022. Biobased plastic: A plausible solution toward carbon neutrality in plastic industry? J. Hazard. Mater. 435: 129037.

Syberg, K., B.C. Almroth, M.O. Fernandez, J. Baztan, M. Bergmann, R.C. Thompson, et al. 2024. Informing the Plastic Treaty negotiations on science-experiences from the Scientists' Coalition for an Effective Plastic Treaty. Microplastics and Nanoplastics. 4: 14.

Tang, K.H.D. 2023. Climate change and plastic pollution: A review of their connections. Tropical Environ. Biol. Technol. 1: 110–120.

Ter Halle, A., L. Ladirat, X. Gendre, D. Goudouneche, C. Pusineri, C. Routaboul, et al. 2016. Understanding the fragmentation pattern of marine plastic debris. Environ. Sci. Technol. 50: 5668–5675.

Tewksbury, J.J., R.B. Huey and C.A. Deutsch. 2008. Putting the heat on tropical animals. Science. 320: 1296–1297.

Thomas, G.O., E. Sautkina, W. Poortinga, E. Wolstenholme and L. Whitmarsh. 2019. The English plastic bag charge changed behavior and increased support for other charges to reduce plastic waste. Front. Psychol. 10: 266.

Tibbetts, J., S. Krause, I. Lynch and G.H. Sambrook Smith. 2018. Abundance, distribution and drivers of microplastic contamination in urban river environments. Water. 10: 1597.

Tumu, K., K. Vorst and G. Curtzwiler. 2023. Global plastic waste recycling and extended producer responsibility laws. J. Environ. Manage. 348: 119242.

UNEP. 2022. End plastic pollution: Towards an international legally binding instrument. Available at https://wedocs.unep.org/bitstream/handle/20.500.11822/39812/OEWG_PP_1_INF_1_UNEA%20resolution.pdf (accessed 10 December 2023).

Van den Oever, M., K. Molenveld, M. van der Zee and H. Bos. 2017. Bio-based and biodegradable plastics: facts and figures: Focus on food packaging in the Netherlands (Report 1722). Wageningen Food & Biobased Research.

van Raamsdonk, L.W.D., M. van der Zande, A.A. Koelmans, R.L.A.P. Hoogenboom, R.J.B. Peters, M.J. Groot, et al. 2020. Current insights into monitoring, bioaccumulation, and potential health effects of microplastics present in the food chain. Foods. 9: 72.

Vanapalli, K.R., H.B. Sharma, V.P. Ranjan, B. Samal, J. Bhattacharya, B.K. Dubey, et al. 2021. Challenges and strategies for effective plastic waste management during and post COVID-19 pandemic. Sci. Total Environ. 750: 141514.

Vegter, A.C., M. Barletta, C. Beck, J. Borrero, H. Burton, M.L. Campbell, et al. 2014. Global research priorities to mitigate plastic pollution impacts on marine wildlife. Endanger. Species Res. 25: 225–247.

Velders, G.J.M., A.R. Ravishankara, M.K. Miller, M.J. Molina, J. Alcamo, J.S. Daniel, et al. 2012. Preserving Montreal Protocol climate benefits by limiting HFCs. Science. 335: 922–923.

Villarrubia-Gómez, P., S.E. Cornell and J. Fabres. 2018. Marine plastic pollution as a planetary boundary threat–The drifting piece in the sustainability puzzle. Marine Policy. 96: 213–220.

Virdin, J., R. Karasik, T. Vegh, A. Pickle, Z. Diana, D. Rittschof, et al. 2020. 20 Years of Government Responses to the Global Plastic Pollution Problem: The Plastics Policy Inventory. NI X 20-05. Durham, NC: Duke University.

Vishwakarma, A. 2020. Unsustainable management of plastic wastes in India: A threat to global warming and climate change. Contemporary Environmental Issues and Challenges in Era of Climate Change. 235–244.

Wabnitz, C and W.J. Nichols. 2010. Plastic pollution: An ocean emergency. Marine Turtle Newsletter. 129: 1.

Walker, T., D. Gramlich and A. Dumont-Bergeron. 2020. The case for a plastic tax: A review of its benefits and disadvantages within a circular economy. Sustainability. 185–211.

Walls, M. 2011. Deposit-refund systems in practice and theory. Resources for the Future Discussion Paper. 11–47.

Wang, N. and Y.-C. Chang. 2014. The development of policy instruments in supporting low-carbon governance in China. Renewable Sustainable Energy Rev. 35: 126–135.

Wang, Y., M. Jiao, T. Li, R. Li and B. Liu. 2023. Role of mangrove forest in interception of microplastics (MPs): challenges, progress and prospects. J. Hazard. Mater. 445: 130636.

White, E.M., S. Clark, C.A. Manire, B. Crawford, S. Wang, J. Locklin, et al. 2018. Ingested micronizing plastic particle compositions and size distributions within stranded post-hatchling sea turtles. Environ. Sci. Technol. 52: 10307–10316.

Wichai-Utcha, N. and O. Chavalparit. 2019. 3Rs Policy and plastic waste management in Thailand. J. Mater. Cycles Waste Manage. 21: 10–22.

Wiesinger, H., Z. Wang and S. Hellweg. 2021. Deep dive into plastic monomers, additives and processing aids. Environ. Sci. Technol. 55: 9339–9351.

Williams, A.T. and N. Rangel-Buitrago. 2019. Marine litter: Solutions for a major environmental problem. J. Coastal Res. 35: 648–663.

Winans, K., A. Kendall and H. Deng. 2017. The history and current applications of the circular economy concept. Renewable Sustainable Energy Rev. 68: 825–833.

Wingfield, S. and M. Lim. 2022. The United Nations Basel Convention's global plastic waste partnership: history, evolution and progress. Microplastic in the Environment: Pattern and Process. 323–331.

Zandalinas, S.I., F.B. Fritschi and R. Mittler. 2021. Global warming, climate change and environmental pollution: Recipe for a multifactorial stress combination disaster. Trends Plant Sci. 26: 588–599.

Zantis, L.J., E.L. Carroll, S.E. Nelms and T. Bosker. 2021. Marine mammals and microplastics: A systematic review and call for standardisation. Environ. Pollut. 269: 116142.

Zeb, A., W. Liu, N. Ali, R. Shi, Q. Wang, J. Wang, et al. 2023. Microplastic pollution in terrestrial ecosystems: Global implications and sustainable solutions. J. Hazard. Mater. 132636.

Zen, I.S., R. Ahamad and W. Omar. 2013. No plastic bag campaign day in Malaysia and the policy implication. Environ. Dev. Sustainability. 15: 1259–1269.

Zhang, C., T. Xu, H. Feng and S. Chen. 2019. Greenhouse gas emissions from landfills: A review and bibliometric analysis. Sustainability. 11: 2282.

Zheng, J. and S. Suh. 2019. Strategies to reduce the global carbon footprint of plastics. Nat. Clim. Change. 9: 374–378.

Index

R

S

T

U

V

W

For Product Safety Concerns and Information please contact our EU representative GPSR@taylorandfrancis.com Taylor & Francis Verlag GmbH, Kaufingerstraße 24, 80331 München, Germany

Role of Cancer Stem Cells in Cancer Biology and Therapy